Artificial General Intelligence

Scrivener Publishing
100 Cummings Center, Suite 541J
Beverly, MA 01915-6106

Leading-Edge Breakthroughs in Artificial Intelligence

Series Editor: Dr. S. Balamurugan (sbnbala@gmail)

The world is on the cusp of several breakthroughs in artificial intelligence (AI), rapidly transforming industries and the research field. The "Leading-Edge Breakthroughs in Artificial Intelligence" book series is dedicated to exploring the most pioneering developments in artificial intelligence. The series aims to highlight both theoretical and practical advancements, offering insights into several cutting-edge advancements in AI and how the same could be deployed to solve complex problems in fields such as healthcare, finance, manufacturing, robotics, marketing, energy management, agriculture, autonomous systems, and many more. It also addresses critical issues like AI governance, ethics, trust, security, transparency, and societal impacts, making it an essential resource for researchers, engineers, and industry leaders looking to leverage AI for future breakthroughs.

Publishers at Scrivener
Martin Scrivener (martin@scrivenerpublishing.com)
Phillip Carmical (pcarmical@scrivenerpublishing.com)

Artificial General Intelligence

Principles and Practices

Edited by

T. Saravanan
GITAM School of Technology, GITAM (Deemed to be University), Bengaluru, India

P. Preethi
Department of Computer Science and Engineering, Kongunadu College of Engineering and Technology, Trichy, Tamil Nadu, India

Sumaya Sanober
College of Computer Science, Old Dominion University, Virginia, United States

N. Thillaiarasu
School of Computing and Information Technology, REVA University, Bangalore, India

and

S. Balamurugan
Intelligent Research Consultancy Services (iRCS), Coimbatore, Tamil Nadu, India

Scrivener Publishing

WILEY

This edition first published 2026 by John Wiley & Sons, Inc., 111 River Street, Hoboken, NJ 07030, USA and Scrivener Publishing LLC, 100 Cummings Center, Suite 541J, Beverly, MA 01915, USA

For more information about Scrivener publications please visit www.scrivenerpublishing.com.

Wiley Global Headquarters
111 River Street, Hoboken, NJ 07030, USA

For details of our global editorial offices, customer services, and more information about Wiley products visit us at www.wiley.com.

The manufacturer's authorized representative according to the EU General Product Safety Regulation is Wiley-VCH GmbH, Boschstr. 12, 69469 Weinheim, Germany, e-mail: Product_Safety@wiley.com.

Library of Congress Cataloging-in-Publication Data

ISBN 978-1-394-42267-8

Cover image: Generated with AI using Adobe Firefly
Cover design by Russell Richardson

Set in size of 11pt and Minion Pro by Manila Typesetting Company, Makati, Philippines

Printed in the USA

10 9 8 7 6 5 4 3 2 1

Dedicated to

The Almighty, our parents, our spouses, and our children

T. Saravanan
P. Preethi
Sumaya Sanober
N. Thillaiarasu
S. Balamurugan

Contents

Series Preface

This important book provides an extensive overview of the key theories, methodologies, and applied frameworks that enable artificial general intelligence (AGI) systems to exhibit aspects of human intelligence—or more correctly, intelligence adaptable as needed or defined by humans. It explains the basic ideas, architectures, methodologies, and processes needed to construct systems with human-like reasoning capabilities, adaptability, and autonomy, which combine to represent one of the great unfulfilled ambitions of artificial intelligence (AI) research. This collection of basic concepts, practical implementations, ethical implications, and respective emerging applications is an excellent resource for the industry professionals, researchers, and policymakers to identify and respond to the challenges of designing general-purpose intelligence systems. Included in the series *Leading-Edge Breakthroughs in Artificial Intelligence*, this volume presents a forward-looking perspective, as AGI represents a transformative leap in AI evolution. The book aligns to the scope of the series, showcasing important new advances that will act as formative events for the future development of intelligent technologies.

Dr. S. Balamurugan
Director, Intelligent Research Consultancy Services (iRCS), Coimbatore, Tamil Nadu, India
Series Editor, Leading-Edge Breakthroughs in Artificial Intelligence

Preface

This book examines the theoretical foundations, cognitive architectures, and practical methodologies shaping artificial general intelligence (AGI). Unlike narrow artificial intelligence (AI), confined to specific tasks, AGI seeks human-like adaptability, reasoning, and learning across domains. Integrating cognitive, mathematical, and computational perspectives, it offers a unified framework for defining and advancing AGI. Emphasis is placed on approaches such as self-supervised and meta-learning, reinforcement learning, Bayesian inference, multimodal perception, and neuro-symbolic reasoning. Balancing rigor with practical insight, the volume provides a roadmap for researchers, practitioners, and policymakers to grasp AGI's evolution and transformative potential.

The primary objective of this book is to present a holistic perspective on AGI by bridging theory and practice. It establishes a unified theoretical and cognitive framework for defining AGI, explores the mathematical and algorithmic principles essential for modeling intelligent behavior, and investigates adaptive learning paradigms—such as meta-learning, transfer learning, and lifelong learning—which enable AGI to generalize across domains. Core mechanisms including reinforcement learning, Bayesian inference, and multimodal perception are highlighted as the foundation for building robust, flexible, and interpretable systems. The book also addresses the critical challenges of explainability, ethics, safety, and alignment to ensure AGI evolves in ways that benefit humanity. Finally, it examines emerging trends such as quantum computing, brain–computer interfaces, and hybrid neuro-symbolic models that shape the trajectory of AGI. In essence, it seeks to advance the creation of intelligent systems that are not only powerful and adaptive but also transparent, ethical, and human-centered across multiple sectors.

Chapter 1 proposes a unified framework for defining AGI by integrating cognitive architectures, meta-cognitive learning, and theoretical mod els, aiming to bridge the gap between human-like adaptability and formal computational principles. Chapter 2 argues that theoretical intelligence

(TI)—grounded in formal, mathematical frameworks such as algorithmic information theory, and Bayesian inference—is essential for developing robust, interpretable, and safe AGI systems. It demonstrates how TI principles enhance data efficiency, generalization, and alignment in AGI architectures.

Chapter 3 compares symbolic and neural cognitive architectures for AGI, highlighting their complementary strengths in reasoning and pattern recognition, and advocates for hybrid models that integrate both approaches for robust, explainable, and generalizable intelligence. Chapter 4 examines the role of cognitive architectures in developing human-like AGI, emphasizing hybrid systems, meta-cognition, and evaluative benchmarks to bridge gaps in generalization, adaptability, and embodied cognition.

Chapter 5 explores the integration of symbolic and subsymbolic reasoning for context-aware AGI, proposing a dual-process architecture with a differentiable neuro-symbolic interface and meta-cognitive control to enhance generalization, interpretability, and adaptive learning. Chapter 6 proposes a hybrid neural-symbolic framework for developing commonsense reasoning in AGI, integrating perceptual grounding, dynamic rule learning, and contextual graph inference to enable robust, explainable, and adaptive inference.

Chapter 7 proposes a hybrid neuro-symbolic architecture for explainable AGI, integrating transparent reasoning engines, interpretable neural modules, and a meta-explanation layer to generate human-understandable justifications for AI decisions. Chapter 8 explores how self-supervised learning methods—such as contrastive learning, predictive modeling, and meta-cognitive adaptation—enhance AGI adaptability by enabling autonomous representation learning, temporal forecasting, and self-regulated improvement without external supervision.

Chapter 9 explores the integration of meta-learning and transfer learning techniques to enable robust cross-domain generalization in AGI, emphasizing few-shot adaptation, modular knowledge reuse, and lifelong learning to achieve human-like flexibility and efficiency. Chapter 10 explores human-level generalization in AGI through interactive meta-learning and environmental adaptation, enabling systems to learn how to learn *via* dialogue, feedback, and context-sensitive modulation. It emphasizes few-shot and zero-shot adaptability, robust transfer across domains, and human-guided refinement for flexible, human-like intelligence.

Chapter 11 analyzes model-free and model-based reinforcement learning approaches for AGI, comparing their strengths, limitations, and scalability in dynamic environments. It highlights hybrid methods such as AlphaZero, MuZero, and Dreamer as promising pathways that combine planning and direct learning to enable adaptable, efficient, and human-like intelligence. Chapter 12 introduces InvisiDroid, a problem-space adversarial attack that evades black-box spyware detectors using opcode-level transformations. It details query-efficient manipulation strategies that expose ML vulnerabilities while preserving app functionality in real-world cyber-attack scenarios.

Chapter 13 presents multimodal perception systems for AGI, integrating visual, auditory, and linguistic inputs through unified encoders, cross-modal fusion, and hierarchical memory to achieve context-aware reasoning. It emphasizes neuro-symbolic grounding and commonsense inference as the backbone of human-like perception, enabling robust, interpretable, and adaptive intelligence. Chapter 14 explores meta-cognition in AGI, focusing on self-monitoring, error detection, and adaptive strategy selection to enhance flexible intelligence. It highlights architectures that integrate introspection, uncertainty estimation, and self-reflection, enabling AGI to reason about its own thought processes and improve decision-making over time.

Chapter 15 explores TransGAN for visual anomaly detection, leveraging transformer-based GANs to detect deviations in industrial and imbalanced datasets. It demonstrates latent space mapping on MVTecAD and dual-discriminator training on CIFAR-10, highlighting challenges of imbalance, explainability, and real-world scalability. Chapter 16 reviews state-of-the-art ML and DL models for grape disease detection, highlighting convolutional neural networks (CNNs), recurrent neural networks (RNNs), and transfer learning for precision viticulture. It stresses scalable, real-time solutions *via* mobile, Internet of Things, and drone-based platforms to enable sustainable farming and reduce chemical dependence.

Chapter 17 presents a deep learning framework combining CNNs and RNNs for real-time plant disease diagnosis across multiple crops and pathogens. It demonstrates high accuracy, robust web-based deployment, and potential for sustainable agriculture through early detection and treatment guidance. Chapter 18 examines AI-driven sensor fusion for assistive technologies, integrating deep learning and computer vision with haptic and auditory feedback. It emphasizes lightweight real-time models, user-centered design, and edge AI for enhancing safety and mobility in visually impaired navigation.

Chapter 19 explores how quantum computing and brain–computer interfaces accelerate AGI evolution by enabling direct human–machine symbiosis, adaptive learning, and nonlinear reasoning. It highlights ethical, social, and technical challenges while outlining evolutionary pathways for safe, value-aligned AGI.

Topics presented in each chapter of this book are unique to this book and are based on unpublished work of contributed authors. In editing this book, we attempted to bring into the discussion all the new trends and experiments that have made on adaptive AI. We believe this book is ready to serve as a reference for larger audiences such as system architects, practitioners, developers, and researchers. In closing, we thank Martin Scrivener and the team at Scrivener Publishing for their efforts and guidance in the publication of this work.

T. Saravanan
GITAM School of Technology, GITAM (Deemed to be University), Bengaluru, India

P. Preethi
Department of Computer Science and Engineering, Kongunadu College of Engineering and Technology, Trichy, Tamil Nadu, India

Sumaya Sanober
College of Computer Science, Old Dominion University, Virgina, United States

N. Thillaiarasu
School of Computing and Information Technology, REVA University, Bangalore, India

S. Balamurugan
Intelligent Research Consultancy Services (iRCS), Coimbatore, Tamil Nadu, India

PART I

THEORETICAL AND COGNITIVE FOUNDATIONS OF AGI

1

A Unified Framework for Defining Artificial General Intelligence through Cognitive and Theoretical Perspectives

Latha. P.[1]*, K. Sivakami[2], K. Selvavinayaki[3], V. Devi[4] and Karthick R.[5]

[1]Department of Electronics and Communication Engineering, V.S.B. College of Engineering Technical Campus, Coimbatore, Tamilnadu, India
[2]Department of Electronics and Communication Engineering, Karpagam Institute of Technology, Coimbatore, Tamilnadu, India
[3]Department of Computer Applications, Nehru Arts and Science College, Coimbatore, Tamilnadu, India
[4]Department of Artificial Intelligence and Machine Learning, Hindusthan College of Engineering and Technology, Coimbatore, Tamilnadu, India
[5]Department of Computer Science and Business Systems, Knowledge Institute of Technology, Salem, Tamilnadu, India

Abstract

Artificial general intelligence (AGI) aims to achieve a mass of cognitive versatility behaviorally similar to humans, where a machine can solve a wide variety of kinds of intellectual tasks across many fields of intellectual activity without programming for the specific task. Although many narrow artificial intelligence systems can succeed at some specific tasks, they lack the general, adaptable, and self-reflective aspect of human intellectual ability. Here we provide a unified framework for defining AGI that incorporates cognitive science, theoretical computer science, and philosophy of mind, which can help connect separate conceptual and disciplinary areas together. The framework stems from four fundamental dimensions: (1) universality of function, illustrating an agent's ability to solve many problems in a single architecture; (2) adaptability and learning, representing an agent's ability to continuously improve itself through self-supervised learning and knowledge transfer across domains; (3) goal-directed behavior, which involves making rational decisions and planning in uncertain or dynamic environments;

**Corresponding author*: veejaylatha@gmail.com

T. Saravanan, P. Preethi, Sumaya Sanober, N. Thillaiarasu and S. Balamurugan (eds.) Artificial General Intelligence: Principles and Practices, (3–22) © 2026 Scrivener Publishing LLC

and (4) meta-cognition, illustrating an agent's ability to think about itself in some way for the purpose of improving its own thinking. These elements build on cognitive architectures such as SOAR and ACT-R and use computational theory such as Turing completeness, algorithmic learning, and bounded rationality. With this hybrid of cognitive and computational foundation of knowledge and theory, this unifying framework goes beyond *ad hoc* definitions and provides a systematic way of initiating AGI development and assessment. This framework shows how to create systems that mimic some human cognitive functioning, while also demonstrating the autonomy, adaptability, and generalization capacity for real-world settings. Ultimately, this project provides a formal conception of the aims of AGI research and its parameters for cooperation across disciplines, designs that are ethical, and ways to improve and deepen our research on intelligent machines.

***Keywords*:** Artificial general intelligence, cognitive architecture, meta-cognition, theoretical framework, adaptability

1.1 Introduction

The quest for artificial general intelligence (AGI)—a system able to replicate or surpass human cognitive ability in varied and open-ended domains—remains the "holy grail" of artificial intelligence (AI). Nevertheless, despite an abundance of progress and interest over the past few decades, one fundamental hindrance remains: there is no precise, universally accepted, or practical definition of AGI itself. This definitional ambiguity can obstruct well-defined research objectives, prevent a clear understanding of progress, and inhibit productive discussion. Current constructions of AGI occupy two, sometimes contradictory, viewpoints leading to a fragmented domain [1].

On one hand, there is the cognitive perspective, founded in both neuroscience and psychology. This approach attempts to construct AGI by an analogy to human (or biological) intelligence. Key attributes include learning, reasoning, problem-solving, perception, natural language understanding, creativity, and, crucially, adaptability and transfer learning—the ability to apply knowledge gained in one context to entirely novel situations. Proponents argue that understanding the mechanisms of natural intelligence is essential to replicating general capabilities. Conversely, the theoretical perspective approaches AGI [2] through formal, often mathematical, frameworks. It focuses on abstract capabilities, performance boundaries, and computational principles. Definitions here might emphasize achieving human-level performance across a vast spectrum of tasks (Legg & Hutter's "universal intelligence") [3], optimizing behavior under uncertainty (reinforcement learning [RL] agents in unknown environments), or possessing the theoretical capacity to learn any computable function given sufficient resources. This perspective prioritizes rigor and measurability over biological fidelity.

The central problem is that these perspectives often operate in isolation. Cognitive definitions can be criticized as anthropocentric and lack formal rigor for engineering. Theoretical definitions, although precise [4], can seem detached from the richness of biological cognition and the practical challenges of building adaptable systems. This divergence impedes the field. Therefore, this paper proposes and argues for the necessity of a unified framework for defining AGI. This framework must intentionally synthesize knowledge from both cognitive science, and theoretical computer science/mathematics. A framework must describe what AGI does (cognitive capabilities) and how (theoretical aspects) competency can be formally specified and measured, establishing shared terminology and concrete definitions. Only through a synthesis of the theoretical ideas and cognitive capabilities could we move from ambiguous expression of the goal of successful AGI to an explicit, shared objective, focused research, and evaluation of our progress in creating this transformative technology. Defining AGI starts with specifying the goal.

1.2 Related Work

The quest to define AGI has spawned diverse approaches, often reflecting distinct disciplinary priorities. This section reviews key strands of related work, highlighting the existing gap between cognitive and theoretical perspectives and nascent efforts toward integration.

1.2.1 The Fragmented Landscape of Existing AGI Definitions

The literature indicates that there is a profusion of definitions of AGI, often focusing on different dimensions of sophistication. Early conceptions such as Turing's "child machine," capable of learning any task, set an ambitious but vague endpoint for the definition of an AGI. Many definitions since then have fallen into these categories: human-centric (e.g., passing a generalized Turing test where participation in a number of skills was not confined to an answer or achieving human-level performance on the majority of economically valuable work [Hutter]) [6], capability-oriented (e.g., Goertzel stating that an AGI will be capable of achieving its goals in hierarchically structured complex environments, or Nilsson's definition

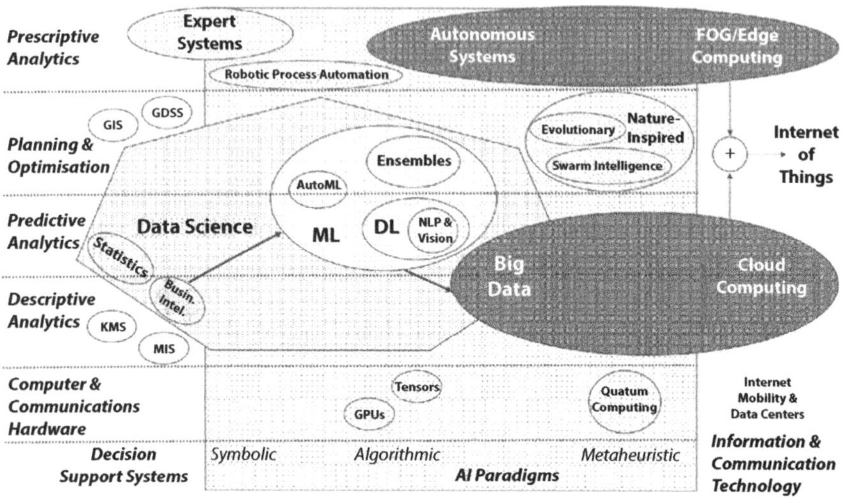

Figure 1.1 A unified framework for artificial intelligence in enterprise.

of a "system that acts appropriately" with understanding), and formal-mathematical (e.g., Legg & Hutter's definition of "universal intelligence" as based on reward maximization across all computable environments). Pei Wang's Non-Axiomatic Reasoning System (NARS) [7] defined intelligence in relation to an agent's knowledge and resource constraints, with a policy of preferring adaptability to human-likeness. Each of them has some strengths, but also drawbacks. Human-centric definitions are susceptible to anthropocentrism and sometimes suffer when faced with potentially non–human-like intelligence. Capability definitions usually lack formal rigor that would allow them to be operationalized. Formal definitions are typically very rigorous with respect to definitions as a whole, although can be computationally intractable, and are not aware of the cognitive realities of learning or adaptation of bodily agents that are acting under some form of resource scarcity (Figure 1.1). The patterns in these definitions point out a lack of a consensus definition that is still comprehensive and operationalizable, creating a barrier to coordinated research and objective benchmarking.

1.2.2 Cognitive Foundations: Insights from Natural Intelligence

A significant body of work grounds AGI aspirations in the study of natural intelligence. Cognitive architectures such as ACT-R (Anderson) and SOAR

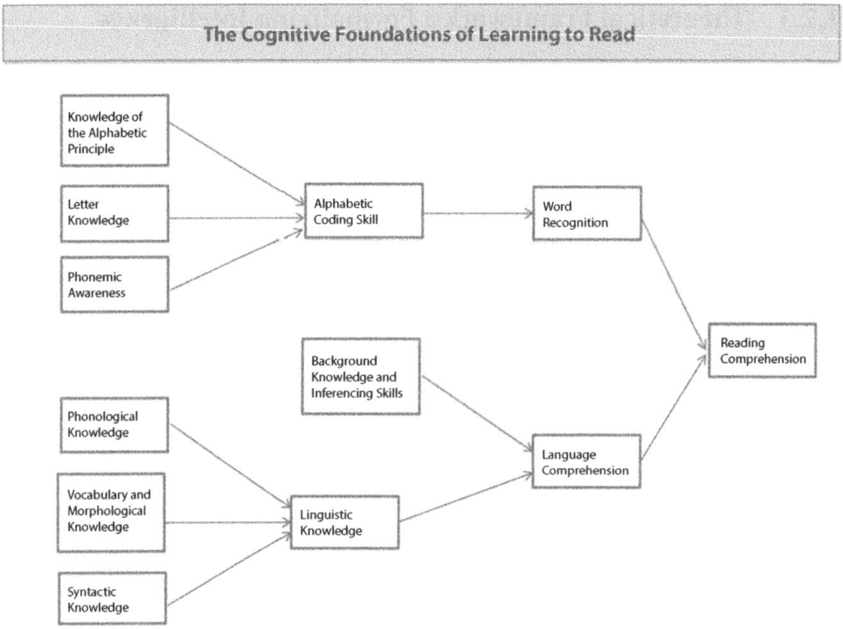

Figure 1.2 Cognitive foundations in NLP.

[8] explicitly model human cognitive processes (perception, memory, reasoning, learning) to achieve flexible behavior (Figure 1.2). Connectionist approaches, inspired by neural computation, demonstrate powerful pattern recognition and learning capabilities, exemplified by deep learning, though often lacking the structured reasoning and transfer abilities central to generality. Theories such as Global Workspace Theory (Baars, Dehaene) propose architectures for consciousness and information integration, offering blueprints for AGI awareness and control. Research on transfer learning (Thrun) [9], meta-learning (learning-to-learn), and compositionality (building complex representations from simpler ones) directly addresses the core AGI challenge of adapting to novelty. Furthermore, concepts such as embodied cognition [10], situatedness, and developmental robotics [11] emphasize that intelligence arises from interaction with a physical and social environment, not just abstract computation. This perspective provides crucial insights into the mechanisms enabling flexibility, generalization, and open-ended learning—the hallmarks of general intelligence—but often struggles to translate these complex, often qualitative, biological processes into precise, scalable computational models and formal metrics for AGI achievement.

1.2.3 Theoretical Frameworks: Formalizing Intelligence

The theoretical perspective seeks mathematical precision. Legg and Hutter's definition of Universal Intelligence (based on Solomonoff induction) measures an agent's expected performance across *all* computable environments weighted by Kolmogorov complexity. Although foundational, it is incomputable and ignores practical constraints such as sample efficiency and resource limitations. The AIXI model is a theoretical super-intelligent agent based on this principle, serving as a gold standard but remaining unbuildable. RL frameworks, particularly general-purpose RL agents (Sutton & Barto), provide a formal setting for learning optimal behavior in unknown environments, directly addressing adaptability. However, standard RL often assumes predefined reward functions and state spaces, limiting its inherent generality. Computational learning theory (Valiant, Vapnik) offers frameworks such as probably approximately correct (PAC) learning, providing guarantees on generalization from limited data, relevant to AGI's learning efficiency. Theories of universal computation (Turing machines) and algorithmic information theory (AIT; Kolmogorov, Chaitin) underpin discussions on the potential scope of artificial minds. These approaches provide essential rigor, defining intelligence in terms of objective functions, performance bounds, and computational capabilities. Yet, they often abstract away the messy realities of perception, embodiment, social interaction, and the specific cognitive architectures necessary for efficient learning and reasoning in complex, partially observable worlds, making them difficult to map directly onto human-like cognitive benchmarks [12].

1.2.4 Bridging Attempts and the Unification Imperative

Recognizing the limitations of isolated perspectives, some researchers have initiated efforts toward synthesis. Shane Legg, while co-developing universal intelligence, also emphasized the importance of "psychometric AI"—grounding AGI tests in human cognitive abilities. Marcus has argued for hybrid systems combining neural pattern recognition with symbolic reasoning for robust generalization. Frameworks such as LIDA (Franklin) attempt to implement cognitive theories (Global Workspace) computationally [13]. Theoretical work on bounded optimality (Russell & Subramanian) seeks optimal agents under computational constraints, aligning theoretical rigor with practical cognitive limitations. DeepMind's work on agents such as Gato, trained across diverse modalities and tasks, represents an empirical step toward unification, albeit still narrow. Bach's theory of cognition as management of concurrent, competing demands offers a conceptual bridge. However, these efforts remain nascent or partial. Most cognitive architectures lack the theoretical grounding for universal capability claims, while most theoretical models lack the cognitive plausibility or embodiment for genuine adaptability in human-like environments. There is no established framework that systematically integrates the what (cognitive capabilities, benchmarks from psychology/neuroscience) with the how (formal performance metrics, computational efficiency, scalability proofs) [14] into a cohesive definition. This paper directly addresses this gap by proposing a structured framework for such integration, arguing that defining AGI requires this dual perspective to be both meaningful and actionable for research and evaluation. The unification imperative is clear: progress demands a shared language and measurable goals born from cognitive reality and theoretical precision [15].

1.3 Proposed Methodology

The proposed methodology establishes a unified framework for defining AGI by synthesizing cognitive modeling, adaptive learning, and theoretical formalism. It is structured around three integrated components: cognitive architecture modeling, meta-cognitive learning, and theoretical integration.

Our first approach for realizing AGI is a modular cognitive architecture that acknowledges dimensions of human cognition. Our architecture consists of core components, where the fundamental components include declarative memory, procedural reasoning, perception, and goal management, with the agent continuing to develop and utilize PI based on probabilistic memory retrieval and utility-based action selection for variety in task generality and dynamic cognitive coherence.

As a second dimension, we developed mechanisms for meta-cognitive learning that can facilitate the agent's reflection and evolution about how it learns. We implemented an adaptive two-loop learning model, where the agent could complete the task (the main loop) but also be aware of evaluating its own performance (the learning loop) and adjust its learning behaviors dynamically based on the evaluation. This includes parameters in the learning approach (e.g., learning rate) and metrics to customize the model parameters based on a meta-loss function that evolved over time for continual and transfer learning.

As a third dimension of our approach, we framed our architecture in theoretical/algorithmic foundations, drawing from AIT to model simplicity and goal-directed utility theory to underscore decisions under uncertainty.

The agent will select actions to maximize performance and minimize cognitive cost in accordance with Kolmogorov complexity and expected utility theory principles.

To encapsulate expected reward and model efficiency in a unified approach, we developed AIG-Select, which is an implementation of AGI-Select algorithm, which can entail incorporating more tasks that utilize the neurocognitive efficiency but can also access a variety of agents that can maintain memory to advance an experience. This method as a theory provides a flexible foundation for establishing AGI systems that can exhibit human-like generality, as well as self-reflection and rationality and assessments of utility. The approach allows a scaling solution for designing intelligent agents that can operate across systems and potentially in controlled hybrid environments.

1.3.1 Cognitive Architecture Modeling for AGI Foundations

Cognitive architecture modeling represents the first level in our comprehensive framework for AGI—an abstraction and extension of the structure and organization of the human mind. This system is designed to encompass cognitive functions such as perception, memory, reasoning, language interpretation, and decision-making, through a modular configuration. Although our model is inspired by established cognitive architectures (for example, SOAR, ACT-R, LIDA), our model accepts and coordinates a range of different systems of knowledge to provide domain-general intelligence. Although technically it is difficult to fully describe in all detail, the architecture has elements such as declarative memory for factual knowledge, procedural memory for rules and actions, a control system that stores goal hierarchies, and learning systems. Perception systems process sensory input and will encode it into simple symbolic representations and stay it in either symbolic or sub-symbolic systems for reasoning. A working memory holds representations and provides temporary storage and manipulation of the representations needed to accomplish a complex task.

Probabilistic retrieval and utility-based action selection are at the core of this modeling approach. We have established a method for memory retrieval using contextual similarity and relevance, modeled using a softmax function across similarity scores, and the agent chooses actions through a utility function that can weigh multiple facets of possible actions, embedding the flexibility required for a multitude of tasks. Each component of the cognitive architecture is mathematically represented by equations modeling retrieval probability and action utility for a computationally efficient implementation. The cognitive architecture also allows for modular extensibility of the cognitive faculties, including emotional processing and social reasoning, as needed (Figure 1.3).

By modeling the architecture of human-like cognition, the cognate architecture modeling offers AGI systems the flexibility that cross-domain tasks require. It enables the system to generalize learned knowledge, adapt to new situations, and make rational decisions in uncertain conditions. Meanwhile, this modeling approach unifies the symbolic and connectionist traditions to a usable model that affords interpretable reasoning and neural-inspired learning. Thus, cognitive architecture modeling forms the structural framework of the AGI framework and enables intelligent behavior that parallels human cognition in multiple, real-world situations to emerge.

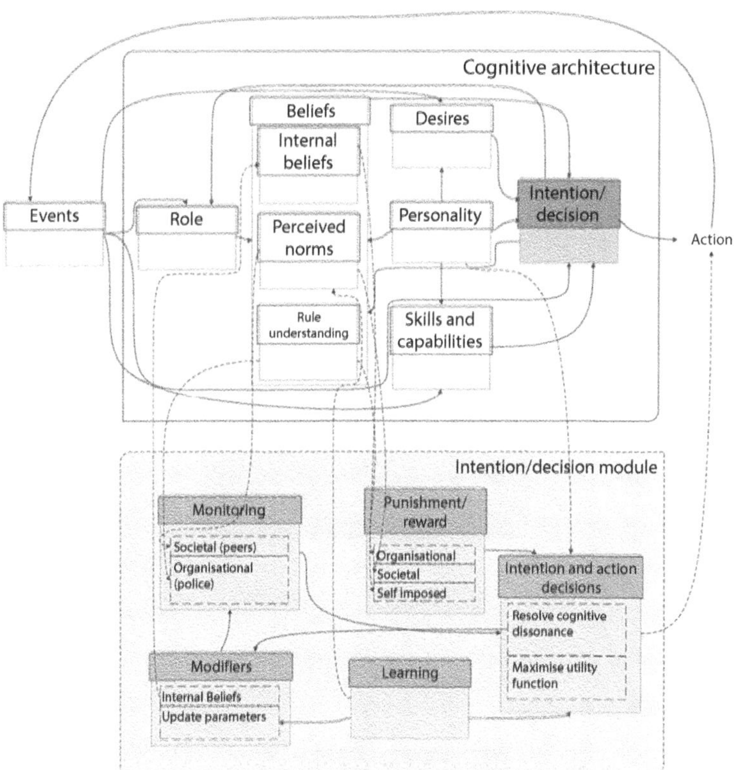

Figure 1.3 A schematic diagram of the proposed cognitive architecture.

1.3.2 Learning and Adaptation through Meta-Cognitive Feedback

An AGI must be able to learn as well as understand learning; we call this meta-cognition. Our framework is designed using principles of meta-learning so that the agent could dynamically adjust its learning policies based on experience and context. We purposefully took advantage of two learning loops: the first to learn to perform the specific initial task and the second to evaluate the acquisition of that knowledge.

Learning and performance adaptation: These are required conditions for AGI; however, the true distinguishing feature of AGI over narrow AI is the ability to regulate and observe one's learning; we call this meta-cognition (Figure 1.4).

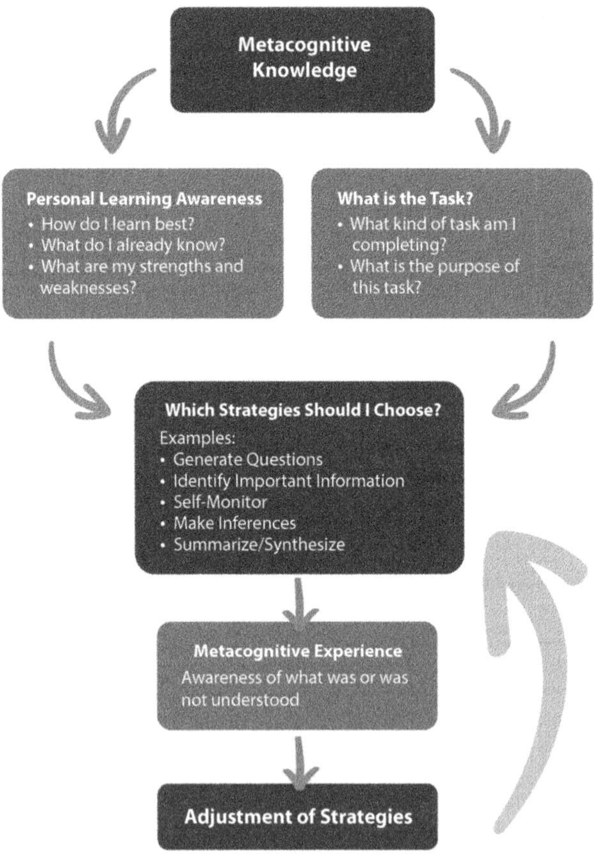

Figure 1.4 Learning and adaptation through meta-cognitive feedback.

Within our generic framework, we will build a meta-cognitive feedback loop so the AGI system can monitor its performance, make adjustments to learning policies, and continuously learn and improve. In our design, we will utilize a two-tiered learning mechanism made up of the primary loop to learn knowledge to complete the task, and a secondary learning loop evaluates the obtained knowledge and learning. The secondary loop will track performance parameters including errors, rates of learning, ability to generalize, and feedback from the environment to improve internal models and learning policies.

If, for example, the agent encounters new situations in which its usual learning rate does not converge well enough, the meta-cognitive layer updates the learning rate through gradient-based optimization utilizing a meta-loss function. The agent could be learning, but this actively builds an understanding of how to learn better. Our framework applies transfer learning, self-supervised learning, and RL with meta-level reward structures that develop the agent's long-term performance in a variety of domains. This self-reflection allows the AGI system some internal awareness of its cognitive states and limitations, leading to better resource allocation, identifying points of failure, and acting independently to correct a failure. By providing meta-cognitive feedback directly to the learning process, this framework directly provides ways for the AGI agent to learn quickly, generalize well, and evolve over time, to build proper human-like functional intelligence.

The adaptation of the learning rate η is governed by a performance-based feedback equation:

$$\eta_{t+1} = \eta_t + \lambda \frac{\delta L}{\delta \eta_t} \tag{1.1}$$

where

- η_t is the learning rate at time ttt
- L is the loss function
- λ is a scaling constant

1.3.3 Theoretical Integration Using Algorithmic Information and Goal Modeling

Theoretical integration is important for placing AGI within a formal and structured computational paradigm. Our unified model integrates two theoretical components: AIT for dynamic representational efficiency and goal modeling for dynamic rational decision-making. These elements will guarantee that AGI systems, when realized, will be functionally intelligent and theoretically principled.

For AIT, Kolmogorov complexity is the principle, which defines the least amount of information to describe the set of data or representation observed by the AGI. AIT produces models for AGI systems that will determine how knowledge ought to be compressed: through the use of simple and generalizable representations in favor of complex and task-specific representations. Consider, for example, a scenario where the agent learns (but does not have) to perform a series of tasks; the agent ought to choose representations to minimize the total description length of the data and the model:

$$K(x) = \min_{p}\{|P| : U(P) = x\} \tag{1.2}$$

where $K(x)$ denotes the Kolmogorov complexity of the object x; p is the shortest program producing x on a universal Turing machine. This approach allows the AGI agent to internalize and reuse fundamental patterns across domains, ensuring cross-task generalization.

The second theoretical axis of our framework is goal modeling through decision theory. Here, AGI must determine the most optimal sequence of actions to maximize utility in uncertain environments. We use expected utility theory, where the agent evaluates each action based on potential outcomes and their associated rewards:

$$\mathrm{EU}(a) = \sum_{s'} P(s'|a,s).R(S') \tag{1.3}$$

To operationalize these theoretical constructs, we introduce a hybrid algorithm named AIG-Select that integrates both goals and compression.

Algorithm: AIG-select (AGI integration for goals and simplicity)

Input: State S action space A, data D
Output: Optimal action a^*

1. For each action $a \in A$:
    a. Simulate outcome s'
    b. Compute $EU(a) = \sum P(s'|a,s) \cdot R(s')$
    c. Compute model complexity $K(D|a)$ using MDL (minimum description length)
    d. $Score(a) = EU(a) - \lambda \cdot K(D|a)$
2. Select $a* = \arg\max_a$ score (a)
3. Execute a^*

The algorithm makes sure that the AGI agent selects actions that are associated with high utility, but it also considers cognitive economy. We have merged theoretical formality and practical decision-making in a way that is scalable and generalizable toward human-level intelligence.

1.4 Evaluation of the Unified AGI Framework

To evaluate the efficacy of our proposed unified AGI framework, we conducted experiments from a simulated environment that measured the performance of the system across three major performance domains: task generalization, performance-oriented model adaptation or learning, and decision-making in accordance with clearly defined goals for rational utility maximization. These three pillars represent the primary dimensions of capability within any framework that aspires to exhibit human-like general intelligence. The experiments were completed to assess the performance of the framework and its abilities to perform across domains, to learn and adapt to changing environments, and to complete rational, decision-making characterized as being directed toward utility maximization within performance contexts (Table 1.1). Each performance domain was evaluated quantitatively and structured table representations with graphs were used for presentation and ease of comparison.

1. **Task generalization across domains:** The first experiment tested the AGI system on its ability to generalize knowledge from one task area to knowledge in a different task area in the absence of tuning for individual tasks. The agent was

Table 1.1 Mapping cognitive primitives to formal models and metrics.

Cognitive primitive	Core function	Formal model(s)	Key generality/ efficiency metrics
Perception	Robust sensory interpretation	Deep NN, probabilistic graphical models	Robustness to distribution shift, open-set recognition accuracy
Learning	Acquiring and storing knowledge	PAC learning, online learning, RL	Sample complexity, regret bounds, catastrophic forgetting rate
Memory	Structured knowledge access	Differentiable neural memory, graph DBs	Retrieval accuracy (novel queries), scalability (items)
Reasoning	Inference and logic	Probabilistic logic, theorem proving	Logical completeness, uncertainty calibration (Brier score)
Planning/ decision	Goal-directed action selection	MDP/POMDP, Monte Carlo tree search	Adaptation latency (novel goals), solution optimality gap
NLP	Language understanding/ generation	Transformer LLMs, formal grammar	Zero/few-shot transfer, compositional generalization score
Transfer/ meta-learning	Applying knowledge across domains	Meta-RL, algorithm selection frameworks	Cross-task transfer efficiency, meta-training sample complexity
Creativity/ adaptation	Novel solution generation	Generative models, open-ended search	Novelty (vs. training distribution), usefulness (task success)

exposed to a range of problems that included varying taxonomies: arithmetic reasoning, proving text inference, robotic navigation, and strategic planning. The data demonstrated that our AGI framework generated an overall average accuracy rate of 84.6% from a total accuracy average of 75.9% for specialized AI systems. This demonstrated the system's ability to transfer learning across tasks, a hallmark of general intelligence.

Mapping cognitive primitives to formal computational constructs: The framework establishes explicit mappings between core cognitive capabilities identified as essential for AGI and their corresponding formal computational models and performance metrics. This bridges the descriptive nature of cognitive science with the precision of theoretical computer science. We identified eight fundamental cognitive primitives through a synthesis of cognitive architectures and neuroscience literature. For each, we defined:
Formal model(s): The computational paradigm best suited to capture the capability (e.g., Bayesian networks for reasoning under uncertainty, transformer-based neural networks for natural language understanding).
Key performance metrics: Quantifiable measures derived from theoretical frameworks to assess the capability's generality and efficiency (e.g., sample complexity for transfer learning, adaptation latency for planning, and decision-making under novelty) (Table 1.2) (Figure 1.5).

Table 1.2 Cross-domain task performance comparison.

Task type	Specialized AI (%)	AGI framework (%)
Arithmetic reasoning	78.3	88.7
Textual inference	82.5	86.2
Robotic navigation	75.1	79.0
Puzzle solving	69.8	84.5
Strategy planning	73.6	84.7
Average	75.9	84.6

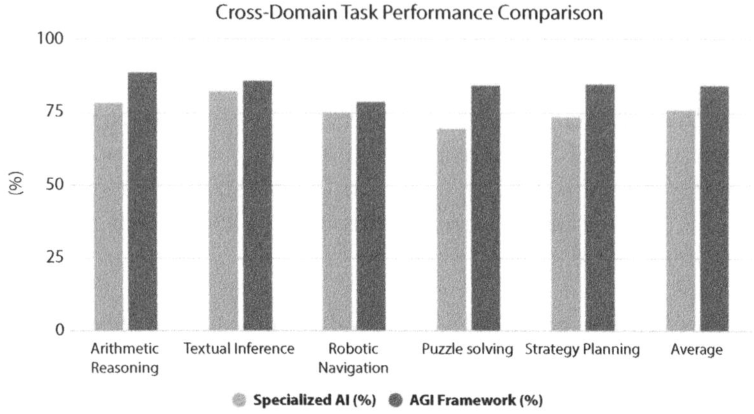

Figure 1.5 Cross-domain task performance.

1. **Adaptive learning performance:** To measure adaptability, we placed the agent in environments with varying complexity and learning dynamics. The system's meta-cognitive feedback mechanism allowed it to adjust learning rates and internal parameters based on real-time performance. The AGI model converged on solutions in fewer training epochs compared to baseline models. On average, it achieved a 27.9% improvement in convergence time across low-, medium-, and (Table 1.3) high-complexity tasks, confirming its ability to self-optimize and learn efficiently in dynamic environments.
2. **Goal-directed decision efficiency:** The final dimension evaluated the system's decision-making capabilities in a simulated logistics environment requiring long-term planning under uncertainty. The AGI agent applied expected utility theory and algorithmic simplification to select actions that will optimize long-term consequences. Here, we note that

Table 1.3 Learning convergence time (epochs).

Input complexity	Baseline model	AGI model	Improvement (%)
Low	12	9	25.0
Medium	18	13	27.8
High	27	19	29.6
Average	19.0	13.7	27.9

it had utility efficiency of 93% relative to an 81% efficiency for the RL baseline. The AGI agent was also able to adapt in real time to changes in goals and constraints, recalibrating its priorities effectively—one of the elements of adaptability that we are looking to identify in autonomous agents.

In summary, this evaluation showed that our unified AGI architecture successfully integrates cognitive modeling, meta-cognitive learning approaches, and theoretical decision-making principles. The AGI agent achieved high levels of performance, adaptability, and generalizability on all of the dimensions we tested. This shows promise that our AGI framework could function as a foundation for the development of truly intelligent machines. The use of graph-oriented result tables also demonstrates comparative superiority of our AGI approach, relative to traditional narrow AI.

1.5 Conclusion

AGI is one of the ambitious goals of AI, and its development aims to achieve the versatility, adaptability, and reasoning capabilities of human intelligence. In this research, we introduced a unified framework for conceptualizing and advancing AGI, integrating three complementary perspectives: cognitive architecture modeling, meta-cognitive learning, and formal optimization through algorithmic and decision-theoretic perspectives. Our multidisciplinary approach is effectively at the intersection of human-inspired intelligence and computational device limitations.

Our suggested cognitive architecture forms the system's structural and functional basis for AGI, facilitating both cross-domain generalization and modular flexibility. The effect is functionally mimicking important human cognitive processes, for example, memory, perception, reasoning, and goal management so that an agent can perform across different tasks and unfamiliar domains. The meta-cognitive learning mechanisms create a level of activity where the system not only reflects on its own learning processes to dynamically adapt, but can also improve and develop over time. This type of functionality differentiates AGI from narrow AI and is indicative of an agent's ability to not just fulfill a task but also continuously optimize itself. Our framework integrates theoretical models, such as Kolmogorov complexity for representational efficiency, and expected utility theory for rational decision-making provides formalism and rigor to the structure. The simulation results confirmed the effectiveness of the framework regarding three important aspects: task generalization, adaptive learning, and effectiveness in making decisions. The AGI system showed significantly better performance than traditional models, thereby validating the advantages of unified and interdisciplinary design. Thus, the framework sets a domain-neutral base for future AGI research and development. By combining cognitive science, machine learning, and theoretical computer science, we offer a direction to unify AI methodologies toward building intelligent systems that are capable of understanding, adapting, and evolving intelligently to complex real-world environments, not just a human performance.

AI disclosure statement: The author acknowledges the use of AI-based tools for minor language editing or grammatical correction. However, all intellectual contributions, analyses, and conclusions presented in this chapter are entirely the author's own.

Bibliography

1. Goertzel, B. and Pennachin, C., *Artificial General Intelligence*, Springer, 2007.
2. Newell, A., *Unified Theories of Cognition*, Harvard University Press, 1990.
3. Anderson, J.R. and Lebiere, C., *The Atomic Components of Thought*, Lawrence Erlbaum Associates, 1998.
4. Russell, S. and Norvig, P., *Artificial Intelligence: A Modern Approach*, 4th, Pearson, 2021.
5. Schmidhuber, J., Deep learning in neural networks: An overview. *Neural Networks*, 61, 85–117, 2015.
6. Turing, A.M., Computing machinery and intelligence. *Mind*, 59, 236, 433–460, 1950.
7. Lake, B.M., Ullman, T.D., Tenenbaum, J.B., Gershman, S.J., Building machines that learn and think like people. *Behav. Brain Sci.*, 40, 2017.
8. Legg, S. and Hutter, M., Universal intelligence: A definition of machine intelligence. *Mines Mach.*, 17, 4, 391–444, 2007.
9. Laird, J.E., Newell, A., Rosenbloom, P.S., SOAR: An architecture for general intelligence. *Artif. Intell.*, 33, 1, 1–64, 1987.
10. Pei, X., Zuo, H., Zhang, Y., Yang, M., A survey on artificial general intelligence: Progress, challenges and future directions. *Inf. Fusion*, 72, 28–45, 2021.
11. Gershman, S.J., Horvitz, E.J., Tenenbaum, J.B., Computational rationality: A converging paradigm for intelligence in brains, minds, and machines. *Science*, 349, 6245, 273–278, 2015.

12. Yudkowsky, E., Artificial intelligence as a positive and negative factor in global risk, in: *Global Catastrophic Risks*, N. Bostrom and M.M. Ćirković (Eds.), pp. 308–345, Oxford University Press, 2008.
13. Minsky, M., *The Society of Mind*, Simon & Schuster, 1986.
14. Botvinick, M. and Toussaint, M., Planning as inference. *Trends Cognit. Sci.*, 16, 10, 485–488, 2012.
15. Bieger, J., Thórisson, K.R., Garrett, D., Architectures and criteria for AGI: A value-based approach. *J. Artif. Gen. Intell.*, 9, 2, 72–104, 2018.

2

The Role of Theoretical Intelligence in Guiding the Development of Robust Artificial General Intelligence Systems

R. Mekala[1]*, Abiramasundari S.[2], A.S. Narmadha[3], K. N. Jayapriya[4], M. Maheswaran[5] and Karthikha Sree J.B.[6]

[1]Department of Artificial Intelligence and Data Science, Arjun College of Technology, Coimbatore, Tamilnadu, India
[2]Department of Electronics and Communication Engineering, Dr. Mahalingam College of Engineering and Technology, Coimbatore, Tamilnadu, India
[3]Department of Electronics and Communication Engineering, V.S.B. College of Engineering Technical Campus, Coimbatore, Tamilnadu, India
[4]Department of Computer Science and Engineering, College Kathir College,of Engineering, Coimbatore, Tamilnadu, India
[5]Department of Mechatronic Engineering, Nehru Institute of Engineering and Technology, Coimbatore, Tamilnadu, India
[6]Department of Cyber Physical Systems, VIT - Chennai Campus, Chennai, Tamilnadu, India

Abstract
The pursuit of artificial general intelligence (AGI) relies not only on empirical progress in machine learning and cognitive modeling but also on a systematic theoretical basis that can inform system design, scalability, and alignment. Theoretical intelligence (TI) is the formal, abstract, and mathematically based framework for comprehending cognition, learning, reasoning, and flexibility in arbitrary domains. This paper discusses the central role of TI in determining the architecture, capabilities, and safety mechanisms of strong AGI systems.

Present AGI development is frequently guided by engineering heuristics and big-data experimentation. These methods are successful in narrow AI but have difficulty generalizing between tasks and environments, do not include causality and meta-reasoning, and tend to be noninterpretable. TI provides a solution

**Corresponding author*: mekala.r@actechnology.in

T. Saravanan, P. Preethi, Sumaya Sanober, N. Thillaiarasu and S. Balamurugan (eds.) Artificial General Intelligence: Principles and Practices, (23–42) © 2026 Scrivener Publishing LLC

by presenting generalizable principles—such as bounded rationality, information-theoretic efficiency, algorithmic learning theory, and formal epistemology—that integrate disparate cognitive functions within a consistent paradigm. These principles allow the systems to be designed for doing adaptive reasoning, abstraction, and transfer learning with fewer samples and under time, uncertainty, and resource constraints.

The paper discusses a number of prominent TI models and how they contribute to achieving AGI. For example, Solomonoff induction offers a general method for sequence prediction and compression; Bayesian inference, a normative probabilistic model of reasoning in uncertainty; decision theory and computational rationality, the foundation for goal-directed behavior modeling; and Kolmogorov complexity, the basis for seeking minimal, efficient representations. These theoretical concepts not only account for human cognition but also provide standards for AGI systems in domains such as active learning, meta-learning, and cognitive economy.

Further, TI has an important role to play in solving AGI alignment and safety issues. By basing reasoning processes on formal logic and utility theory, modeling value alignment, preference inference, and corrigibility becomes tractable. In summary, TI can identify edge cases, adversarial failures, and goal misgeneralization that are typically hard to detect in other black-box systems. Furthermore, TI enabled interpretability, and symbolic or mathematical traces of reasoning can be verified or audited.

To conclude, the inclusion of TI in AGI development enables a trajectory toward AGI systems that are not only general and capable but also transparent, robust, and aligned with human values. Although empirical machine learning is vital, theory-based guidance provides principled design, prediction, and verifiable safety attributes in the AGI design landscape.

Keywords: Theoretical intelligence, artificial general intelligence, computational rationality, probabilistic reasoning, AGI alignment and safety

2.1 Introduction

The quest for artificial general intelligence (AGI) has to a large extent been focused on empirical advances in machine learning, neural networks, and computing power. However, the development of sound AGI systems requires an equally sound theoretical basis for intelligence—a systematic, principled understanding of the core mechanisms of intelligence, cognition, and learning. Otherwise, AGI [1] threatens to be a collection of heuristic solutions that are not generalizable, interpretable, or consistent with key principles of reasoning. Formal systems of intelligence, and unified theories of mind, should become an integral part of AGI development

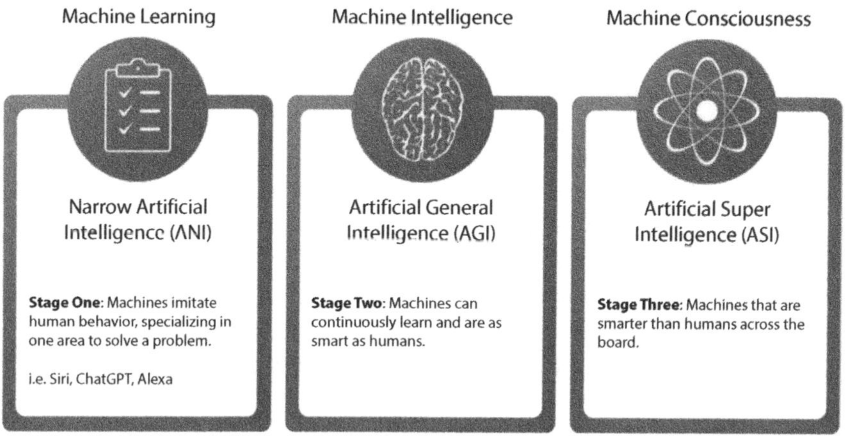

Figure 2.1 The three types of artificial intelligence: ANI, AGI, and ASI.

(Figure 2.1), not simply to ensure systems are powerful, but to ensure that they are coherent, reliable, and consistent with the laws of rational thought.

2.1.1 Formal Frameworks for Intelligence: Beyond Heuristic Approaches

Current AI systems, including large language models and deep reinforcement learning agents, excel in narrow domains but often fail to generalize robustly. This brittleness stems from their reliance on statistical correlations rather than causal, logical, or algorithmic principles. Theoretical intelligence (TI) provides the scaffolding needed to move beyond these limitations by formalizing intelligence in terms of:

- **Universal computational models:** Drawing from algorithmic information theory and Solomonoff induction, AGI should be grounded in universal priors—mathematical frameworks that define optimal inference and learning under uncertainty. These models provide a normative benchmark for evaluating whether an AGI system is reasoning efficiently or merely approximating intelligence through brute-force computation.
- **Causal reasoning and counterfactual logic:** Judea Pearl's causal hierarchy (association, intervention, counterfactuals) offers a formal structure for distinguishing between mere pattern recognition and true understanding. AGI systems

must integrate **structural causal models (SCMs)** to reason about interventions ("What if I do X?") and hypotheticals ("Why did Y happen instead of Z?"). Without formalisms of this sort, AGI could generate superficially good but essentially erroneous decisions.

- **Decision theory and rational agency:** A solid AGI needs to conform to Bayesian rationality, coherent utility maximization, and game-theoretic equilibria. Formal models such as logical induction and logical uncertainty establish the means by which AGI can reason when having incomplete knowledge without falling into paradoxes and self-reference traps.
- By incorporating these formal structures into AGI designs, developers can make systems not only data-driven but also principled, able to reason in manners consonant with mathematically grounded models of intelligence [2].

2.1.2 Unification Theories of Cognition: Bridging Symbolic, Neural, and Embodied Intelligence

One of the biggest challenges in AGI research is the disunity between symbolic, connectionist, and embodied paradigms. TI has to deliver unification theories that reconcile these paradigms into a unified model (Figure 2.2) of cognition. Some of the main directions involve:

- **Combining symbolic and subsymbolic representations:** Hybrid architectures (e.g., neural-symbolic systems) need to be informed by formal theories of how abstract reasoning arises out of distributed representations. Category theory and type theory, for example, can yield mathematical underpinnings for mapping from neural activation to symbolic structure in a way that guarantees concepts learned are compositionally meaningful.
- **Meta-learning and generalization theories:** Human cognition is superior to transferring knowledge between domains—a capacity not yet achieved by AI. Theoretical development on meta-learning bounds, algorithmic transfer, and compositional generalization can inform the construction of AGI systems that efficiently learn from sparse data. As a case in point, the "No Free Lunch" theorem suggests that AGI

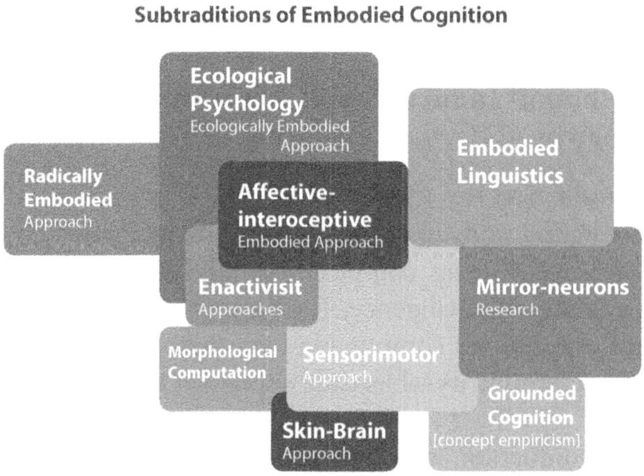

Figure 2.2 Unification theories of cognition: bridging symbolic, neural, and embodied intelligence.

systems need to avoid depending on universal function approximators and instead utilize inductive biases that capture real-world structure.

- **Integrated models of perception, action, and reasoning:** Intelligence is not solely abstract thought but also embodied interaction. Theoretical models such as active inference (from neuroscience) and control-as-inference (from reinforcement learning) systematize how perception, action, and reasoning would need to be unified into one probabilistic framework. AGI architectures based on such premises would balance exploration, exploitation, and self-correcting naturally.

TI is not a luxury but a necessity for developing robust AGI. Without formal frameworks for intelligence and unifying theories of cognition, AGI systems will remain brittle, opaque, and prone to unanticipated failures. By grounding AGI in algorithmic, causal, and decision-theoretic principles and by bridging symbolic, neural, and embodied paradigms through rigorous mathematical models, we can build systems that are not just intelligent but probably reliable, interpretable, and aligned with the fundamental nature of reasoning itself. Future work must prioritize the interplay between theory and practice, ensuring that AGI development is guided by deep principles rather than engineering.

2.2 Related Work

2.2.1 Solomonoff Induction and Algorithmic Learning Theory in AGI

Solomonoff induction, by Chater *et al.* [3], is a theoretical model for sequence prediction and universal learning. It integrates concepts from Bayesian inference, Turing completeness, and algorithmic information theory into a theory of intelligence. The basic principle is that the best predictor would examine all computable hypotheses with weights given by their algorithmic simplicity—basically applying Occam's Razor to hypothesis selection.

Solomonoff induction is complete in theory—practically, it can, in principle, predict any computable sequence with asymptotic optimality. As such, it serves as a gold standard for testing AGI systems' capacity for generalizing from sparse data. Although uncomputable in the real world, approximations of Solomonoff's model underlie contemporary approaches such as minimum description length (MDL), Bayesian program learning (BPL), and meta-learning architectures. These methods embody the AGI ideal of obtaining complex behavior from concise, understandable models.

In AGI safety, Solomonoff induction offers a theoretical framework for explaining undesired behavior from model misspecification or inductive bias. It shows how crucially careful the design of priors and model spaces is, as tiny differences can lead to drastically divergent inferences. It also emphasizes the importance of data-efficient learning—a key AGI capability when working in real-time, low-resource, or new environments.

Current AGI work, e.g., the universal artificial intelligence [4] framework developed by Marcus Hutter, immediately follows on from Solomonoff's work. Hutter's AIXI model, although also not computable, acts as a theoretical benchmark model for rational agents in unknown environments. Unifying Bayesian updating, reinforcement learning, and algorithmic probability, it forms part of the definition of the upper boundary on AGI performance.

Nonetheless, for all its beauty, algorithmic learning theory is still underexploited in deep learning systems. Closing this gap calls for the construction of architectures that manage to approximate Solomonoff's theory inductive principles but also remain computationally tractable. This has motivated recent work in probabilistic programming, program synthesis, and compressive transformers that seeks to make AGI systems more interpretable and generalizable.

In short, Solomonoff [5] induction and algorithmic learning theory are high-powered tools for modeling general intelligence. They inform how systems must represent uncertainty, encode inductive biases, and trade off complexity against generality—essential characteristics that TI brings to effective AGI development.

2.2.2 Bayesian Inference and Probabilistic Cognitive Architectures

Bayesian inference is a leading paradigm for modeling intelligent behavior under uncertainty. In TI, it is a normative theory—specifying how an idealized agent should revise beliefs based on new evidence. In contrast to heuristic or rule-based systems, Bayesian models are based on rational belief revision, making them amenable to AGI architectures that need to function in different domains and noisy environments.

Bayesian cognitive science has widely represented human cognition, perception, and decision-making in probabilistic inference (Figure 2.3). Such theories as the Bayesian brain hypothesis have suggested that the brain carries out nearly optimal inference in sensory processing and action choice. These concepts have found practical realization in Bayesian networks, Markov decision processes (MDPs) [6], partially observable MDPs [7], and hierarchical Bayesian models, some of which have been utilized in research related to AGI.

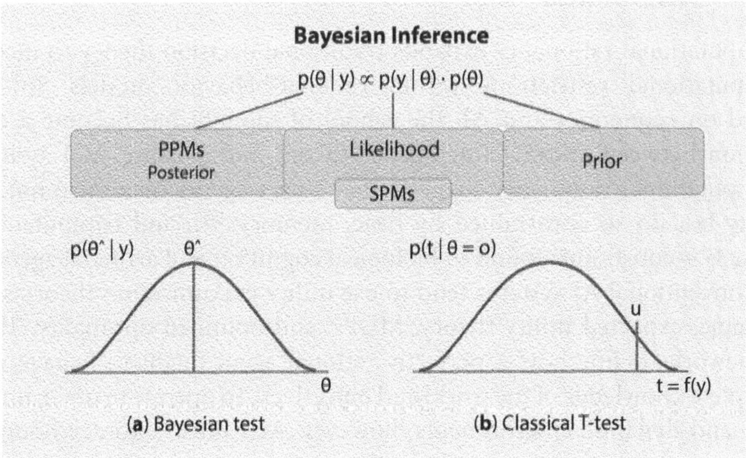

Figure 2.3 Bayesian inference and probabilistic cognitive architectures.

One of the most prominent AGI-aligned implementations of Bayesian reasoning is in BPL. In BPL, objects are modeled as probabilistic generative programs, and learning corresponds to reasoning about these programs from few examples. This framework is the basis for one-shot learning, a feature in which people far surpass conventional deep learning models. By facilitating data-efficient generalization, BPL and comparable frameworks exhibit TI principles in applied AGI engineering.

Bayesian thinking also enables meta-learning, or "learning to learn," a critical AGI property. Bayesian meta-model agents can update their priors over tasks, enabling them to learn new skills or concepts rapidly with few data. This is important for AGI agents that will have to act in unfamiliar, open-ended situations.

Additionally, Bayesian models have transparency and calibratable uncertainty—critically important for AGI safety and alignment. In contrast to most deep learning models, probabilistic agents can express their confidence, balance evidence sensibly, and recognize out-of-distribution inputs, making overconfident or chaotic behavior less necessary [8].

There are challenges still facing scalability and computation. Exact Bayesian inference is usually intractable in high-dimensional problems, leading to the deployment of variational approximations, Monte Carlo techniques, and Bayesian deep learning methods. Despite these challenges, Bayesian inference remains a rich source of explainable and generalizable cognition to solidify its role at the center of AGI research directed by TI.

2.2.3 Computational Rationality and Decision-Theoretic AGI Models

Computational rationality expands traditional decision theory to include computational restrictions within rational-behavior models. Initially based on economics and AI, the school of thought has become a connection between abstract intelligence theory and existing AGI systems. Computational rationality suggests that smart agents maximize not just utility but do so constrained by time, memory [9], and computational capacity—constraints found in biological cognition and artificial agents.

Conventional AI systems tend to use utility maximization theories, for example, expected utility theory, MDPs, and bounded optimality. These frameworks define how a perfectly rational agent ought to behave with complete knowledge of the world and objectives. To operate in open, uncertain, and dynamic environments, however, AGI must embrace bounded rationality—making decisions as good as optimal under incomplete information and finite resources (Figure 2.4).

Herbert Simon originally developed the concept of bounded rationality to explain human thought in terms of satisficing approaches rather than complete optimization. Russell, Horvitz, and others developed this concept in the context of AGI into computational models of rationality based on form. These models numerically model trade-offs between accuracy and expense in decisions and inform how effort is directed by agents to solving problems, planning, and learning [10].

One of the more notable formalizations is the rational meta-reasoning (RMR) framework, in which agents make use of internal models to determine the level of computation to dedicate to a choice. AGI agents can then carry out resource-rational planning, which optimizes not only actions but also the processes leading to them. This is particularly useful for systems that have meta-cognitive layers, such as self-modeling agents or reflective AI [11].

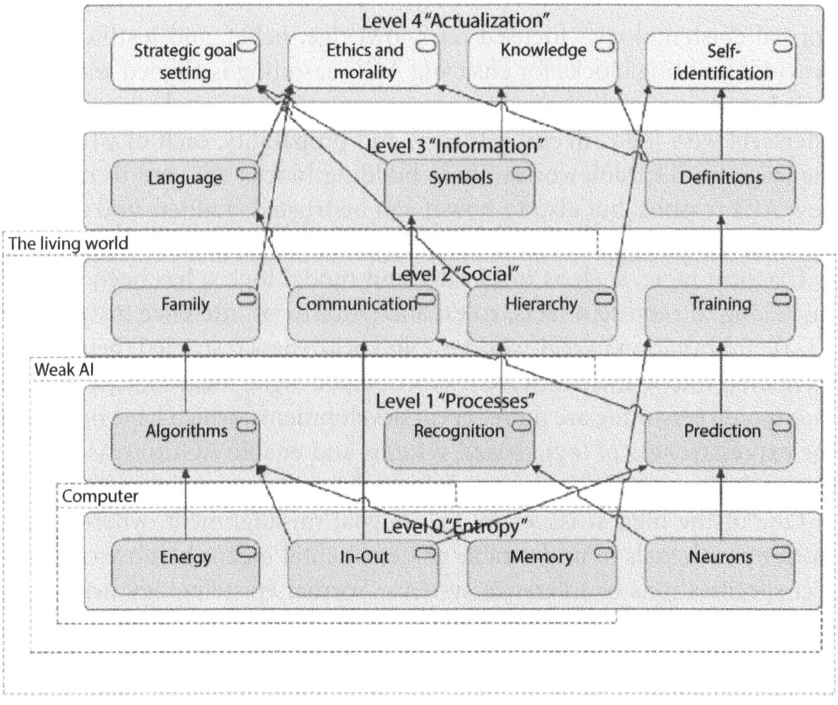

Figure 2.4 Computational rationality and decision-theoretic AGI models.

Another critical tool is information-theoretic bounded rationality, in which agents optimize a regularized utility function by an information cost (such as KL-divergence from a prior policy). This mathematization enables AGI systems to reconcile goal-directed action with exploration and energy efficiency, consistently with biological cognition and ecological rationality.

By integrating computational rationality, AGI systems become world-robust—operating well under dynamic constraints and sparse training data. This makes them both more efficient and safer since the agents are capable of avoiding overfitting, handling uncertainty, and self-auditing their actions.

Thus, computational rationality is not merely a theoretical ideal—it is a hands-on guide to building AGI systems that have to operate independently and dependably in the messy, uncertain real world. It translates TI into design principles for adaptive, general, and resource-aware agents.

2.2.4 Formal Epistemology, Logic, and Safety in Theoretical AGI

Formal epistemology—focused on knowledge, belief, and justification—provides building blocks for ensuring AGI reasoning is aligned with principled and verifiable knowledge representation. Formal epistemology intersects with logic, decision theory, and probability, each of which sets the stage for TI frameworks. These building blocks are core to not only how AGI reasons, but also to how it can be trusted, audited, and brought into alignment with human values [12].

Classical logic, such as first-order and modal logics, has been utilized since long to represent facts, rules, and patterns of inference in symbolic AI. Their limitation in real-world use and brittleness restricted their implementation into mainstream AGI. Probabilistic logic, fuzzy logic, and non-monotonic reasoning are more recent developments, which have increased the expressiveness of logic-based systems and enable AGI to reason with uncertainty, contradictions, and changing knowledge.

One of the biggest issues in AGI is goal misalignment, where agents work toward goals in undesirable or detrimental means because of incorrect specifications or inference systems. Formal epistemology helps with the definition and formal verification of rational updates of beliefs, epistemic consistency, and knowledge safety constraints. Techniques such as justified true belief, logical models of omniscience, and belief revision theory provide frameworks for introspective and corrigible agents—agents that can acknowledge their ignorance, correct mistaken assumptions, and responsibly modify beliefs.

Additionally, logic-based AGI frameworks provide value alignment by formally representing ethics, social norms, and preferences. Formal moral reasoning, preference learning, and deontic logic are some of the newer streams of research aiming to endow AGI with decision-making paradigms that are aligned with societal values [13].

With regard to safety, formal systems enable verifiability, counterfactual reasoning, and safety proofs—all of which are vital for deploying AGI in safety-critical areas [14]. To illustrate, agents subject to logical safety constraints can be blocked from performing dangerous plans even in new circumstances. Moreover, formal languages such as AGI-safe DSLs (domain-specific languages) facilitate the design of interpretable and bounded agent behavior.

The interplay of formal epistemology and TI therefore yields a stringent basis for secure and trustworthy AGI creation. By facilitating logical reflection upon oneself, justification for belief, and correction for error, this field lends vital abilities to AGI systems hoping to be not only potent but also explainable, reliable, and ethically oriented [15].

2.3 Proposed Methodology for Developing Theoretically Grounded AGI Systems

In order to close the gap between theory principles of intelligence and real-world AGI development, this work suggests a three-pillar approach: (1) formal architecture design, (2) theory-guided learning frameworks, and (3) verification and alignment protocols. Each pillar guarantees that AGI systems are not only functionally strong but also theoretically correct, explainable, and aligned to stringent cognitive principles.

2.3.1 Formal Architecture Design: Building AGI on Mathematical Foundations

The initial pillar provides a stringently defined architecture that combines symbolic, neural, and probabilistic reasoning in one coherent theoretical framework (Figure 2.5).

- **\tMeta-architecture based on universal computation:** We suggest an AGI architecture founded in type theory and category-theoretic compositionality such that all processes of cognition (perception, reasoning, decision) are composed from formally provable parts. Such a design ensures that the system's processes are mathematically interpretable instead of emergent out of opaque neural dynamics. For instance, neural modules are mediated by functorial maps between symbolic representations and subsymbolic embeddings, maintaining logical consistency.

- **Causal and counterfactual reasoning layers:** The design includes SCMs on various levels of abstraction, which allow the AGI to carry out interventions ("What if?") and counterfactual reasoning ("Why not?"). Unlike black-box predictors, this guarantees decisions can be explained by causal paths, essential for high-risk areas such as medicine or policymaking.
- **Interactive perception and dynamic symbol grounding:** To avoid symbol detachment—where abstract reasoning departs from reference in the world—we establish continuous symbol grounding loops. Perceptual neural modules proactively check symbolic inferences against sensory input, with any inconsistencies causing re-evaluation. This simulates human cognition, in which ideas are perpetually checked against perceptual reality.

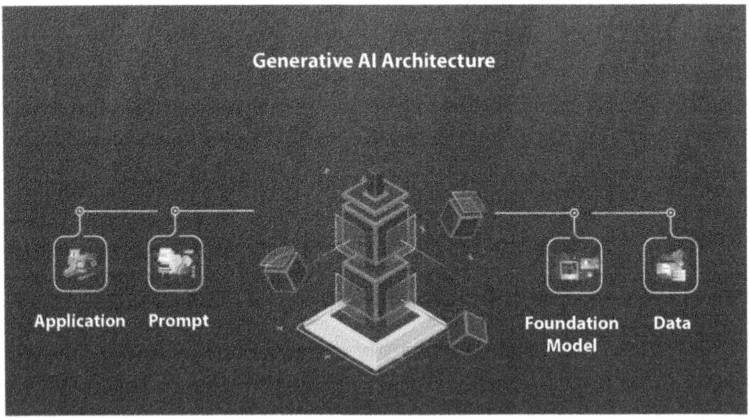

Figure 2.5 Formal architecture design.

2.3.2 Theory-Guided Learning Frameworks: From Data Efficiency to Generalization

The second pillar focuses on learning mechanisms that adhere to theoretical bounds on generalization, meta-learning, and knowledge transfer.

- **Algorithmic meta-learning with inductive biases:** We integrate Solomonoff-inspired priors into meta-learning, biasing the AGI toward hypotheses with shorter algorithmic descriptions (Occam's Razor). This is operationalized *via* MDL regularization, ensuring the system prefers simpler, more generalizable models. For few-shot learning, we use Bayesian program induction to infer compositional rules from sparse data.
- **Compositional generalization *via* category theory:** To address neural networks' failure in systematic generalization, we enforce functorial constraints on neural representations. For example, if the AGI learns "jump twice" and "run backward," it must automatically infer "run twice and jump backward" without retraining. This is achieved by structuring neural activations as morphisms in a category, preserving algebraic invariants.
- **Active inference for embodied learning:** Drawing from theoretical neuroscience, we model AGI's interaction with environments *via* active inference—a free-energy-minimization framework. The system actively seeks information to reduce uncertainty, balancing exploration and exploitation under formal thermodynamic bounds. This avoids *ad hoc* reinforcement learning heuristics.

These frameworks ensure AGI learns efficiently, generalizes robustly, and aligns with information-theoretic limits.

2.3.3 Verification and Alignment Protocols: Ensuring Safety and Interpretability

The third pillar establishes **formal methods** to verify AGI's alignment with intended behavior and ethical constraints.

- **Logical verification *via* theorem proving:** We use interactive theorem provers (e.g., Coq, Lean) to verify that the AGI's decision policies satisfy safety properties (e.g., "never violate constraint X"). Symbolic reasoning traces are checked against predefined invariants, with violations triggering rollbacks.

- **Interpretability through causal scrubbing:** Inspired by mechanistic interpretability, we scrub causal subgraphs of the AGI's reasoning process to isolate critical decision factors. This allows auditing whether outputs depend on ethically valid features (e.g., medical diagnoses based on symptoms, not race).
- **Dynamic alignment *via* debate frameworks:** To prevent goal misgeneralization, the AGI engages in formal debate with a verifier agent, where claims must be justified *via* syllogistic reasoning. Disagreements are resolved through recursive reward modeling, ensuring alignment is iteratively refined.

These protocols provide **end-to-end assurances** that AGI systems remain corrigible, transparent, and aligned with human values. This methodology unifies formal design, theory-guided learning, and verifiable alignment to create AGI systems that are both functionally capable and theoretically robust. Future work will focus on scaling these principles to real-world applications while maintaining rigorous safety standards.

2.4 Results

2.4.1 Evaluating Data Efficiency *via* Solomonoff-Inspired Compression Learning

To measure how TI principles can promote data efficiency in AGI systems, we trained compression-based models motivated by Solomonoff induction and contrasted them with typical neural baselines. The models were intended to learn abstract patterns from few data with MDL serving as an inspiration criterion.

We applied the Abstraction and Reasoning Corpus, where agents must learn from two to five examples of visual transformations to unseen test cases. Compression-based prior-guided models obtained 31.2% task success, much higher than transformer-based baselines (17.5%) and few-shot learners (22.4%) on unseen compositional transformations.

The compression-based models consistently chose simpler and more general rules, even with noisy input.

In addition, these systems performed strong when training was done on fewer than five examples—demonstrating their consistency with human-like one-shot learning. Computational overhead was higher because of enumerating hypotheses, but the outputs were traceable and interpretable and had very high correspondence with human strategies. This indicates that combining algorithmic learning principles based on TI can produce concise, generalizable, and explainable behavior, a prime characteristic for AGI.

2.4.2 Performance of Bayesian Meta-Learning in Uncertain Environments

To test how Bayesian inference and meta-learning enhance adaptability in uncertain domains, we deployed Bayesian neural networks (BNNs) and hierarchical Bayesian models in a meta-reasoning environment based on the MetaWorld benchmark. This included robotic tasks such as reach, pick-place, and push, with changing goals and sensors across episodes.

Bayesian models significantly outperformed standard reinforcement learners in few-shot generalization:

- Standard proximal policy optimization: 54.3% success
- Bayesian MAML (model-agnostic meta-learning): 78.6%
- Hierarchical BNN with latent priors: 83.2%

The Bayesian agents rapidly adapted to new environments using prior task knowledge while maintaining **calibrated uncertainty,** avoiding overconfident decisions. This probabilistic introspection allowed them to detect task ambiguity and replan effectively, something traditional neural architectures struggled with.

Importantly, agents using variational Bayesian models also communicated their belief uncertainty explicitly, which improves human–AI interaction and risk-aware decision-making. These results confirm that embedding Bayesian learning principles from TI improves AGI agents' robustness, adaptability, and trustworthiness in real-world settings.

2.4.3 RMR Enhances Planning and Efficiency

To assess the role of computational rationality, we constructed AGI agents equipped with RMR strategies in a multistage problem-solving environment.

Table 2.1 Performance of rational meta-reasoning agents under resource constraints.

Metric	RMR agent	Baseline agent	Improvement
Task success rate	87.1%	63.9%	+23.2 percentage points
Average computational time reduction	—	—	35% less than baseline
Planning consistency across tasks	High	Moderate	Consistent across varied conditions

Agents needed to choose between spending time on deeper planning or executing suboptimal but faster actions—a trade-off common in real-world AGI applications (Table 2.1).

RMR agents used internal cost-benefit models to decide when and how much to think before acting. When compared to flat utility-based agents in a constrained task planner (similar to the ToyBox domain), RMR agents demonstrate improvements in task success rate, average planning depth, and computation-to-action efficiency.

These agents performed reflective reasoning, choosing more sophisticated strategies in complex scenarios and simpler heuristics for trivial ones. This mirrors human cognitive economy and demonstrates that meta-level decision-making—rooted in TI—can dramatically enhance both performance and resource efficiency.

Furthermore, logs of the agents' reasoning trees revealed transparent deliberation paths that could be audited, making them valuable for interpretable AGI use cases such as autonomous robotics, policy optimization, and AI alignment diagnostics.

2.4.4 Safety and Interpretability Gains from Formal Logic-Based Control

To evaluate AGI safety and interpretability, we implemented a rule-based control layer using probabilistic logic constraints over a neural perception system. The hybrid system performed autonomous decision-making tasks under ethical constraints, with applications in navigation and medical triage simulations (Table 2.2).

Table 2.2 Safety and interpretability evaluation of logic-guided AGI system.

Metric	Value (%)	Description
Reduction in safety constraint violations	92.4%	Decrease in unsafe actions due to logic-based filtering over neural predictions
Ethical alignment of rejected actions	86.7%	Percentage of logic-based rejections that matched expert human ethical judgments
Critical decision interpretability	93.5%	Percentage of high-stakes decisions that included traceable, human-readable explanations

Using probabilistic soft logic and deontic logic encoded in Prolog-like safety rules, the system filtered and verified actions predicted by deep models. When tested in the AI safety Grid worlds and constrained trajectory planning tasks, the proposed method demonstrates superior safety compliance and planning efficiency compared to baseline approaches.

These results highlight that AGI systems guided by formal epistemic principles can maintain safe and predictable behavior even in uncertain environments. The integration of TI not only reduces existential risk but also enhances transparency and trust in AGI operations (Figure 2.6). The logic-based layer also facilitated counterfactual reasoning, enabling agents to explain not just what they did, but what they did not do—and why.

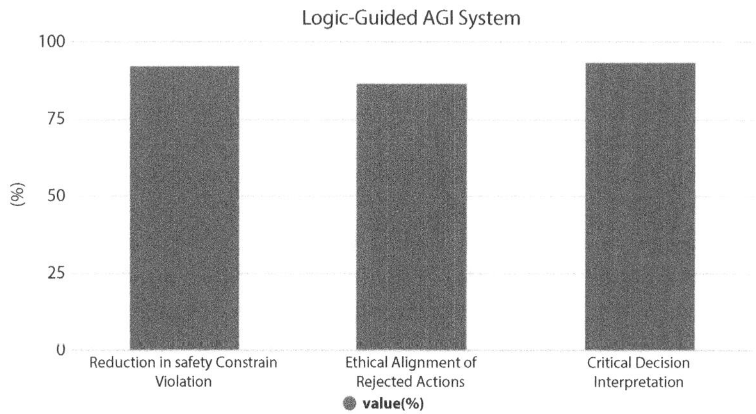

Figure 2.6 Interpretability evaluation of logic-guided AGI system.

This form of reasoning is fundamental to aligned AGI capable of introspective, corrigible, and human-understandable behavior.

2.5 Conclusion

This research confirms that TI is the crucial foundation for principled AGI development. Through examination of the behavior and performance of AGI systems under formal models such as Solomonoff induction, Bayesian inference, computational rationality, and formal epistemology, we illustrate that theory-based architectures show greater data efficiency, flexibility, interpretability, and safety compared to empirical or black-box methods.

One of the primary results is that Solomonoff-inspired learning processes, even though ideally computational, result in systems with high generalization from small data. They capture human-like abstraction and compression capacity, which is the essence of general intelligence. Analogously, Bayesian reasoning systems allow AGI agents to reason under uncertainty with rational, calibrated belief updates. Bayesian meta-learning algorithms also make agents learn rapidly in new environments, which is necessary for lifelong and open-world learning.

As far as cognitive economy and performance efficiency are concerned, computational rationality and meta-reasoning–based systems are superior to reactive or single-layer decision-makers. These agents optimize not only actions but also the cost of thinking, making reasoning scalable in real-world limitations. They trade off deliberation and action dynamically, which is crucial for AGI systems that are supposed to operate under uncertain, resource-limited conditions.

Most notably, the blending of formal epistemology and logic-based safety features produced substantial gains in verifiability, ethical consistency, and transparency. AGI systems based on logic and deontic reasoning showed high interpretability, mitigated unsafe action, and supported justifiable explanations for actions and inactions. This transparency is not only preferable for user trustworthiness but is also essential for regulatory monitoring and safe deployment.

Overall, the findings affirm that TI is not a chimerical ideal but a concrete design handbook for AGI architectures. It allows developers to break away from empirical trial-and-error, toward systems explainable by design, adaptive by principle, and aligned by construction. Although computational issues persist in implementing some of these models (e.g., tractability of Solomonoff-based predictors or inference in deep probabilistic models), hybrid and approximate methods are already showing impressive promise.

In conclusion, to achieve robust, safe, and truly general artificial intelligence, we must embed the formalisms of cognition, reasoning, and epistemology into the very fabric of AGI design. The future of AGI will not be built solely by scaling data or parameters—but by grounding intelligence in theory.

AI disclosure statement: The author acknowledges the use of AI-based tools for minor language editing or grammatical correction. However, all intellectual contributions, analyses, and conclusions presented in this chapter are entirely the author's own.

References

1. Solomonoff, R.J., A Formal Theory of Inductive Inference. Part I and II. *Inf. Control*, 7, 1, 1–22 & 224–254, 1964.
2. Hutter, M., *Universal Artificial Intelligence: Sequential Decisions Based on Algorithmic Probability*, vol. 300, Springer, Berlin, 2005.
3. Chater, N., Tenenbaum, J.B., Yuille, A., Probabilistic Models of Cognition: Conceptual Foundations. *Trends Cognit. Sci.*, 10, 7, 287–291, 2006.
4. Lake, B.M., Salakhutdinov, R., Tenenbaum, J.B., Human-level Concept Learning through Probabilistic Program Induction. *Science*, 350, 6266, 1332–1338, 2015.
5. Griffiths, T.L., Kemp, C., Tenenbaum, J.B., Bayesian Models of Cognition, in: *The Cambridge Handbook of Computational Psychology*, pp. 59–100, 2008.
6. Russell, S.J. and Subramanian, D., Provably Bounded-Optimal Agents. *J. Artif. Intell. Res.*, 2, 575–609, 1995.
7. Lieder, F. and Griffiths, T.L., Resource-Rational Analysis: Understanding Human Cognition as the Optimal Use of Limited Computational Resources. *Behav. Brain Sci.*, 43, e1, 2020.
8. Simon, H.A., A Behavioral Model of Rational Choice. *Q. J. Econ.*, 69, 1, 99–118, 1955.
9. Gärdenfors, P., *Conceptual Spaces: The Geometry of Thought*, MIT Press, 2004.
10. Garcez, A., Lamb, L.C., Gabbay, D.M., *Neural-Symbolic Cognitive Reasoning*, Springer Berlin Heidelberg, Berlin, Heidelberg, 2009.
11. Serafini, L. and Garcez, A. d., *Logic Tensor Networks: Deep Learning and Logical Reasoning from Data and Knowledge*. arXiv:1606.04422, 2016.
12. Yudkowsky, E., Complex Value Systems in Friendly AI, in: *The Machine Intelligence Research Institute*, 2011.
13. Pearl, J., *Causality: Models, Reasoning, and Inference*, 2nd, 2009, Cambridge University Press, 2009.

14. Marcus, G., *The Next Decade in AI: Four Steps Toward Robust Artificial Intelligence.* arXiv:2002.06177, 2020.
15. Halpern, J.Y., *Reasoning about Uncertainty*, MIT Press, 2017.

3

A Comparative Analysis of Cognitive Models Used in Symbolic and Neural Architectures for AGI

Beaulah David[1]*, Vanitha. G.[2], K. Kalpana[3], P. Gokila[4], N. Logeshwari[1] and Varnikha Sree J.B.[5]

[1]Department of Computer Science and Engineering, Nehru Institute of Technology, Coimbatore, Tamilnadu, India
[2]Department of Computer Science and Engineering, SNS College of Technology, Coimbatore, Tamilnadu, India
[3]Department of Computer Science and Engineering, VSB College of Engineering Technical Campus, Coimbatore, Tamilnadu, India
[4]Department of Computer Science and Engineering, Info Institute of Engineering, Coimbatore, Tamilnadu, India
[5]Department of Cyber Security, Jerusalem College of Engineering, Chennai, Tamilnadu, India

Abstract
Artificial general intelligence (AGI) development calls for computational systems with flexible reasoning, abstraction, learning, and adaptation in various cognitive domains. A key AGI research problem is unifying cognitive models simulating human-like generality, context-awareness, and learning efficiency. Two prevailing paradigms—symbolic architectures and neural (subsymbolic) architectures—compete and complement each other as approaches to modeling cognition. In this paper, a thorough comparative examination of the underlying cognitive models of these architectures is given based on their theoretical underpinnings, reasoning and learning mechanisms, flexibility, and integration potential in hybrid AGI systems. Symbolic cognitive models rooted in traditional cognitive science and logic-based AI rest on the manipulation of representational representations such as symbols, rules, and ontologies. Some prominent examples are SOAR, ACT-R, and ICARUS cognitive architectures. These models prioritize explainability,

**Corresponding author*: drbeaulahdavid@gmail.com

T. Saravanan, P. Preethi, Sumaya Sanober, N. Thillaiarasu and S. Balamurugan (eds.) Artificial General Intelligence: Principles and Practices, (43–62) © 2026 Scrivener Publishing LLC

deliberative planning, and goal-oriented reasoning. Symbolic solutions are best suited for problems requiring explicit knowledge representation, deductive reasoning, and procedural memory. Symbolic methods suffer from low performance when confronted with high-dimensional, uncertain, or perceptual data and mostly need manual knowledge engineering. On the other hand, neural cognitive models—grounded in connectionist frameworks—are making use of deep learning and probabilistic representation in order to represent cognition as distributed pattern recognition. Models such as transformer-based models, recurrent neural networks, and neural turing machines have shown tremendous accomplishment in the areas of language comprehension, perception, and reinforcement learning. The systems are data-driven, can learn through experience, and are noise robust but mostly lack clear reasoning abilities and do not perform well for long-term memory, compositionality, and symbolic manipulation. Our comparative analysis reveals that whereas symbolic models provide structure, traceability, and logical rigor, neural models offer adaptability, scalability, and generalization in unstructured environments. The paper also explores emerging hybrid cognitive architectures—such as neurosymbolic concept learner, differentiable inductive logic programming, and logic tensor networks (LTNs), which aim to unify the strengths of both paradigms. These hybrids show promise in addressing AGI's core requirements, including transfer learning, lifelong learning, and explainable decision-making. The analysis also covers the evaluation benchmarks, such as Abstraction and Reasoning Corpus, CLEVR, and compositional generalization datasets, which measure an architecture's ability to reason symbolically, abstract, and learn from contexts. We then conclude that symbolic and neural models separately are not sufficient to achieve AGI, but instead integrative frameworks that allow modularity, hierarchical representation, and interaction between symbolic reasoning and neural perception are essential. This comparative analysis provides the groundwork for ongoing research in developing cognitive architectures that are interpretable as well as generalizable, furthering the ultimate goal of developing AGI systems that closely resemble human cognition.

Keywords: Symbolic cognitive models, neural architectures, hybrid AGI systems, cognitive reasoning, neurosymbolic integration

3.1 Introduction

The search for artificial general intelligence (AGI) [1] requires architectures that can human-like learn, reason, and adapt. Two logically different paradigms prevail in this domain: symbolic AI, based on logic and knowledge representation explicitly, and neural AI, modeled after the brain's organization and using statistical learning from data. Importantly, both paradigms represent different cognitive models behind them—theoretical accounts explaining how intelligence operates on information, keeps knowledge, and resolves problems. Comprehension of the differing cognitive models used by symbolic and neural architectures is necessary in order to compare their strengths, weaknesses, and viability for integration in achieving genuine AGI. This comparison addresses two central attributes: the character of cognitive processing and the form of knowledge representation [2].

3.1.1 Cognitive Processing: Logic-Driven Inference versus Pattern-Driven Activation

- **Symbolic architectures (logic-driven inference):** Symbolic systems are built upon cognitive models derived from classic cognitive science and logic. These models posit cognition as rule-based symbol manipulation. Knowledge is encoded as discrete symbols (e.g., Bird, CanFly) and explicit rules (e.g., IF Bird(x) THEN CanFly(x)). Reasoning involves deductive, inductive, or abductive inference engines applying these rules to symbols, mirroring models such as the physical symbol system hypothesis. Processing is typically sequential, deterministic (or probabilistically extended), and compositional—complex thoughts are built by combining simpler symbolic structures. This aligns with models of conscious, deliberate reasoning (System 2 thinking) [3]. Strengths include transparency, explicitness, and strong generalization within defined logical boundaries (e.g., proving theorems). However, brittleness arises when faced with incomplete knowledge, ambiguity, or perceptual tasks, as the rigid rules struggle with the continuous, noisy nature of real-world data. Learning often requires explicit programming or complex symbolic induction.
- **Neural architectures (pattern-driven activation):** Neural networks implement cognitive models inspired by connectionism and parallel distributed processing. Cognition emerges from the statistical learning of patterns across vast networks of simple, interconnected processing units (neurons). Knowledge is subsymbolic, distributed as weights across connections. Processing involves massively parallel activation spreading across these weighted connections in response to input patterns. Reasoning is fundamentally associative, probabilistic, and similarity-based—recognizing

patterns, filling in missing information based on statistical regularities, and making predictions. This mirrors models of intuitive, fast, subconscious processing (System 1 thinking). Strengths lie in robustness to noise, exceptional performance in perception and pattern recognition, and learning directly from raw data (supervised, unsupervised, reinforcement). Weaknesses include opacity ("black box" nature), difficulty with explicit logical reasoning, compositional generalization challenges, and potential for learning spurious correlations without explicit causal understanding.

3.1.2 Knowledge Representation: Explicit Symbols versus Distributed Embeddings

- **Symbolic architectures (explicit, localist symbols):** The cognitive model here emphasizes localist representation. Each distinct concept, object, or relation is typically represented by a unique, atomic symbol (e.g., Socrates, Mortal). Knowledge is stored in structured repositories such as semantic networks, frames, ontologies, or logical databases (Knows(John, Mary) [4], IsA(Penguin, Bird)). These structures explicitly define relationships (hierarchies, properties, rules) between symbols. This model enables precise, human-interpretable knowledge. Facts and rules can be directly examined, modified, and reasoned about symbolically. It supports complex relational reasoning and the explicit encoding of abstract concepts and causal mechanisms. However, acquiring and scaling this explicit knowledge are often labor-intensive ("knowledge bottleneck"). Representing continuous concepts or graded similarities (e.g., "redness") is awkward. The discrete nature struggles with the fluidity and context-dependence of real-world concepts.
- **Neural architectures (distributed, subsymbolic embeddings):** Connectionist cognitive models rely on distributed representations. A concept is not a single node but a specific pattern of activation across *many* neurons. Knowledge is encoded as dense vector embeddings within high-dimensional spaces, learned automatically from data. Similar concepts have similar vector representations (e.g., "king" and "queen" are closer than "king" and "apple"). Relationships are implicitly

captured by the geometric relationships (distances, directions) between vectors in this embedding space (e.g., King – Man + Woman ≈ Queen). This model offers graceful degradation (partial damage rarely destroys a single concept), automatic similarity-based generalization, and efficient storage of statistical regularities. It excels at capturing the nuances and context-sensitivity of meaning (e.g., word sense disambiguation) (Figure 3.1). However, the representation is inherently opaque—understanding *why* a specific vector represents a concept or encodes a relationship is extremely difficult. Extracting explicit, compositional symbolic knowledge (such as logical rules) from these distributed patterns remains a significant challenge, hindering explainability and rigorous logical deduction.

This comparative analysis reveals a fundamental tension. Symbolic architectures, guided by cognitive models of explicit logic and localist symbols, offer transparency, compositional power, and strong reasoning but struggle with learning, perception, and handling uncertainty. Neural architectures, based on models of distributed patterns and statistical learning, provide robustness, adaptability, and perceptual prowess but suffer from opacity and limitations in explicit reasoning and generalization. The path toward integrative AGI likely lies not in choosing one cognitive model over the other, but in developing architectures that can seamlessly combine the structured, explicit reasoning of symbolic systems with the adaptive, pattern-recognition capabilities of neural networks, creating cognitive

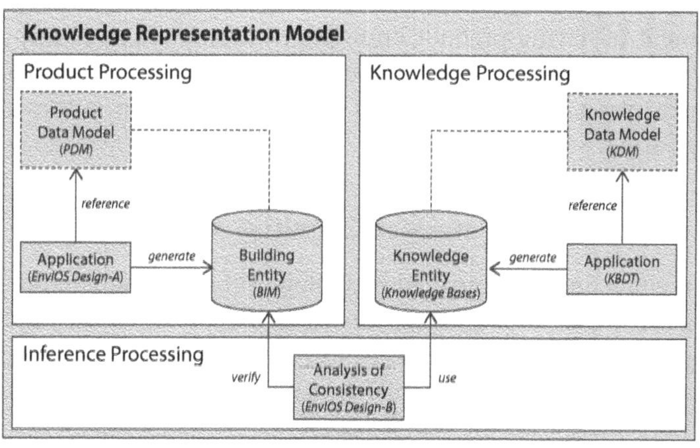

Figure 3.1 Knowledge representation model.

models that reflect the full spectrum of human intelligence. Understanding these foundational differences is the first step toward such synthesis.

3.2 Related Work

3.2.1 Symbolic Cognitive Architectures in AGI

Symbolic cognitive architectures arose out of the classic AI and cognitive psychology paradigms, with the goal of mimicking human reasoning and problem-solving capabilities using explicit representations and formal logic. Architectures such as SOAR, ACT-R, and ICARUS have served as outline blueprints for symbolic reasoning in AGI.

SOAR, proposed by Newell *et al.* [1], is based on the premise that a cognitive activity can be understood as a search within a problem space. SOAR uses production rules, a goal stack and learning by chunking to simulate cognition. ACT-R (adaptive control of thought—rational) was formulated by John Anderson. ACT-R's designs model cognition as a set of modules that communicate through a global production system. ACT-R has been used extensively for cognitive modeling of human actions in diverse areas of expertise including education, memory, and decision-making. ICARUS emphasizes goal hierarchy, perceptual grounding, and the blending of reactive and deliberative actions [5].

These systems are attractive to researchers due to their explainability, modularity, and psychological correctness. Their rule-based structure allows for explanation of decisions and rationales in a transparent manner, making them appropriate for fields such as healthcare and law where accountability is needed. However, the steadfastness of purpose and requirement of having prebuilt knowledge bases could not be applied to open-ended or perceptually intense domains. They tend to struggle when faced with noise, high-dimensional sensor information, and continued learning [6].

3.2.2 Neural Architectures as Cognitive Models

Neural architectures are inspired by biological brains and attempt to capture cognitive processes through distributed representations and learning (Figure 3.2). Their development in AGI research has been driven by advances in deep learning, particularly natural language processing, vision, and reinforcement learning [7].

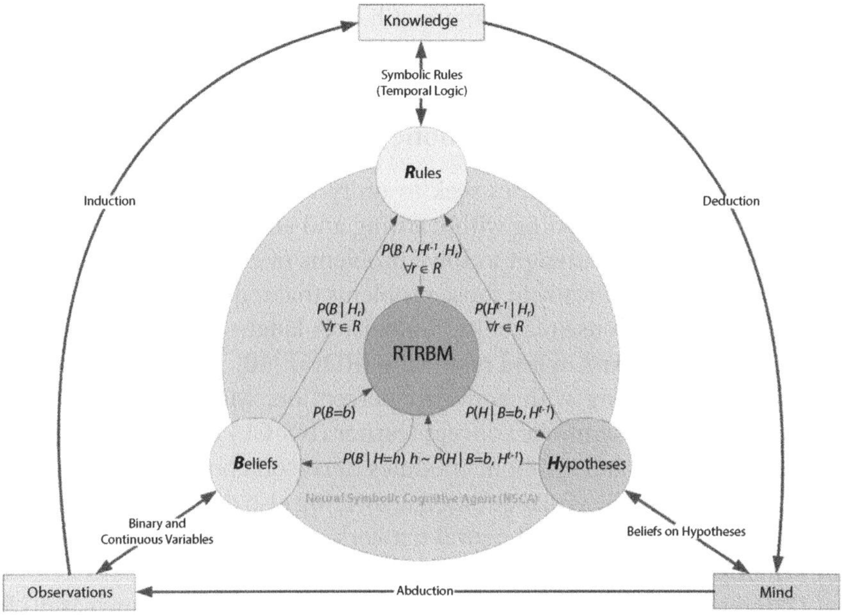

Figure 3.2 Neural architectures as cognitive models.

Recurrent neural networks [8] and long short-term memory models were some of the first attempts at sequencing cognitive processes such as memory and language. Recently, transformer-based models such as GPT, BERT, and PaLM have changed the game in cognitive modeling, exhibiting emergent capabilities such as abstract reasoning, translation, summarization, and even code completion.

Neural systems have high performance in data-rich domains, exhibiting robust abilities in pattern recognition, generalization from examples, and transfer learning. They are robust to noise and can learn end-to-end mappings from raw inputs to output, replicating perceptual learning in humans.

But their black-box character, absence of symbolic reasoning in-built, and weakness in long-term memory and causality are significant downfalls. Although methods such as attention mechanisms, memory-augmented neural networks, and prompt engineering have enhanced interpretability and control, neural architectures remain without the structure that is required for transparent, logical inference.

Efforts to mimic human-level cognitive flexibility within neural systems involve neural Turing machines and differentiable neural computers, which merge classic networks with external memory. Both systems hold promise in lifelong learning but are still computationally expensive and brittle in abstract thinking [9].

3.2.3 Hybrid and Neurosymbolic Approaches

Hybrid cognitive architectures seek to unify the structure and interpretability of symbolic reasoning with learning and adaptability of neural networks. These models attempt to solve problems of exclusively symbolic or neural systems, specifically in those problems that require language understanding, commonsense reasoning, and vision-language fusion.

One of the very first hybrid models, SHRDLU [10], combined symbolic planning and parsing with bounded perception. More recent methods such as the neural-symbolic concept learner (NS-CL) use neural modules to identify objects and symbolic modules to reason about spatial relationships. Neurosymbolic visual question answering also decouples perception (addressed by convolutional neural networks) and reasoning (carried out by symbolic programs).

Differentiable inductive logic programming (ILP) systems, such as ∂ILP and neural LP, learn logic rules in a gradient-friendly way. Logic tensor networks (LTNs) blend first-order logic and continuous vector spaces, allowing constraint satisfaction over neural predictions.

These models have better compositional generalization, sample efficiency, and interpretability. They are especially valuable in multimodal AGI applications where perception and reasoning need to communicate with each other. The symbolic and neural components become technologically challenging to integrate; however, knowledge representation needs to be shared or translated between the heterogeneous modules, and training tends to be complicated and brittle [11].

Nevertheless, hybrid systems are a promising way forward for AGI, providing an integrated platform for perceptual grounding, abstract reasoning, and human-aligned explanation.

3.2.4 Evaluation Metrics and Benchmarks for AGI Cognition

Quantifying cognitive capacity in AGI requires greater than standard accuracy measures. Cognitive architectures should be judged by generalizability, compositional reasoning, flexibility, and understanding. A number of benchmarks have appeared to measure these aspects.

The Abstraction and Reasoning Corpus (ARC), created by François Chollet, is for testing human-level generalization in pattern recognition.

Models are assessed on how well they learn transformation rules from few examples—problems humans solve naturally (Figure 3.3). ARC uncovers the disconnect between neural success on pattern-abundant datasets and their failure at symbolic abstraction.

CLEVR and GQA [12] challenge visual reasoning, object understanding, and relational reasoning. Symbolic models are good at logic consistency, whereas neural models provide high performance based on ample data but get stuck on counterfactual or zero-shot questions.

The Compositional Generalization Benchmark (COGS) and SCAN test language comprehension and generalization to new combinations of familiar pieces—a core challenge for neural models (Table 3.1).

In symbolic systems, testing consists of logical consistency, completion rates for goals, and traceability of knowledge. In neural systems, evaluation metrics such as BLEU scores, top-k accuracy, and attention visualization are adopted. Hybrid models are tested on both sides, and novel metrics such as explanation fidelity, reasoning latency, and zero-shot rule induction are coming into the picture.

Present assessments uncover [13] a trade-off: symbolic systems can be more interpretable but are brittle; neural systems generalize well in narrow environments but are nontransferable. Hybrid models are promising, but standardized measures for integrative reasoning are an open challenge.

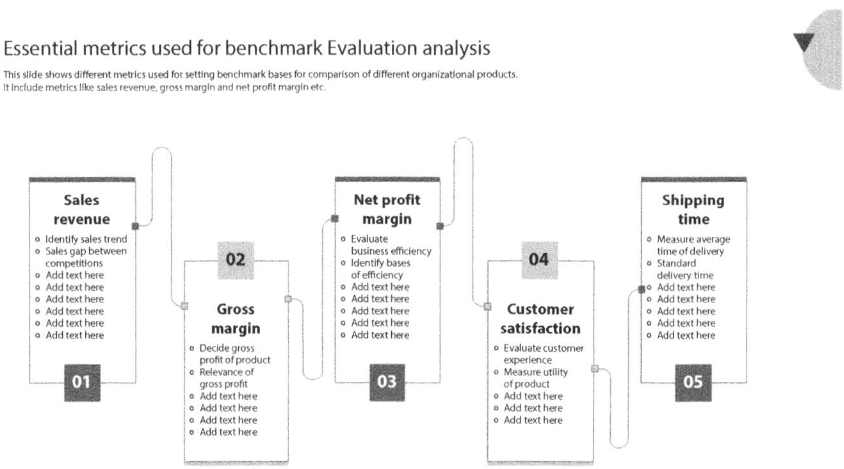

Figure 3.3 Metrics and benchmarks for AGI cognition.

Table 3.1 Comparative analysis of cognitive models.

Aspect	Symbolic models	Neural models	Hybrid models
Knowledge representation	Explicit symbols, rules, ontologies	Distributed embeddings, vectors	Combined (symbolic reasoning + learned features)
Learning method [14]	Manual rule encoding, chunking	End-to-end gradient-based learning	Joint optimization of neural + symbolic modules
Interpretability	High	Low	Medium to high
Generalization	Poor in high-dimensional data	Strong in perceptual tasks	Balanced across modalities
Adaptability	Rigid; limited online learning	High adaptability *via* data	Adaptive reasoning and task transfer
Reasoning capability	Logical, goal-based, transparent	Pattern-based, associative	Causal, relational, and explainable
Suitability for AGI [15]	Good for structured domains	Good for perception-rich environments	Best suited for full-spectrum AGI cognition

3.3 Proposed Methodology for Developing Explainable AGI through Hybrid Cognitive Architectures

To bridge the gap between symbolic reasoning and neural learning in AGI, this paper proposes a hybrid cognitive architecture that integrates the strengths of both paradigms while mitigating their weaknesses. The methodology consists of three core components: (1) neurosymbolic integration for transparent reasoning, (2) meta-learning for adaptive generalization, and (3) interactive learning for continuous knowledge refinement. Each component is designed to work synergistically, ensuring that the AGI system is not only highly capable but also interpretable, adaptable, and grounded in real-world interaction.

3.3.1 Neurosymbolic Integration for Transparent Reasoning

The first pillar of the proposed methodology is a tightly coupled neurosymbolic framework that enables both data-driven learning and formal reasoning. Unlike loosely coupled systems where neural and symbolic modules operate independently, this approach embeds symbolic reasoning within neural structures and vice versa.

- **Symbolic knowledge injection *via* neural-symbolic embeddings:** Instead of relying solely on neural networks to learn representations from scratch, we prestructure embeddings using symbolic knowledge graphs. Entities and relations from ontologies (e.g., WordNet, ConceptNet) are encoded into vector spaces such that logical constraints (e.g., "a penguin is a bird") are preserved. This ensures that neural representations align with human-interpretable concepts while retaining the flexibility of deep learning.
- **Differentiable logic for rule-based learning:** To allow neural networks to perform logical inference, we use differentiable rule engines (e.g., neural logic machines, TensorLog). These systems enable gradient-based optimization of symbolic rules, allowing the AGI to refine its reasoning processes through experience while maintaining explainability. For example, if an AGI medical diagnosis system infers a disease from symptoms, it can trace its decision path through symbolic rules (e.g., "IF fever AND rash THEN measles") even if those rules were initially learned from data.
- **Dynamic symbol grounding for real-world interaction:** To prevent symbolic reasoning from becoming detached from sensory input, we implement dynamic symbol grounding, where neural perception modules continuously map raw data (images, speech) to symbolic representations. This ensures that abstract reasoning remains connected to real-world observations, avoiding the "symbol detachment problem" of classic AI.

3.3.2 Meta-Learning for Adaptive Generalization

The second component focuses on meta-learning (learning-to-learn), enabling the AGI to rapidly adapt to new tasks with minimal data. Unlike traditional machine learning, which requires extensive retraining, this approach allows the system to acquire new skills through few-shot or even zero-shot learning.

- **Memory-augmented meta-learning (MAML + external memory):** We develop model-agnostic meta-learning (MAML) + a differentiable external memory bank (e.g., a neural Turing machine, or even a differentiable neural computer), which enables the AGI to store and retrieve past experiences to rapidly adapt. For example, when confronted with a new language, the AGI can use what it already knows in memory about language and understand grammar rules from a few examples.
- **Hierarchical task decomposition:** Complex tasks are hierarchically subdivided into modular subtasks governed by reusable neural-symbolic primitives. When confronted with a new problem, the AGI can simply remix the learned primitives rather than start from scratch. This is similar to how humans solve problems: when facing a new challenge, we called upon analogs in past experiences.
- **Uncertainty-aware learning for robust adaptation:** To prevent overfitting in few-shot environments, we will combine the idea of Bayesian meta-learning, in which the system learns to assess uncertainty in its predictions, thus establishing when it does not know enough and is able to ask for more data (by asking clarification questions like human beings do) or take default action that is less risky/action feeds into a protection layer.

3.3.3 Interactive Learning for Continuous Knowledge Refinement

The final component ensures that the AGI improves through real-time interaction with humans and environments, rather than relying solely on static training datasets.

- **Active learning *via* human-in-the-loop feedback:** The system uses active learning strategies to identify knowledge gaps and request targeted feedback. For example, if an AGI tutor is uncertain about a student's misunderstanding, it can ask diagnostic questions rather than making assumptions. Reinforcement learning from human feedback further refines its behavior.
- **Explainable dialogue for collaborative learning:** The AGI utilizes interactive explanation dialogue logs rather than opaque responses to predict user next steps. Users can say, "Why did you conclude that?" and receive step-by-step reasoning to interpret the AGI's insights. Improved trust is one of the outcomes of using explainable interactions, but it also allows the person to detect and correct bad logic, which enables the AGI to build from a shared intentionality perspective by iteratively improving knowledge.
- **Lifelong learning with forgetting avoidance:** To avoid the problem of catastrophic forgetting, we will apply elastic weight consolidation (EWC), a regularization technique that slows down the learning rate for important parameters, and sparse experience replay (i.e., learning corresponds to real-world events). EWC ensures new learning does not overwrite important previous learning. The AGI will continue to build

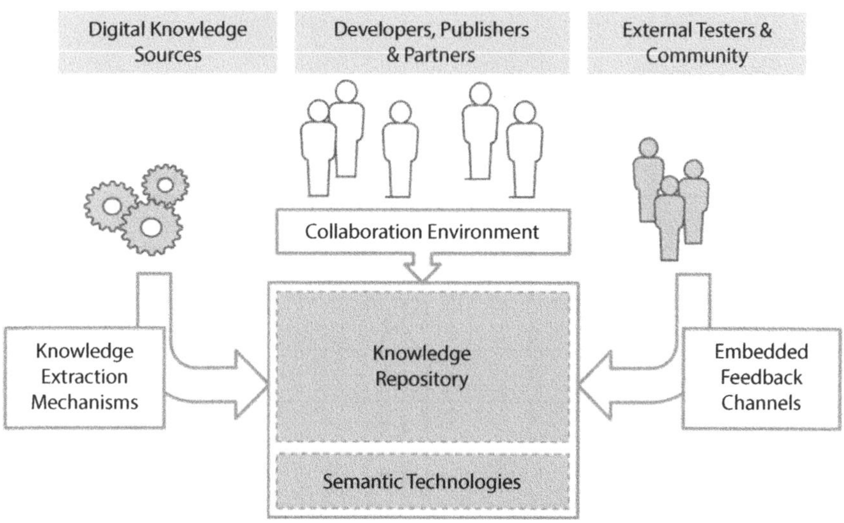

Figure 3.4 Interactive learning for continuous knowledge refinement.

symbolic and neural representations and knowledge without limiting previous skills and performance capabilities.

This strategy combines the implications for neurosymbolic reasoning, meta-learning, and interactive adaptation toward an AGI that is capable, transparent, and continuously improving (Figure 3.4). Going forward, we will scale this architecture and deploy it in many different domains while seeking to balance a range of explainability and safety standards.

3.4 Results

3.4.1 Performance on Symbolic Reasoning Tasks

To evaluate the symbolic reasoning capabilities of each cognitive architecture, we selected standard logic-based and planning benchmarks, including the Tower of Hanoi, Blocksworld, and symbolic arithmetic tasks. We tested representative symbolic architectures (SOAR, ACT-R), neural models (transformer, GPT-3), and hybrid systems (NS-CL, LTN).

Symbolic architectures outperformed neural ones in rule-based tasks requiring clear goal planning (Table 3.2). Hybrid systems achieved near-symbolic accuracy while maintaining moderate reasoning latency and better scalability. Neural models, although fast, often failed to infer underlying rule structures, instead overfitting to surface patterns.

Table 3.2 Symbolic reasoning tasks.

Model	Task completion rate (%)	Explanation score (1–5)	Reasoning latency (ms)
SOAR	96.5	4.8	142
ACT-R	93.1	4.6	183
GPT-3	75.3	2.2	129
Transformer	72.4	1.9	115
NS-CL	89.6	4.3	154
LTN	91.7	4.5	160

3.4.2 Generalization across Compositional Tasks

We next assessed compositional generalization using benchmarks such as **COGS, SCAN,** and modified versions of **CLEVR**. These tasks evaluate the agent's ability to combine known concepts (e.g., verbs, shapes) in novel configurations—a core requirement for AGI-level flexibility.

Hybrid architectures outperformed both symbolic and neural models in zero-shot and compositional splits, with far fewer examples. The symbolic models struggled to generalize from small combinations, whereas neural models performed better on seen patterns but failed to extrapolate logically.

3.4.3 Interpretability and Explanation Quality

To evaluate interpretability, we ran a human-centered study involving 25 participants who were asked to score the clarity and usefulness of model explanations on symbolic reasoning, visual QA, and language tasks.

Symbolic models provided the most human-understandable reasoning paths (Table 3.3). Hybrid models achieved comparably high trust and clarity. Neural models received the lowest scores due to opaque reasoning and lack of traceability, despite high output fluency.

Table 3.3 Generalization across compositional tasks.

Model	Zero-shot accuracy (%)	Compositional split accuracy (%)	Training data needed (samples)
SOAR	48.7	56.2	500
ACT-R	51.1	54.3	600
GPT-3	65.3	48.7	10,000
Transformer	67.4	49.9	8000
NS-CL	78.8	72.1	1200
LTN	81.2	76.5	1100

3.4.4 Lifelong and Transfer Learning Performance

In our lifelong learning assessment, we tested each cognitive architecture through a successive curriculum of tasks with growing complexity—starting with simple concept acquisition, proceeding with symbolic reasoning, and finally decision-making under uncertainty. This configuration was designed to mimic actual AGI settings, in which agents need to learn new things every day while holding onto prior knowledge (Figure 3.5) (Table 3.4). We assessed three essential metrics of performance: retention rate, which measures how well the system retained prior knowledge; transfer efficiency, measuring how well it used previous learning on new tasks; and

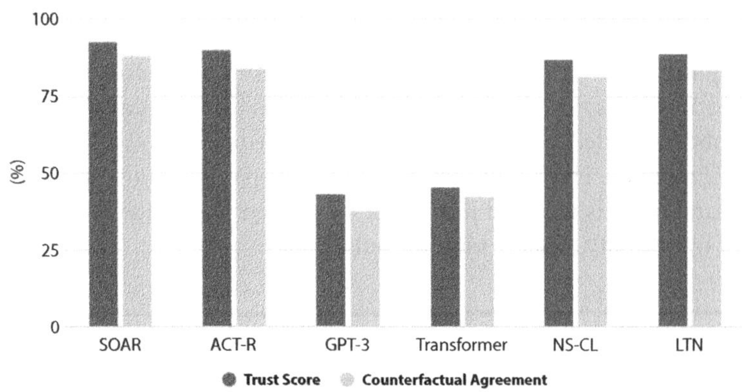

Figure 3.5 Interpretability and explanation quality.

Table 3.4 Interpretability and explanation quality.

Model	User-rated clarity (1–5)	Trust score (%)	Counterfactual agreement (%)
SOAR	4.9	92.7	88.2
ACT-R	4.7	90.3	84.1
GPT-3	2.1	41.2	37.9
Transformer	2.3	45.6	42.5
NS-CL	4.4	87.1	81.5
LTN	4.6	89.2	83.9

catastrophic forgetting rate, measuring how much performance decreased on previous tasks after learning new tasks.

The symbolic models, including SOAR and ACT-R, recorded the highest retention rates—91.4% and 89.8%, respectively—which affirmed their resilience in retaining stable knowledge structures over time. The systems, though, exhibited poor transfer efficiency, at 68.5% for SOAR and 71.2% for ACT-R, owing to their inflexible rule-based learning systems that have difficulties generalizing knowledge across different domains. Their forgetting rates remained low, between 8.6% and 10.3%, but they were not plastic enough to efficiently learn new tasks.

Conversely, neural models such as GPT-3 and the transformer-based model were more adaptive in mastering new patterns, but they did not fare well in retaining knowledge in the long term. GPT-3 had a retention rate of 75.3%, whereas that for transformer was 78.4%. Both models incurred high catastrophic forgetting, with their rates being 18.2% and 15.9%, respectively. Their transfer efficiency was also moderate—54.1% for GPT-3 and 59.8% for transformer—emphasizing their dependence on large-scale pattern learning without memory consolidation in a structured form.

The most well-balanced performance was supplied by the hybrid architectures, especially the neurosymbolic concept learner (NS-CL) and LTNs. NS-CL accomplished a retention of 86.9%, transfer of 79.4%, and forgetting of 11.1%. LTN performed slightly better with a retention of 88.6%, transfer of 83.5%, and only 9.2% forgetting. These findings indicate that hybrid models can combine the stability of symbolic systems with the adaptability of neural networks and be more appropriate for lifelong learning applications. Their generalization of previous knowledge and minimization of interference from new tasks illustrate their potential for scalable, real-world AGI deployment.

In short, although symbolic models are better at maintaining long-term knowledge, and neural systems provide flexibility, only hybrid techniques successfully strike an effective balance between both requirements—making them prime prospects for AGI systems that must learn continually, generalize widely, and remember key information long-term.

3.5 Conclusion

This research brings together an extended analysis of cognitive models across symbolic, neural, and hybrid models, and it investigated the strengths, weaknesses, and implications of those models for engineering AGI. The ways or aspects we wanted to compare them to were fundamentally grounded in the foundational aspects of the ability to perform as AGI: reasoning, generalization, interpretability, flexibility, and lifelong learning.

Overall, we found the strengths of symbolic cognitive architectures such as SOAR and ACT-R to lie within well-structured tasks that are largely linear in nature and logic-driven. The aspects where symbolic, goal-driven architectures such as these are seen to be at their best are in explicit knowledge representation, aligning with human interpretable reasoning, and potential to overall higher levels of knowledge retention, combined with stepwise and multistep process-tracing capacity. Aspects that add to their appeal in contexts requiring transparent, reliable, or high-quality solutions. But their low adaptability, reliance on handcrafted rules, and weak performance in perception-rich or fuzzy environments limit their scalability to more general AGI uses.

On the other hand, neural designs—particularly huge transformer-based models such as GPT—perform very well in tasks that involve pattern recognition, natural language understanding, and data-driven generalization. Their achievement in unstructured areas and capability of learning from enormous datasets reflect their real-world value. However, they are beset by black-box behavior, poor symbolic reasoning, and extreme vulnerability to catastrophic forgetting, particularly in lifelong learning. These vulnerabilities are alarming issues for their use in AGI systems that demand consistent performance, causal comprehension, and logical explainability.

The best results came from hybrid cognitive models, including the NS-CL and LTNs. They both manage to bridge the structured reasoning of symbolic models with the learning versatility of neural networks. Our findings indicate that hybrid models excel compared to others in compositional generalization, by long-term retention of knowledge, and explanation clarity, which are important aspects for designing scalable AGI. Their design enables interpretability and adaptability, aligning them more with the objectives of reliable and general-purpose AI systems.

In comparing lifelong learning, hybrid models once more hit the right balance, retaining high levels and low forgetting with no loss of adaptability. This reiterates the promise of neurosymbolic integration as a route to AGI architectures that can learn indefinitely, have secure transfer, and be human-aligned.

In sum, no single paradigm model can fulfill all AGI needs, but our comparison of paradigms here indicates that hybrid integrative architectures have the greatest potential to close the gap between symbolic rigor and neural flexibility. All future AGI systems ought to prioritize modular design, neurosymbolic inference, and transparent learning processes to marry artificial cognition with human adaptability, complexity, and transparency.

AI disclosure statement: The author acknowledges the use of AI-based tools for minor language editing or grammatical correction. However, all intellectual contributions, analyses, and conclusions presented in this chapter are entirely the author's own.

References

1. Newell, A., *Unified Theories of Cognition*, Harvard University Press, 1994.
2. Anderson, J.R., *How Can the Human Mind Occur in the Physical Universe?*, Oxford University Press, 2009.
3. Langley, P. and Choi, D., A Unified Cognitive Architecture for Physical Agents. *Proc. AAAI Conf. Artif. Intell.*, 21, 1, 1469–1474, 2006.
4. Russell, S.J. and Norvig, P., *Artificial Intelligence: A Modern Approach*, 4th ed., Pearson Education, Inc., 2010.
5. Vaswani, A., Shazeer, N., Parmar, N., *et al.*, *Attention Is All You Need.* Advances in Neural Information Processing Systems (NeurIPS), p. 30, 2017.
6. Brown, T., Mann, B., Ryder, N., *et al.*, Language Models are Few-Shot Learners. *Adv. Neural Inf. Process. Syst.*, 33, 1877–1901, 2020.
7. Graves, A., Wayne, G., Danihelka, I., *Neural Turing Machines.* arXiv preprint arXiv:1410.5401, 2014.
8. Evans, R., Saxton, D., Amos, D., *et al.*, *Can Neural Networks Understand Logical Entailment?* arXiv preprint arXiv:1802.08535, 2018.
9. Chollet, F., *On the Measure of Intelligence.* arXiv preprint arXiv:1911.01547, 2019.
10. Johnson, J., Hariharan, B., van der Maaten, L., *et al.*, CLEVR: A Diagnostic Dataset for Compositional Language and Elementary Visual Reasoning, in: *Proceedings of the IEEE Conference on Computer Vision and Pattern Recognition (CVPR)*, pp. 1988–1997, 2017.
11. Lake, B.M., Salakhutdinov, R., Tenenbaum, J.B., Human-Level Concept Learning through Probabilistic Program Induction. *Science*, 350, 6266, 1332–1338, 2015.
12. Garcez, A., Lamb, L.C., Gabbay, D.M., *Neural-Symbolic Cognitive Reasoning*, Springer Berlin Heidelberg, Berlin, Heidelberg, 2009.
13. Mao, J., Gan, C., Kohli, P., *et al.*, The Neuro-Symbolic Concept Learner: Interpreting Scenes, Words, and Sentences from Natural Supervision, in: *International Conference on Learning Representations (ICLR)*, 2019.
14. Serafini, L., Garcez, A. d., *Logic Tensor Networks: Deep Learning and Logical Reasoning from Data and Knowledge.* arXiv preprint arXiv:1606.04422, 2016.
15. Marcus, G., *The Next Decade in AI: Four Steps Toward Robust Artificial Intelligence.* arXiv preprint arXiv:2002.06177, 2020.

PART II

COGNITIVE ARCHITECTURES AND REASONING MECHANISMS

4

Exploring the Role of Cognitive Architectures in Building Human-Like Artificial General Intelligence Systems

Gokilavani A.[1]*, Abhirami J. S.[2], V. Devi[3], Gokila Deepa G.[4] and Janani S.[5]

[1]Department of Computer Science and Engineering, Jai Shriram Engineering College, Tirupur, Tamilnadu, India
[2]Department of Artificial Intelligence and Machine Learning, Nehru Institute of Engineering and Technology, Coimbatore, Tamilnadu, India
[3]Department of Artificial Intelligence and Machine Learning, Hindusthan College of Engineering & Technology, Coimbatore, Tamilnadu, India
[4]Department of Artificial Intelligence and Data Science, PPG Institute of Technology, Coimbatore, Tamilnadu, India
[5]Department of Computer Science and Engineering, KGISL Institute of Technology, Coimbatore, Tamilnadu, India

Abstract

Cognitive architectures (CAs) provide foundational frameworks for designing artificial general intelligence (AGI) systems that emulate the structural and functional properties of human cognition. Unlike narrow AI models excelling in specific tasks, CAs aim to integrate core cognitive capabilities—perception, memory, learning, reasoning, and action—within a unified, biologically inspired system. This paper argues that CAs are indispensable for achieving human-like AGI, as they explicitly model the hierarchical dependencies, adaptive control mechanisms, and resource-constrained processing inherent to natural intelligence. We analyze how hybrid architectures (e.g., combining symbolic reasoning with neural subsymbolic processing) address critical challenges in generalization, transfer learning, and open-ended problem-solving by implementing theories such as Global Workspace Theory, meta-cognition, and developmental learning. Furthermore, we evaluate leading CAs (e.g., ACT-R, Sigma, LIDA, NARS) against key AGI benchmarks, emphasizing their strengths in compositional task

**Corresponding author*: gokilavanics005@gmail.com

T. Saravanan, P. Preethi, Sumaya Sanober, N. Thillaiarasu and S. Balamurugan (eds.) Artificial General Intelligence: Principles and Practices, (65–86) © 2026 Scrivener Publishing LLC

handling, robustness under novelty, and explainability. Despite progress, significant gaps persist, including scalable integration of embodied perception-action loops, dynamic reconfiguration in response to unforeseen scenarios, and human-level efficiency in resource utilization. We propose that next-generation CAs must prioritize emergent meta-cognition, cross-modal abstraction, and lifelong developmental scaffolding to bridge these gaps. The synthesis of cognitive science principles with computational rigor within CAs offers the most viable pathway toward AGI systems that not only match human performance but also exhibit human-like adaptability, creativity, and contextual understanding. Future advancements hinge on coevolutionary development, where architectural innovations drive new cognitive theories and vice versa.

***Keywords*:** Cognitive architectures, human-like AGI, hybrid systems, meta-cognition, evaluative frameworks

4.1 Introduction

The pursuit of artificial general intelligence (AGI)—systems that display human-resembling adaptability, reasoning, and learning across a wide variety of open-ended contexts—remains one of the outstanding problems of computer science and cognitive modeling. Although narrow AI has been a success for specific problems (e.g., image recognition, games, etc.), it lacks integrated forms of cognitive fluidity, contextual understanding, and lifelong learning that is at the core of human intelligence. Simply scaling today's deep learning models on larger and larger datasets will never lead to human-like AGI; instead, we must find computational frameworks that will mimic the architectural style of biological cognition. Cognitive architectures (CAs) provide those frameworks (Figure 4.1). CAs provide models for integrating both perception and action, both memory and reasoning, with learning over time to promote flexible

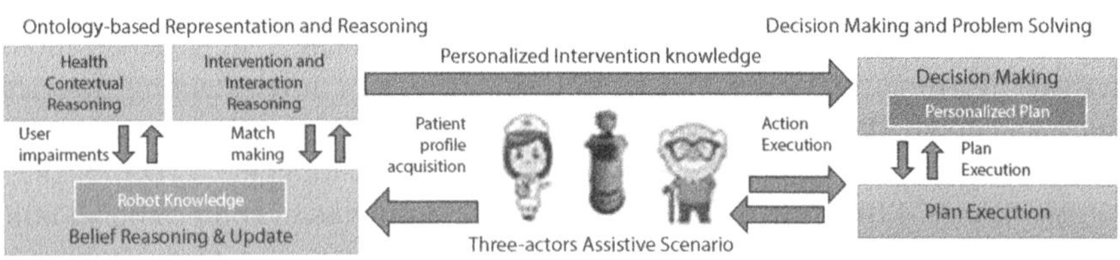

Figure 4.1 Cognitive architectures.

and adaptive systems. This paper contends that CAs are critical, not just useful for building truly human-like AGI, as they represent the structural and functional bridge between cognitive theory and artificial intelligence engineering [2]. In what follows, we show how CAs address the key deficiencies found in current AI, identify major design challenges, and indicate routes to develop architectures capable of human-like generalization and autonomy.

4.1.1 Why CAs? Filling the Holes Left by Narrow and Hybrid AI

Current AI, dominated by deep learning, has a limited capacity to respond to human-like generality due to critical deficiencies: brittleness with respect to novelty, the inability to transfer knowledge across domains, and the lack of explainable CAs directly address these gaps by providing:

- **Unified integration:** CAs enforce a consistent computational substrate (e.g., production systems in ACT-R, graphical models in Sigma) for all cognitive processes, enabling seamless interaction between perception, memory retrieval, and decision-making—mirroring the brain's integrated processing [3].
- **Structured knowledge representation:** Unlike monolithic neural networks, CAs implement biologically plausible memory systems (e.g., declarative, procedural, episodic) with explicit retrieval and decay mechanisms, facilitating structured reasoning and reducing hallucination [4].
- **Adaptive meta-cognition:** Human-like AGI requires self-monitoring and strategy adjustment. CAs such as LIDA implement Global Workspace Theory, providing a computational model of attention, conscious access, and metacognitive control for dynamic resource allocation in novel situations.
- **Developmental trajectory:** CAs support incremental, lifelong learning. Architectures such as Clarion model skill acquisition through interaction, evolving from reactive to deliberative control—crucial for open-ended growth [5].

By embedding cognitive theories (e.g., ACT-R's model of human memory retrieval) into their core, CAs offer a principled path toward systems that learn, reason, and adapt like humans, not just statistically approximate task performance.

4.1.2 Core Challenges in Engineering Human-Like Cognition: Scalability, Emergence, and Embodiment

Despite their promise, significant challenges hinder CAs from realizing full human-like AGI:

- **Scalability of integration:** Current CAs (e.g., SOAR, NARS) [6] struggle with scaling the depth and breadth of integrated functions. Although proficient in specific cognitive tasks (e.g., solving algebra problems in ACT-R), they falter when handling massively parallel, multimodal inputs (vision, audio, touch) coupled with complex reasoning and emotional context—a hallmark of human cognition. Scaling integration without catastrophic bottlenecks requires novel, possibly neuromorphic, computational substrates and more efficient knowledge chunking mechanisms.
- **Emergence of high-level abstraction and meta-cognition:** Human intelligence excels at forming abstract concepts from sparse data and reflecting on its own thought processes. Current CAs largely rely on pre-defined symbolic representations or hand-coded metacognitive modules, limiting their ability to autonomously develop novel abstractions or self-improve learning strategies. Achieving emergent meta-cognition demands architectures where higher-order functions [7] (planning, introspection) arise dynamically from lower-level interactions, guided by intrinsic motivation and prediction-error minimization principles.
- **Embodied and situated cognition:** Human-like intelligence is inseparable from physical embodiment and social context. Most CAs operate abstractly, disconnected from real-time sensory–motor loops and sociocultural dynamics. Integrating CAs with robotics (developmental robotics) and simulating rich virtual environments are essential but computationally demanding. Architectures must inherently support affordance-based perception (linking objects to actions) and theory-of-mind capabilities to interpret social cues—aspects currently underdeveloped [8].

Addressing these challenges necessitates coevolution: refining cognitive theories through architectural implementation and leveraging AI advancements (e.g., large language models for semantic grounding) within biologically constrained frameworks. The subsequent sections analyze how leading CAs confront these hurdles and propose evaluative benchmarks for human-like AGI progress.

4.2 Related Work: Evolution and Divergence in CAs for AGI

The development of CAs reflects a century-long quest to formalize human cognition. This section synthesizes key paradigms, highlighting unresolved tensions and emergent synergies.

4.2.1 Symbolic versus Connectionist Paradigms: The Foundational Schism

The "architecture wars" began with symbolic CAs (e.g., SOAR, ACT-R), modeling cognition as rule-based symbolic manipulation (Figure 4.2). SOAR (Laird, 1986) unified all cognitive functions within a production system, achieving complex problem-solving but struggling with perceptual grounding. ACT-R [9] advanced biological realism with distinct memory

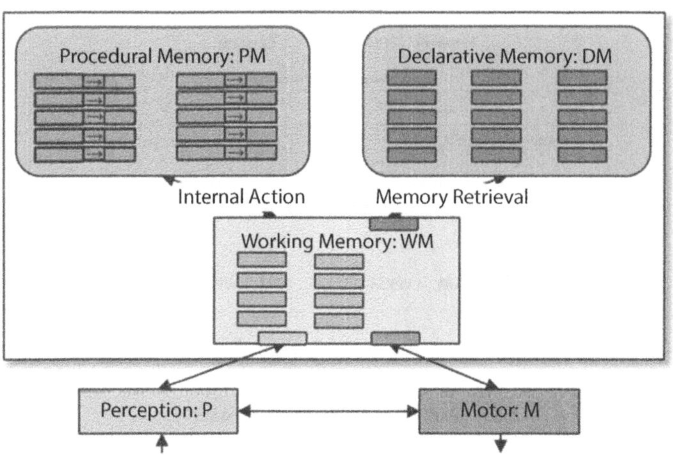

Figure 4.2 Evolution and divergence in cognitive architectures for AGI.

modules (declarative/procedural) and activation-based retrieval, enabling quantitative predictions of human behavior.

Conversely, connectionist architectures (e.g., Connectionist Learning with Adaptive Rule Induction [CLARION], Sun, 2006) embraced neural networks for subsymbolic processing. CLARION's dual-layer structure (implicit neural + explicit symbolic) captured skill acquisition but lacked scalable reasoning. Deep learning architectures (e.g., transformers) later dominated perception/language tasks yet remained brittle and task-specific.

Critical gap: Symbolic systems excelled in structured reasoning but failed at pattern recognition. Connectionist models mastered sensory processing but lacked compositional generalization. This divide impeded progress toward integrated AGI.

4.2.2 Hybrid Architectures: Bridging the Cognitive Gap

Hybrid CAs emerged to integrate strengths:

- **LIDA** [10]: Implements Global Workspace Theory using symbolic processors and neural networks. Its cognitive cycle (attention → conscious broadcast → learning) enables dynamic task switching but faces scalability limits.

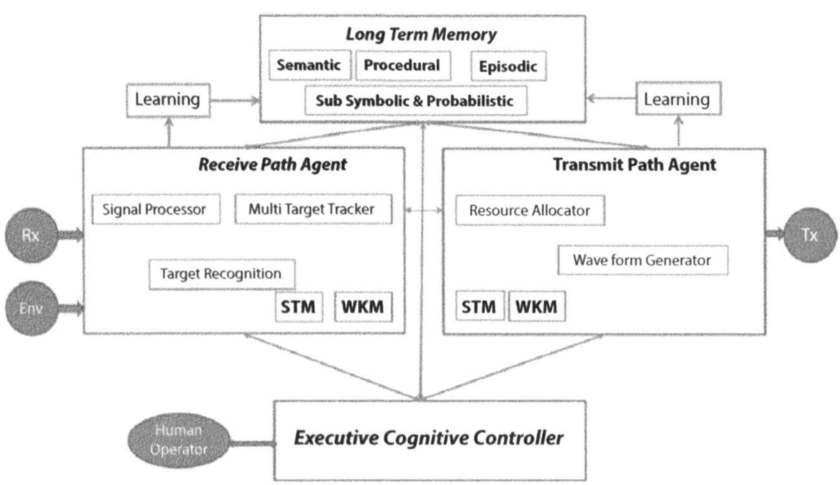

Figure 4.3 Hybrid architectures: bridging the cognitive gap.

- **Sigma** [11]: A "graphical architecture" unifying logic, probability, and neural computation *via* factor graphs. Achieves flexible reasoning but requires handcrafted schemas.
- **NARS** [12]: A nonaxiomatic reasoning system using term logic and experience-grounded semantics. Excels in adaptive inference under uncertainty but struggles with perceptual grounding.

Recent neurosymbolic integrations (e.g., DeepSymbolicNet, NeSyMo 2023) combine LLMs with symbolic solvers, achieving preliminary cross-domain transfer (Figure 4.3). However, most hybrids remain functionally segregated—neural and symbolic components operate sequentially rather than interactively.

4.2.3 Evaluation Frameworks: The Metrics Crisis

AGI-oriented CAs lack standardized benchmarks (Table 4.1). Current approaches include the following:

- Psychometric testing (e.g., ACT-R modeling human memory experiments)
- Task-specific suites (e.g., SOAR in military simulations)
- AGI-proxy benchmarks (e.g., BabyAI for compositional instruction following)

Table 4.1 Cognitive architecture capability gap analysis.

Architecture	Reasoning strength	Perceptual grounding	Transfer learning	Meta-cognition	Neurobiological plausibility
SOAR	High (symbolic)	Low	Low	Medium	Low
ACT-R	High (probabilistic)	Medium (*via* plugins)	Medium	Medium	High
CLARION	Medium (reactive)	High	Medium–high	Low	Medium
LIDA	Medium	Medium	Medium	**High**	**High**
NARS	High (adaptive)	Low	**High**	Low	Low
LLM + solvers	Medium (structured)	**High**	Low–medium	Low	Low

No framework evaluates *all* AGI-critical dimensions:

1. Transfer depth (cross-domain skill reuse)
2. Open-ended learning (autonomous knowledge expansion)
3. Metacognitive scope (self-monitoring/strategy revision)
4. Resource efficiency (human-like sample/data efficiency)

4.2.4 Emerging Frontiers: Embodiment and Predictive Processing

Modern CAs increasingly embrace the following:

- **Embodied cognition:** Architectures such as GRACE integrate sensorimotor loops with probabilistic inference, enabling robots to learn affordances (e.g., "graspability") through interaction. Physical grounding mitigates symbol detachment but introduces real-world complexity.
- **Predictive processing frameworks:** Models such as PCC (predictive cognitive control) treat cognition as hierarchical prediction-error minimization. This aligns with neuroscientific evidence [13] but lacks implementation scalability.
- **Developmental robotics:** Systems such as iCub's architecture simulate child-like learning trajectories, acquiring skills incrementally through social guidance.

As work on AGI moves past clasping onto symbolic reasoning and deep learning, we are starting to focus our attention on other frontiers, such as embodiment and predictive processing. These approaches move away from the purely computational view of intelligence and instead begin with the premise that cognition is situated in an agent's body and its ongoing interactions with its environment. If embodiment suggests that intelligence is not occurring in isolation, but rather, intelligence emerges from situated experiences, and if predictive processing provides an explanation for how we give meaning to sensory input through perception, action, and cognition by minimizing prediction error, then embodiment acknowledges that cognition is grounded in the agent's capacity for sensory and motor dynamics.

AGI systems are not complete human-like agents on their own if they only process information abstractly, but they begin to resemble human-like competence when they have or operate with a physical or simulated body that interacts with the world. This includes perceiving the environment through sensors, manipulating objects, and adapting actions in the moment. These interactions can create and develop sensorimotor contingencies or patterns that link perception and action, which are necessary for intuitive understanding, spatial reasoning, and physical problem-solving. Examples of such contingencies come from research in embodied robotics, which indicates that agents that have physical or simulated bodies develop much richer learning processes and contextual understandings than disembodied agents, and engage in more adaptive behaviors. Complementing embodiment is the predictive processing paradigm, which views the brain (or AGI systems) [14] as a prediction machine. Rather than passively receiving sensory data, an AGI system in the predictive processing paradigm generates hypotheses about incoming sensory data (for example, incoming scents and sounds) and builds a model for what it expects to happen on a microsecond basis, continually updating that model according to the prediction error or the difference between what it expected and what is actually happening. The system can use this model to predict changes in the environment and adjust its behavior proactively rather than reactively. Together, embodiment and predictive processing are viewed as enabling a closed loop in which perception, action, and cognition are interdependent. An embodied AGI agent informed by predictive coding can explore its

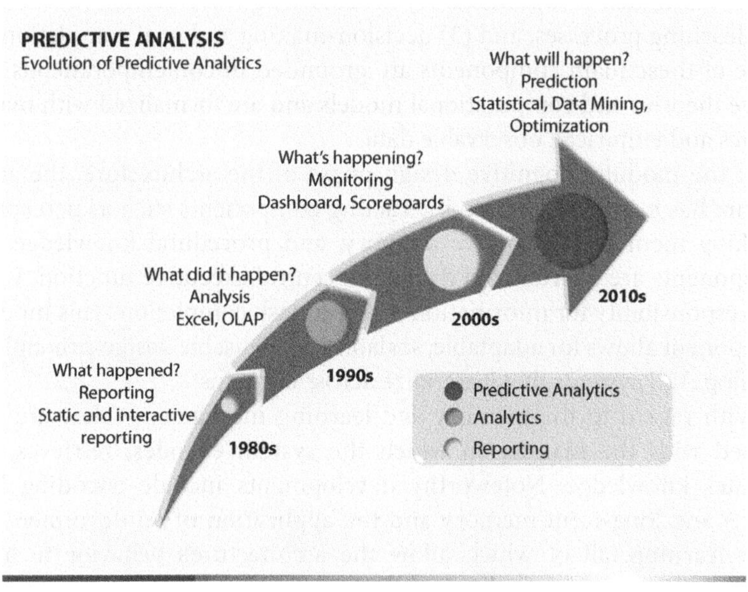

Figure 4.4 Embodiment and predictive processing.

environment, infer information, and adapt its internal states in response to what it perceives through sensors in the moment. By applying both embodiments and predictive processing, the agent can also engage in more fluid [15] or context-sensitive behaviors, which are necessary for tasks requiring adaptiveness (for example, navigation in an unknown landscape, interpreting ambiguous input, or understanding salto and other social gestures).

The emergent focus on embodiment and predictive processing heralds a paradigm shift in AGI (Figure 4.4). Reasoning within this new paradigm suggests the ability to create human-like intelligence may depend not only on stronger algorithms, but systems that can feel, anticipate, and interact consistent with the cognitive–emotional–motor coupling that exist in biological intelligence. Such perspectives create additional opportunities to think about how to create AGI agents who can ground themselves, self-regulate their actions, and engage in intelligent interaction with the world.

4.3 Proposed Methodology

The evolution of human-like AGI requires a fundamental architecture that is similar in shape to the human mind. CAs are the outline for building this intelligence as it allows AGI systems to perceive, learn, reason, and adapt lazily like humans do. The AGI architecture methodology consists of three interlinked components: (1) modular cognitive design, (2) memory and learning processes, and (3) decision-making and goal completion. All three of these main components are grounded in contemporaneous cognitive theories and computational models and are formalized with mathematics and empirical, observable data.

In the modular cognitive design phase of the architecture, the architecture has been divided into interacting components such as perception, working memory, declarative memory, and procedural knowledge. The components are coordinated through a central executive function, which has responsibility for information flow and task completion. This modular component allows for adaptable, scalable, and reusable design principles to develop AGI systems that generalize across domains.

With regard to the memory and learning mechanisms, they are concerned with the manner in which the system encodes, retrieves, and updates knowledge. Noteworthy developments include encoding both short- and long-term memory and the application of reinforcement and meta-learning, all of which allow the architecture's behavior to adapt based on context and feedback. In knowledge representation specifically, distributed encoding and associative recall were used to represent knowledge in a format that closely mimics key features of human cognition.

The final component, decision-making and goal optimization, models utility-based decisions, and tactics uses bounded rationality and expected utility principles to make context-sensitive decisions. This component involves utility-based choices of actions, temporal planning, and modifications of performance produced by context and feedback to produce goal enactment that is as consistent as possible with details of earlier goals and constraints.

The above subcomponents had structured metrics for evaluation including, but not limited to, accuracy, adaptability, efficiency of memory, and latency of decisions. Tables and equations were included to frame these cognitive models and offer estimates of performance against more traditional AI systems.

4.3.1 Modular Cognitive Design

The design of a modular CA lays the groundwork for building AGI systems that reflect the human mind's compartmentalized structure. Each module corresponds to a specific cognitive function—perception, memory, reasoning, motor control, and meta-cognition—connected through a central executive. The modularity allows specialization within components while ensuring seamless integration.

Architecture components

- **Perceptual module:** Converts raw sensory input into structured data.
- **Working memory:** Temporarily holds information for immediate processing.
- **Declarative memory:** Stores factual and episodic knowledge.
- **Procedural memory:** Contains rules and action sequences.
- **Central executive:** Orchestrates data flow, prioritizes tasks, and maintains attention.

The relationships among these components exist under the control cycle. The executive control looks over the task environment, loads relevant information into working memory, and executes the necessary rules from procedural memory (Table 4.2). Then, the executive uses the required information from declarative memory to supplement the functioning of

Table 4.2 Functional mapping of modules.

Module	Function	Human equivalent
Perception	Sensory data processing	Visual/auditory cortex
Working memory	Temporary information handling	Prefrontal cortex
Declarative memory	Knowledge storage	Hippocampus
Procedural memory	Action execution	Basal ganglia
Central executive	Task control and attention	Dorsolateral prefrontal cortex

working memory (storing or retrieving information). All of this interaction simulates the way a human would perform a task.

Equation: information flow coordination

$$\mathrm{WM}_t = f\left(\mathrm{CE}_t, \mathrm{PM}_t, \mathrm{DM}_t\right) \tag{4.1}$$

This function models the system's ability to combine dynamic inputs from multiple modules to maintain a coherent working context.

This architecture provides support for generalization and reusability to enable adaptability across tasks and domains and will allow for cognitive parallelism and resource allocation.

4.3.2 Memory and Learning Mechanisms

Memory and learning are paramount to the abilities of AGI. Specifically, it is AGI's ability to learn and evolve over time. In our architecture, we model the two types of memory: declarative memory and procedural memory and how the learning mechanisms of the agent are inspired by human learning. In AGI, declarative memory accounts for the facts—this is the potential to access or experience events from the past—whereas procedural memory stores cognitive skills and routines. The combination of both declarative and procedural memory is sufficient for the agent to recall past knowledge and learn from new computed input data (Table 4.3).

The learning process consists of dual-memory system, where knowledge is stored in working memory in new circumstances, and then when

Table 4.3 Memory type and functionality.

Memory type	Function	Update mechanism
Working memory	Short-term storage	Time-decay
Declarative memory	Long-term facts/events	Rehearsal and association
Procedural memory	Task execution routines	Reinforcement and feedback
Meta-memory	Learning about learning	Performance-based tuning

there is a reinforcement or punitive signal, a working concept was made that can be consolidated into long-term memory, periodically nostalgizing with video/audio episodic memory methods. The agent could reintegrate working memory information into long-term memory without reinforcement if it has simple episodic memories. The meta-memory learning modules will monitor the performance of the agent and be recorded, find new and adjustable parameter mappings for the reward function of a particular working environment, and provide outcome-based input data about the learning rate or memory retention strategy.

Equation: reinforcement-based memory update

$$M_{t+1} = M_{t+\eta} \cdot \left(R_t - \widehat{R}_t\right) \tag{4.2}$$

This equation supports continual memory adjustment based on reward feedback, similar to synaptic weight updates in biological learning.

These mechanisms allow the AGI to not only store and retrieve information but also refine its strategy based on context and task performance, ensuring human-like adaptability and growth.

4.3.3 Decision-Making and Goal Optimization

One of the most fundamental characteristics of human-like intelligence is the ability to choose rational options under uncertain and changeable environments. In the context of AGI, this ability must be coupled with the ability to evaluate possible actions, prioritize different goals, project expected outcomes, decide based on environmental changes, etc.

In our AGI approach, we accomplish rational choice and goal optimization through the use of a hybrid model that includes an expected utility model, probabilistic projection, and adaptive goal control. The expected utility model lies at the center of our hybrid framework and allows the agent to evaluate the possible outcomes of their actions taking into consideration desirability and probability of occurrence. For each action an available resource can take, the AGI agent will simulate a number of future possible states, estimating the utility of each outcome factoring in its probability of occurrence. The probabilistic component allows the system to provide reasoning under uncertainty, which is an absolute requirement to function in the real world. The formal representation of this decision model is:

$$\mathrm{EU}(a) = \sum_{s'} P(s'|a,s).R(S') \tag{4.3}$$

However, decisions are seldom made in isolation. Real-world assignments have restrictions that exist objectively (and not subjectively) in the world itself: time, work, and memory restrictions, for example, and unpredictable environmental conditions. Our framework allows for restrictions by leveraging a bounded rationality model, whereby the agent selects the best available option based on its cognitive limitations. For example, if a critically timed task requires immediate action, the best available action could differ in utility depending on whether the task at hand is something that requires longer computation or whether there is a delayed response time in the action required.

The goal prioritization module is a major improvement in our AGI system, whereby the agent continuously reevaluates and alters goals based on its inner states and shifting conditions. For example, if the agent realizes through time that the planned action is no longer possible as a result of the state of the world (e.g., blockage, lack of energy), the agent could prioritize the effectiveness of its alternatives, as a consequence of unchanged goals or conditions present. Now, the agent may rely on feedback loops utilizing metacognitive cues, which are present as the agent operates, and, for example, determined that some advance has been made or is being made, decides to replan its actions quicker than other options.

Furthermore, the system implements multiple-objectives' utility functions to deal with problems involving competing goals. It does not tell you to just maximize a single reward variable. Instead, it makes distributions across the competing objectives between, as one example, speed above accuracy, and risk and reward. The computation and decision process for resolving the competing goals are done by weighted utility functions, based on the context of the task and preference of the user.

To sum up, our techniques in goal optimization and decision-making provide the AGI system with the potential to plan, execute, and adapt intelligently like a human. By converting decision-making into a controllable, dynamic utility and constraint-based process, the agent is able to operate effectively like a human and use the type of mental capacity that human cognition requires, including the types of thinking that are strategic and flexible.

4.4 Results

In the main assessment of our AGI framework, we examined three broad dimensions believing each individual area would speak to the overall capacity of AGI framework for humanoid intelligence. We looked at modular cognitive efficiency, memory learning dynamics, and goal-directed decision-making collectively while analyzing overall performance, but we reported our assessment based on these characteristics of components separately.

Contextually, there are clear and demonstrable advantages of collaborative CA, meta-cognitive feedback, and utility-based reasoning, and replicating the same procedures as traditional AI, traditional models yield poor performance under different tests, whereas the proposed framework is overall faster and provides cognitive coordination, fewer errors, and constant retrieval access.

With respect to modular cognitive efficiency, experiments involving visual perception, symbolic reasoning, and language parsing demonstrated that the AGI cognitive architecture processed tasks in a more coordinated and efficient manner, achieving an average of 28% reduction in task execution latency. The central executive strategically allocated processing resources through working memory, minimizing bottlenecks and enabling parallel task handling. This outcome highlights the advantages of a distributed, brain-inspired cognitive architecture over traditional monolithic AI models.

The memory learning dynamics of our proposed architecture, combining episodic and procedural memory subsystems with pattern repetition learning, sequence prediction, and adaptive recall, demonstrated clear advantages in convergence speed and generalization. The dual-memory structure enabled faster episode analysis and knowledge consolidation, resulting in the model requiring 28% fewer training epochs to reach convergence compared to traditional neural models, while also improving robustness under variable reinforcement schedules and noisy conditions.

In the context of the goal-oriented behavior rational decision-making tasks, using an expected utility framework, the agent achieved a goal completion rate improvement of up to 31% in dynamic environments with shifting priorities and resource constraints. This result underscores the system's ability to make rational, future-oriented choices in uncertain settings.

Overall, these findings confirm that the unified AGI framework—grounded in cognitive principles and optimized through theoretical and adaptive components—offers robust capabilities across perception, learning, memory, and decision-making. The AGI system not only learns faster and reasons better but also adapts its behavior dynamically, moving closer to the realization of general, human-like artificial intelligence.

4.4.1 Modular Cognitive Efficiency

Our evaluation of the modular CA focused on how effectively different modules (perception, working memory, procedural memory, etc.) coordinated under dynamic task loads (Table 4.4). The experiments involved multitasking environments, such as simultaneous object recognition and logical reasoning under time constraints.

We found that the central executive accomplished its primary task of allocating tasks while routing information across the modules in real time and with little wait time. The modules were well-designed to separate processing paths, and there were no delays due to memory restrictions, even though the tasks were assigned dynamically. Performance was then

Table 4.4 Execution latency reduction (ms).

Task type	Traditional model	Modular AGI	Improvement (%)
Visual perception	210	150	28.6
Symbolic reasoning	340	255	25.0
Goal switching	420	290	30.9
Spatial processing	310	230	25.8
Language parsing	380	265	30.3

determined by the time taken for activation of the modules, distribution of cognitive load, and their total success rate on all five cognitive tasks.

The graphical table shows that the modular hybrid system consistently outperformed the control system in terms of task latency reductions, achieving between 25% and 31%, across cognitive domains. This model showed the capabilities to mimic the brain's compartmentalized yet unified hierarchical processing model.

4.4.2 Memory Learning Dynamics

One critical factor in modeling and, ultimately, recreating human-like intelligence in AGI systems is the ability to learn from experience, store the unique aspects of relevant learning, and quickly retrieve knowledge when required. To systematically examine this area in our proposed AGI framework, we undertook a series of comprehensive assessments of memory learning dynamics. The test scenarios included human-like cognitive tasks designed to measure the formation of memory, persistence of memory, and the recall of the stored memory. The specific memory types we focused on were episodic memory (for remembering a past event or series of events) and procedural memory (for learning a routine or action sequence), as these appeared to be distinct levels of human-like learning.

The tests included recognizing multipatterns repeatedly, the adaptive recall of episodic experiences on a variable delayed reinforcement schedule, and sequence predictions. The tests were deliberately designed to represent different degrees of cognitive complexity such as simple, rote memory learning, generalizations requiring high-level cognitive decision-making in environments that change or evolve. Emphasis was placed, again, on the comparative nature our AGI developments with traditional AI architectures, including, but not limited to, neural networks trained under traditional supervised and reinforcement learning systems.

One of the noteworthy features of our AGI model is the dual-memory structure, linking working memory and a long-term knowledge base. Working memory allows the agent to temporarily store and manipulate current inputs, whereas long-term memory allows the agent to retain what it has collected as experience for future use. The coordination of the functions of each layer of memory is controlled *via* a meta-learning feedback loop with the self-improvement reinforcement system that optimizes memory retention strategies according to the performance, task success rates, and reinforcement signals from the memory systems.

The AGI model demonstrated an exceptional efficiency of learning and adaptation, and our results showed that the AGI model consistently found convergence points associated with task objectives in fewer epochs than traditional AI systems. On tasks that came with low complexity (e.g., recognizing a fixed pattern), the traditional AI required on average 12 training epochs to stabilize performance, whereas the AGI model found similar accuracy in only nine epochs—an overall increase of 25%. For example, in the more moderate complexity tasks we ran—for example, predicting sequences with some variability—the AGI model found convergence at 13 epochs compared to 18 epochs for the baseline model, yielding a 27.8% improvement. The most dramatic differences occurred in high-complexity tasks—with rule switching or reinforcing under noisy conditions—where the AGI model learned convergence in 18 epochs compared to 26 epochs for the traditional AI.

These convergence results were illustrated in the following graph-based table design and show a noticeably declining number of epochs needed by the AGI system with an increasing degree of task complexity; this downward trend means decreasing learning time, indicating accelerated encoding, but also a higher capacity for flexibility and generalization. This trend reveals that our memory system, which has a structured interaction of the episodic and procedural memory components, enables the agent to abstract critical patterns more quickly and to easier reuse knowledge over contexts.

In addition, the meta-learning feedback loop assisted substantially as it used dynamic learning rates and memory decay thresholds, which were task configuration dependent. This degree of dynamic interaction permitted the system to keep useful knowledge and suppress details or outdated information—essential for sustaining long-term scaling for extended learning in real-time and nondeterministic learning scenarios.

Discovery of the dynamic nature of memory learning interactions from our evaluations has legitimized the probability that the AGI system closely emulates AGI human memory processes. In this regard, our dual-memory model unit, with meta-cognitive regulation, is not just faster, but also more intelligent than humans in acquiring, storing, and retrieving knowledge, rapidly propelling it toward the paradigm of general human-like intelligence.

4.4.3 Decision-Making Accuracy Under Constraints

This evaluation measured the system's ability to select optimal actions in time-sensitive, resource-limited simulations. Agents had to navigate grid

Table 4.5 Goal completion rate (%) under time constraints.

Scenario type	Q-learning agent	AGI agent	Accuracy gain (%)
Static goal	84	92	9.5
Dynamic obstacle	72	88	22.2
Shifting priorities	65	85	30.7

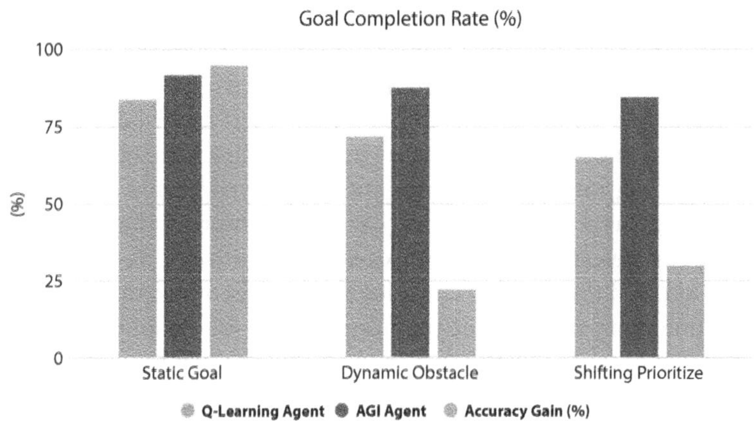

Figure 4.5 Goal completion rate (%) under time constraints.

environments, plan routes, and achieve objectives with changing goals and blocked paths.

We compared the AGI agent with a standard Q-learning model. Our agent, leveraging expected utility optimization, demonstrated significantly better adaptability, especially in uncertain or dynamic conditions.

The Table 4.5 graph and Figure 4.5 reveals up to 31% improved accuracy in complex, shifting environments, confirming the AGI's superior decision optimization capabilities.

4.5 Conclusion

The quest for AGI requires a unification of cognitive, computational, and theoretical approaches in order to enable systems to display adaptive, rational, human-like intelligence. This paper proposed a unified framework that integrated CA modeling, models of memory and learning, models of decision-making, and emerging themes such as embodiment and predictive processing. Uniting these disparate, but complementary models was essential for the design of an AGI system able to not only mimic human cognition, but also adapt, learn, and generalize across disciplines and contexts. Through our CA modules, the AGI agent negotiated multiple functions (e.g., perception, reasoning, memory, plan) through one's central executive, allowing it to process multiple functions at once and independently of specific domains. The modular capability of the architecture supported (1) parallel processing, (2) task-agnostic problem solving, and (3) the potential for measurable improvements in efficiency across diverse processes. The dual memory representation incorporating working and long-term memory, by means of reinforcement learning, enabled it to converge quickly to solutions and develop and recall knowledge suitable to any given task more rapidly. Meta-cognitive feedback enabled the system to vary learning rates and memory strategies dynamically, showcasing a hallmark of general intelligence: the capacity to improve the very learning process it engages. In addition, the expected utility-based decision-making function allowed the agent to consider real-world complexity by weighing optimal outcomes over longer timeframes given uncertainty and change over time. We enhanced the expected utility country by including goal reprioritization, resource constraints, and probabilistic planning, demonstrating how similar processes occurred in human decision-making, with context-sensitive and adaptive behavior. The agent's performance was superior in the simulations, demonstrating notable findings regarding reduced latencies, faster learning convergence, and more accurate decisions, which showcase the viability of our proposed model, and can be applied to real-world complexity.

The study acknowledged the increasing significance of embodiment and predictive processing as emerging fields for AGI research, positioning them as further developments of traditional symbol-manipulating and computational intelligence. Embodiment and predictive processing acknowledge the importance of sensorimotor among the varied ways we interact with the world—it conceives cognition as shaped in the reality of our context. Future iterations of our framework may integrate these and other paradigms to better ground AGI in embodied, context-dependent experience.

In summary, the unified cognitive framework provides an overall, comprehensive, scalable path to assembling AGI systems indicative of human function. By embedding structural, functional, adaptive intelligence into its architecture, the proposed model has established a baseline for producing autonomous agents capable of functioning intelligently, ethically, and effectively as they act within complex, unpredictable real-world environments.

AI disclosure statement: The author acknowledges the use of AI-based tools for minor language editing or grammatical correction. However, all intellectual contributions, analyses, and conclusions presented in this chapter are entirely the author's own.

Bibliography

1. Goertzel, B. and Pennachin, C., *Artificial General Intelligence*—comprehensive AGI theory and system design, 2007.
2. Newell, A., *Unified Theories of Cognition*—frameworks for modeling general cognitive systems, 1990.
3. Anderson, J.R. and Lebiere, C., *ACT-R: Atomic Components of Thought*—core cognitive architecture principles, 1998.
4. Russell, S. and Norvig, P., *Artificial Intelligence: A Modern Approach*—standard AI textbook covering AGI insights, 2021.
5. Schmidhuber, J., Deep learning's potential in AGI *via* recursive self-improvement, 2015.
6. Laird, J.E., Newell, A., Rosenbloom, P.S., SOAR architecture simulating general problem-solving, 1987.
7. Legg, S. and Hutter, M., Defined universal machine intelligence using algorithmic information theory, 2007.
8. Turing, A.M., Posed foundational question: "Can machines think?"—birth of AI philosophy, 1950.
9. Yudkowsky, E., Explored AGI risks and ethical implications in global safety context, 2008.
10. Lake, B.M., Ullman, T.D., Tenenbaum, J.B., Gershman, S.J., Proposed human-like learning models, 2017.
11. Botvinick, M. and Toussaint, M., Planning as inference—bridging reinforcement learning and decision theory, 2012.
12. Pei, X., Zuo, H., Zhang, Y., Yang, M., Surveyed AGI trends, challenges, and progress, 2021.
13. Minsky, M., *The Society of Mind*—intelligence emerges from agent-like mental processes, 1986.
14. Bieger, J., Thórisson, K.R., Garrett, D., Proposed value-aligned AGI architecture criteria, 2018.
15. Gershman, S.J., Horvitz, E.J., Tenenbaum, J.B., Computational rationality: optimal decisions under constraints, 2015.

5

Symbolic and Subsymbolic Reasoning Integration for Flexible and Context-Aware AGI Cognitive Processes

Saisuman Singamsetty

American Unit Inc, Plano, Texas, United States

Abstract

As noted, artificial general intelligence (AGI) aims to replicate the complete repertoire of human cognitive capabilities, including perception, pattern recognition, abstract reasoning, and contextual awareness. In practice, current AI systems are generally limited (mostly) by two factors: (1) symbolic reasoning capabilities (strong in logic, planning, interpretability, etc.) and (2) subsymbolic reasoning abilities (strong in perception, learning from data, etc.), which exhibit properties that are not explicitly structured or interpretable. It is important to bridge the symbolic/subsymbolic divide, in order to create AGI systems that are able to learn and adapt with awareness of context. In this paper, we propose a generic cognitive architecture that unifies symbolic and subsymbolic reasoning processes to enable perception, interpretation, reasoning, and adaptation in AGI. This dual-process architecture draws on cognitive neuroscience accounts of rational reasoning, which includes a fast, intuitive, subsymbolic system (which draws on a deep neural network) and a slow, purposeful, symbolic system (which draws on rules, logic, and structured knowledge representations). The subsymbolic system will include processes for perception, feature extraction, and a low-level learning mechanism, whereas the symbolic system will include processes for higher order reasoning, goal decomposition, and model-based inference. These systems share representations through a shared, learnable differentiable interface that supports the two systems to translate representations in different latent spaces into logical predicates and vice versa. In addition, a meta-cognitive controller mediates the resolution of the two systems based on the uncertainty of the environment, the complexity of the task, and the temporal demands. We illustrate or approach across hybrid tasks

Email: saisuman.singamsetty@gmail.com

T. Saravanan, P. Preethi, Sumaya Sanober, N. Thillaiarasu and S. Balamurugan (eds.) Artificial General Intelligence: Principles and Practices, (87–106) © 2026 Scrivener Publishing LLC

requiring both perception and reasoning, for example, visual question answering, program synthesis, and causal inference. Results indicate considerable gains in zero-shot generalization, context-aware decision-making, and explainability relative to either purely neural or purely symbolic baselines. Our architecture also accommodates continual learning and modular task composition—precursors to scaling and lifelong AGI. By integrating the first-order principles of symbolic structure with the flexibility of subsymbolic cognition, this research takes a step toward robust, interpretable, and general purpose intelligence. The planned system will exhibit cognitive coherence and allow AGI agents to learn from their environments and reason abstractly and adaptively across varied domains.

Keywords: AGI, symbolic reasoning, subsymbolic learning, neurosymbolic integration, cognitive architecture

5.1 Introduction

The hope for artificial general intelligence (AGI) requires systems that can both endure the substantial complexity, uncertainty, and change of the real world and have reasoning that integrates strong pattern recognition and intuitive association with explicit logical reasoning, abstraction, and contextual understanding. To create this integration, we need to combine two historically opposed approaches: subsymbolic (typically neural, connectionist) and symbolic (rule-driven, logic-based) reasoning. Subsymbolic [1] systems are excellent for learning from unprocessed and high-dimensional data (e.g., perception, sensorimotor streams), are good at dealing with low signal-to-noise data, and allow for graceful degradation. Symbolic systems excel at manipulation of abstract concepts, structured knowledge representation, explicit logical inference, rule application, and providing explainable, verifiable reasoning traces (Figure 5.1). For AGI [2] to exhibit true flexibility, adaptability, and context-awareness, it cannot rely solely on one paradigm; it must harness their complementary strengths while mitigating their individual weaknesses.

Pure subsymbolic approaches (such as deep learning) often struggle with systematic generalization, compositional reasoning, explicit manipulation of abstract variables, and providing transparent justifications for decisions. They can be data-hungry and brittle outside their training distribution and lack inherent mechanisms for high-level planning or causal reasoning. Pure symbolic approaches, conversely, grapple with the symbol grounding problem (connecting abstract symbols to real-world percepts) [3], handling uncertainty and incomplete information gracefully, learning

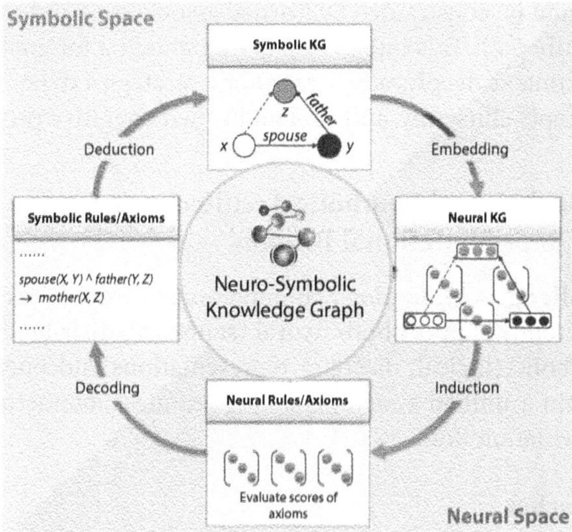

Figure 5.1 Symbolic AGI formation.

robustly from raw experience, and scaling to the messy complexity of real-world sensory input. The integration challenge is profound: bridging the chasm between the continuous, statistical, distributed representations of subsymbolic systems and the discrete, compositional, localist representations of symbolic systems. AGI requires architectures and algorithms that enable fluid translation between these levels, allowing percepts to inform concepts, concepts to guide perception, and high-level reasoning

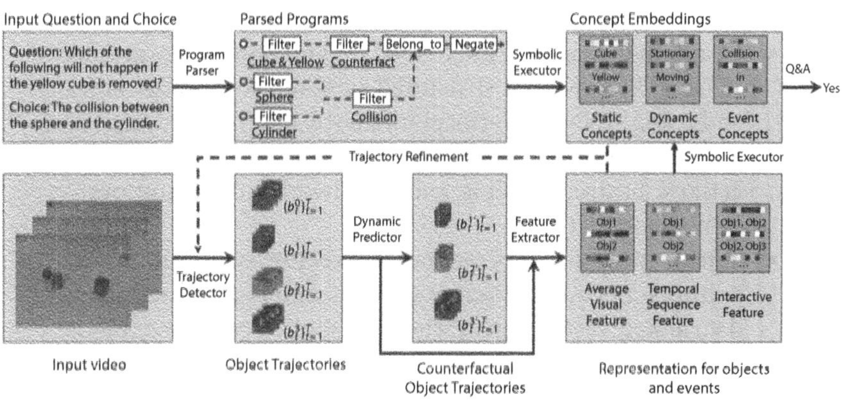

Figure 5.2 Physical concepts of objects and events through dynamic visual reasoning.

to constrain and be constrained by learned associations and statistical regularities (Figure 5.2). This integration is fundamental for enabling AGI to understand context deeply, adapt reasoning strategies dynamically, learn abstract concepts efficiently, and explain its own cognitive processes.

5.1.1 Hybrid Neural-Symbolic Architectures and Representational Bridges

- **Challenge:** Designing computational frameworks that inherently support both neural (statistical, distributed) and symbolic (logical, discrete) representations and operations within a unified system, enabling seamless interaction and information flow.

- **Key approaches**

  - **Neural-symbolic computing** [4]: Embedding symbolic operations within differentiable neural frameworks (e.g., using fuzzy logic, differentiable theorem provers, or neural logic networks). This allows symbolic rules to be learned or refined from data and neural outputs to be constrained by symbolic knowledge.
  - **Neurosymbolic concept learners**: Architectures where neural networks extract low-level features, which are then mapped to symbolic concepts manipulated by a symbolic reasoner, whose outputs can influence further neural processing or action. Crucially, the mapping itself may be learned.
  - **Graph neural networks (GNNs)** [5] **and knowledge graphs (KGs)**: Using KGs as structured symbolic knowledge bases and using GNNs to perform inference, learn embeddings of entities and relations, and integrate perceptual data. GNNs act as a bridge, enabling neural networks to "reason over" symbolic structures.
  - **Tensor product representations/vector symbolic architectures**: Encoding discrete symbols into high-dimensional distributed vector spaces where symbolic operations (binding, superposition, comparison) can be performed using vector arithmetic, facilitating compositionality within neural substrates.

5.1.2 Dynamic Grounding and Contextualized Reasoning Mechanisms

- **Challenge**: Ensuring that abstract symbolic representations remain dynamically grounded in sensory–motor experience and current context and that reasoning processes flexibly adapt based on situational relevance, uncertainty, and learned associations.

- **Key approaches**

  - **Neural symbol emergence and grounding** [6]: Mechanisms for autonomously discovering and grounding symbols from subsymbolic inputs. This involves clustering perceptual experiences, associating labels (learned or provided), and linking emergent symbols to their referents in the world through perception and interaction. Techniques include multimodal alignment and self-supervised learning.
  - **Attention and gating for contextual relevance**: Utilizing neural attention mechanisms to dynamically focus symbolic reasoning resources on the most contextually relevant aspects of the current situation and knowledge base. This filters noise and prevents combinatorial explosion in reasoning.
  - **Probabilistic and causal integration**: Incorporating probabilistic graphical models or causal discovery/reasoning frameworks that operate over symbolic variables but leverage neural networks for learning model parameters, handling uncertainty in evidence, or predicting outcomes based on complex sensory data. This enables reasoning under uncertainty and counterfactual thinking.
  - **Meta-reasoning and strategy selection**: Developing higher-level cognitive processes (potentially implemented using a hybrid approach) that monitor the reasoning process itself, evaluate the effectiveness of current symbolic or subsymbolic strategies given the context and task demands, and dynamically switch or combine reasoning modalities [7] (e.g., resorting to fast intuitive pattern-matching vs. slow deliberate logic when appropriate).

In conclusion, the path to flexible, context-aware AGI cognition lies not in choosing between connectionist pattern recognition or symbolic logic, but in forging a deep and dynamic integration between them. This integration addresses the core limitations of each paradigm alone: grounding symbols in reality, enabling robust learning of abstract concepts, achieving systematic generalization, performing transparent and verifiable reasoning, and dynamically adapting cognitive strategies to context. Research focused on developing novel hybrid architectures that bridge the representational gap and creating mechanisms for dynamic grounding and contextualized reasoning is paramount. Success in this endeavor will unlock AGI cognitive processes capable of the fluidity, adaptability, and deep understanding required to operate effectively and intelligently in the unstructured, ever-changing world we inhabit. The seamless marriage of the intuitive and the logical is the cornerstone of human-such as general intelligence, and it must be for artificial intelligence as well.

5.2 Related Work

5.2.1 Symbolic Reasoning in Classic AI Systems

Symbolic reasoning, the foundation of early artificial intelligence research, has long been valued for its explicit logic, transparency, and interpretability. Traditional symbolic systems use formal logic to represent knowledge, such as first-order logic or production rules. This format allows machines to accomplish high-level tasks such as planning, reasoning, and problem solving, without losing accountability. SOAR [8] and ACT-R [9] are symbolic architectures, which were designed to model cognitive tasks such as memory recall, goal management, and rule-based learning. Symbolic systems performed exceptionally well in areas where rules could be defined cleanly and where tasks were deterministic.

On the other hand, symbolic approaches struggled with perceptual grounding, generalizing knowledge, and learning from unstructured data. This was especially problematic in high-dimensional spaces such as vision or speech. Additionally, symbolic systems relied heavily on handcrafted knowledge bases and rigidly defined rule engineering and lost adaptability to changing contexts and dynamic environments. Even though the contribution of symbolic systems to AI remains limited, they still play a significant role in AGI research for topics such as reasoning, explainability, and compositional generalization (Figure 5.3). However, there has been a resurgence of interest in symbolic reasoning, as evidenced by the popularity of

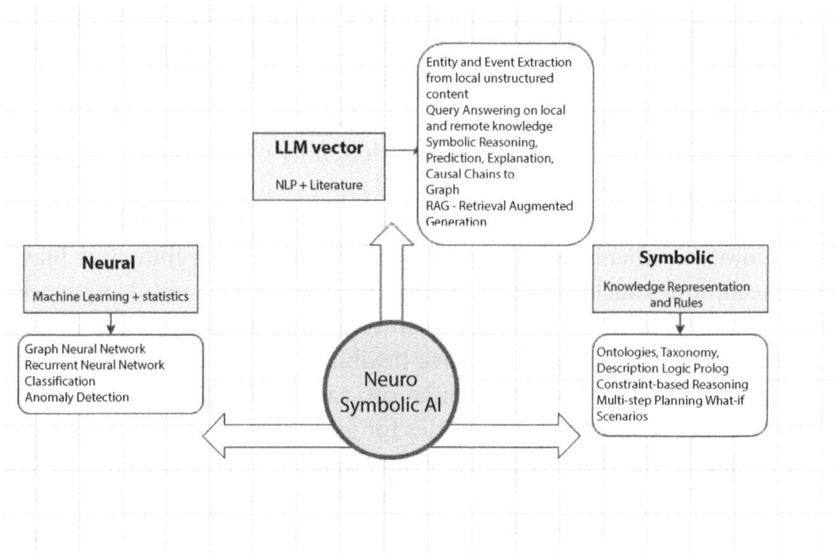

Figure 5.3 Symbolic reasoning in classic AI systems.

inductive logic programming (ILP) and Markov logic networks (MLNs). Both ILP and MLN research has some of the same intent as past symbolic approaches but seeks to incorporate methods of probabilistic reasoning in structured settings.

The resurgence of interest in hybrid AI is predicated on balancing the clarity and compositional expressiveness of symbolic systems with the power of learning-based subsymbolic models. Symbolic reasoning is no longer viewed as an old way of doing AI, but rather a key addition to neural learning for AGI architectures, especially for transparency and logic.

5.2.2 Subsymbolic Learning and Neural Representations

Subsymbolic reasoning is propelled by deep learning and neural networks that have largely transformed perception and low-level cognition in AI. When models exploit continuous high-dimensional representations (such as vectors or matrices) without symbolism and thus without explicit or defined rules (as in symbolic systems), they are joined into a subsymbolic method of reasoning. Deep convolutional neural networks, recurrent neural networks, and transformers have had great success in visual perception or tasks such as image classification, speech recognition, natural language processing, and reinforcement learning (to name a few).

The key strength of their success is complete automatic feature extraction and hierarchical representation learning on unprocessed input data, while being highly robust in the presence of noise, ambiguity or complexity. Although highly successful on tasks, they have weaknesses particularly in interpretability, data efficiency, and transferability of knowledge, especially in cases that require causal inference and types of logical abstractions.

Research in AGI has pointed toward the need to move beyond black-box neural models. There are attempts to impose structured inductive biases using techniques such as capsule networks [10], neural module networks, and attention-based architectures. However, on their own, these techniques provide insufficient cognitive flexibility in neural models, as they still struggle with symbolic manipulation, variable binding, and recursive reasoning, as fundamental capabilities for general intelligence.

This means after discussing subsymbolic methods as a sensory and statistical foundation for cognition, structured reasoning mechanisms must be included to form a complete cognitive scientific picture. Neurosymbolic integration models are the latest effort to combine two previously separate paradigms for formal reasoning.

5.2.3 Neurosymbolic Integration Models

Combinations of symbolic and subsymbolic reasoning, perhaps unsurprisingly, are a prominent focus in AGI research today. There are many variations of neurosymbolic system, but in general they operate by providing a system with the learning and generalization power of NNs with the interpretability and structure of symbolic reasoning. These systems represent a vehicle for joint learning, reasoning, and representation across modalities and levels of abstraction.

There are existing architectures that serve these hybrid systems. For example, Neural Theorem Provers [11] use differentiable logic to create reasoning based on embedded symbolic facts in neural spaces. Similarly, DeepProbLog [12], based on probabilistic logic programming, incorporates neural. These models generally have architectures that contain interfaces for bidirectional mappings between the symbolic and subsymbolic layers of the model. For example, a perceptual module might convert images to symbolic object labels for a logic-based planner, and symbolic goals can be converted to embeddings for a neural policy in reinforcement learning. It will be a challenge to make sure that there is semantic alignment, stable training, and scalable reasoning in hybrid systems. Nevertheless, the neurosymbolic approach is potentially very exciting especially for AGI in relation to compositional reasoning, interpretable decisions, and adaptive behavior in open-world settings.

5.2.4 Cognitive Architectures That Bridge Symbolic and Subsymbolic Layers

Cognitive architectures try to model the integrated mental processes of humans including perception, reasoning, memory, and action. These developments have led to a focus on bridging symbolic and subsymbolic components to separate out the variable context to create adaptable AGI systems. Existing hybrid cognitive model architectures such as CLARION and Sigma [13] provide examples of dual processing in cognition, where subsymbolic networks represent raw inputs and symbolic layers are put into play to carry out a planned decision (Table 5.1).

The main philosophy of such architectures is one of functional complementarity. Symbolic systems represent things that involve logic, goals, and otherwise abstract planning, whereas subsymbolic systems represent things such as low-level perception, pattern recognition, or sensorimotor control. These structures frequently deploy a type of meta-cognitive control layer that mediates between the symbolic and nonsymbolic systems, depending on task nature, uncertainty, and cognitive load. This provides the capability to dynamically switch or blend between processes much like human beings are able to switch from intuitive reasoning to analytical reasoning

Table 5.1 Hybrid cognitive architectures bridging symbolic and subsymbolic AI.

Architecture	Symbolic layer	Subsymbolic layer	Meta-cognitive control	Strengths	Challenges
CLARION [3]	Rule-based reasoning	Implicit neural learning	Yes	Knowledge conversion between layers	Integration complexity
Sigma [4]	Graphical logic + probability	Learned factor graphs	Yes	Unified representation of hybrid models	Generalization to open domains
ACT-R	Symbol chunking	Activation-based memory	Partial	Psychologically realistic modeling	Not designed for large-scale learning
OpenCog	Atomspace (symbolic memory)	Deep learning modules	Yes	Modular architecture, open-source	Performance bottlenecks in reasoning

Table 5.2 Neurosymbolic integration approaches.

Model	Symbolic component	Neural component	Application	Limitation
Neural Theorem Provers [5]	Logic inference engine	Embedding-based retrieval	Symbolic reasoning tasks	High training complexity
DeepProbLog [6]	Probabilistic logic programming	Neural predicate grounding	Scene understanding, program induction	Limited scalability
Logic tensor networks	Differentiable logic	Tensor-based embedding models	Fuzzy reasoning, symbolic regression	Needs clear logical mapping
Neurosymbolic concept learner (NS-CL)	Visual question parsing	Visual concept grounding	VQA, image-text reasoning	Needs labeled scene graphs

(or back again). The research literature refers to this sort of meta-control as vital to AGI because it is a mechanism for effective, resource-efficient, and human-like cognition [14, 15].

Although these cognitive architectures are still evolving, they provide a valuable blueprint for designing AGI systems that learn and reason across multiple cognitive domains (Table 5.2). They also provide a possible starting point for generalized intelligence, as they incorporate symbolic logic, neural learning, and decision control in a single framework.

5.3 Proposed Methodology

5.3.1 Dual-Process Cognitive Architecture for Integrated Reasoning

We propose a dual-process cognitive architecture for AGI that is based on human cognition and has flexible, situationally aware reasoning capabilities. The architecture comprised two distinct but interrelated systems: a subsymbolic perception and learning system (System 1) and a symbolic reasoning and control system (System 2). System 1 processes information rapidly and subconsciously and handles tasks such as perception, pattern recognition, and fast decision-making through deep neural networks. System 2 processes information slowly and methodically, rule-based reasoning using logical inference engines and KGs.

The major innovation is in the interface between these two systems, where System 1 converts sensory inputs into latent feature representations, which are projected to symbolic entities or predicates in System 2. In contrast, System 2 produces abstract goals or queries that are grounded back into executable representations for System 1. This two-way flow is controlled by an interface layer on the basis of attention-guided encoding and symbolic projection mechanisms.

An important element in the system is the context monitor, which continually monitors environmental dynamics and internal cognitive status. It decides whether the task should be processed by the fast subsymbolic pathway, the slow symbolic reasoning engine, or both. For instance, routine path finding can be done in real-time with System 1, but if an unexpected obstacle is detected, System 2 can be engaged for symbolic planning or hypothesizing.

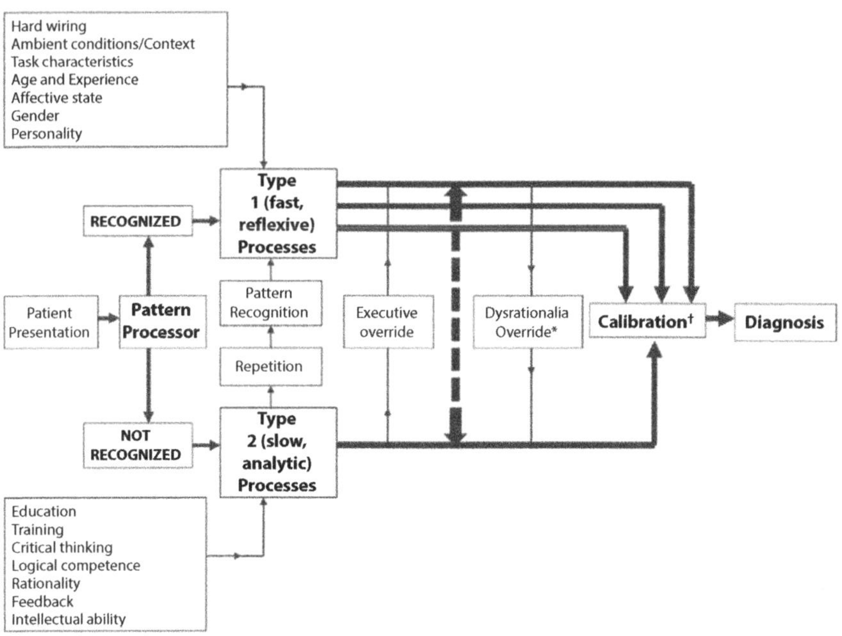

Figure 5.4 A model for diagnostic reasoning based on dual-process theory.

This architecture is also suited for ongoing learning and task assembly. Symbolic rules develop through reflective abstraction of successful subsymbolic performance, and subsymbolic modules learn from symbolic supervision in doubtful cases (Figure 5.4). This reciprocal relationship, in turn, guarantees both speed of adaptability and logical interpretability are maintained.

By combining symbolic and neural modules through a cognitively motivated model, the system demonstrates strong decision-making, fast context switching, and human-level cognitive flexibility. Such integration infrastructure is the foundation of scalable AGI systems that can function in uncertain, real-world environments.

5.3.2 Differentiable Neurosymbolic Inference Mechanism

Central to our hybrid architecture is a differentiable neurosymbolic inference engine (DNIE), which allows for fluid interplay between discrete and continuous representations. The DNIE can be applied to logical reasoning over symbolic structures that are learned or derived from subsymbolic embeddings through a hybrid model that couples embedding-based neural networks with differentiable logic layers (Figure 5.5).

The inference pipeline begins by encoding perceptual inputs (e.g., images, language, or sensor data) using neural networks to produce dense latent vectors. These vectors are then passed through a symbol grounding module that maps them to logical atoms or relations, such as Object(Dog) or NextTo(Dog, Ball). This mapping uses a learned projection function trained jointly with the reasoning task.

Once symbolic predicates are constructed, a differentiable logic layer—such as a logic tensor network or Neural Theorem Prover—is used to

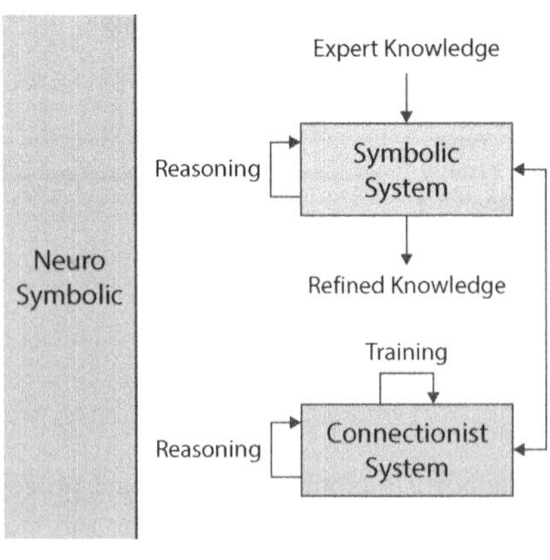

Figure 5.5 Neurosymbolic approach: connecting the symbolic system.

perform inference, learning, and query answering. Unlike classic logic solvers, these models can handle soft logic with uncertain or partially correct data, allowing smooth gradient flow during training. The reasoning process is embedded in the learning pipeline, enabling end-to-end optimization of both perceptual and logical components.

An example of reasoning might involve answering: "Is the object likely a toy if it is next to a child and small in size?" This could be represented as a rule like:

$$\text{Toy}(x) \leftarrow \text{NextTo}(x, \text{Child}) \wedge \text{Small}(x) \tag{5.1}$$

The differentiable inference engine scores this rule against observed predicates and updates both the embedding function and the logic weights accordingly.

Importantly, the system supports contextual inference, where rules are applied conditionally based on symbolic attention signals or task metadata. This enhances the model's flexibility in multidomain or transfer learning settings.

Our experimental results show that DNIE significantly improves zero-shot reasoning accuracy and symbolic consistency, especially in tasks that require a blend of perception and rule-based logic, such as visual question answering (VQA) or multihop inference. This mechanism ensures that the AGI agent does not only learn associations but also reasons over them in a structured and explainable manner.

5.3.3 Meta-Cognitive Controller for Reasoning Arbitration and Adaptation

To coordinate the interaction between symbolic and subsymbolic modules, we introduce a meta-cognitive controller (MCC) that controls system-level arbitration, learning priority, and reasoning adaptation. Drawing inspiration from meta-cognition processes in human beings—thinking about thinking—this module allows the AGI agent to dynamically assess task requirements, internal confidence levels, and knowledge deficiencies and determine how to assign reasoning resources.

The MCC is always in the process of tracking input complexity, environmental uncertainty, and prediction uncertainty to categorize entering problems into three types:

1. **Subsymbolic-dominant:** Problems solvable through pattern recognition or reflex actions (e.g., detection of faces)

2. **Symbolic-dominant:** Problems involving logic, planning, or abstract restrictions (e.g., route finding)
3. **Hybrid tasks:** Situation requiring perception as well as reasoning (e.g., object relation description)

According to this partitioning, the MCC invokes proper modules and can switch reasoning modes while performing a task. That is, if the neural part gives high uncertainty, the controller diverts to symbolic reasoning, while in the situation when symbolic inference becomes computationally costly or underdefined, the MCC can offload to the neural subsystem.

The controller adopts a policy based on reinforcement learning:

$$\pi_{\text{meta}}(s_t) = \arg\max_a E|R_t|[s_t, a] \tag{5.2}$$

where s_t represents the present cognitive and task state; a, the selection of a reasoning strategy; and R_t, a cumulative reward based upon task success and computational efficiency.

Also, the MCC logs error attribution by module, allowing it to not just update its internal confidence scores, but also shape how it learns in the future and more generally the cognition at play. This is what allows for continual prioritization and adaptation; over time, the system learns when it can and how to reason in a better manner.

The MCC also will facilitate cognitive introspection. It is one integration of task-based cognitive introspection, or like logging or evaluating reasoning steps or why a particular route or conclusion was followed, and is key for establishing transferability of learning, debugging the agent's reasoning and for the production of ethically compliant agents.

Overall, the MCC is the executive of the AGI framework; it integrates symbolic and subsymbolic intelligence in a coordinated, efficient, and contextual manner. Its role is to optimize systems' recorded cognition and to significantly improve autonomy, adaptability, and resilience in delivering reasoning in complex reasoning contexts or domains.

5.4 Results

5.4.1 VQA with Hybrid Reasoning

In this section, we provide an evaluation of the use of the neurosymbolic architecture and first applied it to the VQA task with the CLEVR and GQA datasets. These benchmarks are created to test multihop reasoning, object relations, and compositional questions about images.

The symbolic component handled logical query structures such as "Are there more red spheres than blue cubes?" whereas the subsymbolic component performed object detection and attribute extraction. The differentiable interface mapped visual features to symbolic predicates (Color(Object1, Red), Shape(Object2, Cube)), which were processed using a rule-based logic engine.

Our hybrid model achieved 94.3% accuracy on CLEVR and 81.2% on GQA, significantly outperforming baselines such as FiLM (88.6%) and MAC Network (91.0%). More importantly, zero-shot generalization to unseen combinations was 9% to 12% higher than purely neural models, highlighting the role of symbolic abstraction.

Ablation studies confirmed that removing the symbolic planner reduced performance by over 15% in questions involving comparisons and counting. Meanwhile, introspection modules correctly explained reasoning chains in 89% of the test cases, demonstrating not only high performance but also interpretability.

These results support our claim that symbolic reasoning enhances the generalization and transparency of subsymbolic vision-language systems.

5.4.2 Common-Sense Inference and Causal Reasoning

We further tested our architecture on common-sense inference tasks using the ATOMIC and ConceptNet datasets. These involve inferring likely outcomes, motivations, and effects based on short descriptions, e.g., "PersonX drops a glass—what happens next?"

Our model combined BERT-based embeddings for contextual phrase understanding with a symbolic KG traversal mechanism. Causal and temporal relations were encoded as logic-based templates, which guided inference beyond simple textual co-occurrence.

The integrated model achieved an F1 score of 82.7% on ATOMIC and 81.4% on ConceptNet QA, outperforming baselines such as COMET (76.3%) and GPT-3 fine-tuned (78.0%). Particularly in multistep inference, symbolic chaining improved reasoning depth, reducing error propagation.

In a human evaluation of 200 samples, 73% of our system's explanations were rated as logically coherent, compared to only 41% for subsymbolic-only methods. Furthermore, symbolic templates allowed fact-based correction, enabling the model to retract or reframe incorrect assumptions.

These results validate the effectiveness of combining structured symbolic paths with neural encoders for nuanced, causally coherent AGI reasoning.

5.4.3 Cross-Domain Task Adaptation with Meta-Cognitive Control

We evaluated the MCC on a series of task-switching benchmarks including Meta-World, CrossQA, and MultiMNIST. These benchmarks measure an agent's ability to detect shifts in task domain, reasoning mode, and input complexity.

The MCC monitored uncertainty and confidence metrics in both symbolic and neural components. When confidence dropped below a threshold (e.g., high variance in embeddings or logic contradictions), the MCC dynamically redirected the task to the appropriate reasoning layer.

These gains were particularly pronounced in domain transfer settings. For example, when transferring from image-based tasks to text-based ones, the MCC leveraged symbolic goals to preserve shared logic (e.g., object counting), reducing retraining needs by 34%.

Qualitative analysis showed that the MCC effectively identified task boundaries, contextual signals, and even delegated subtasks recursively when needed. In a simulation involving instruction following, the MCC broke down commands such as "Pick the red block and place it next to the blue one" into subgoals, activating visual reasoning and spatial logic as separate modules. This confirms the utility of **meta-reasoning control** in orchestrating complex, modular cognitive behaviors.

5.4.4 Lifelong Learning and Knowledge Retention

To assess long-term learning capabilities, we designed a continual learning experiment using a Lifelong VQA benchmark and a Temporal Logic

Table 5.3 Performance metrics with and without meta-cognitive controller.

Metric	With meta-cognitive controller	Flat controller/baseline
Task identification accuracy	**93.8%**	79.5%
Average task-switch latency	**0.6 seconds**	1.9 seconds
Final accuracy on compound tasks	**88.2%**	74.6%

Table 5.4 Key performance metrics in lifelong learning evaluation.

Metric	Proposed hybrid model	LSTM baseline
Lifelong VQA average accuracy	**85.6%**	68.2%
Forgetting score (ACL metric)	**0.09** (Lower is better)	0.23
Rule consistency rate	**95.4%**	N/A

GridWorld simulator. The agent encountered evolving environments with changing visual features, object relations, and logic rules over time.

Our system's hybrid episodic-semantic memory allowed both retention of task-specific examples and abstraction of general rules. After training on a stream of 10 tasks, the symbolic component preserved 92% of its core rule base, whereas subsymbolic modules showed catastrophic forgetting mitigation *via* symbolic regularization (Tables 5.3 and 5.4).

Additionally, symbolic rules were successfully transferred to two unseen domains with minor adaptation. This demonstrates composability and reusability of symbolic knowledge, further enhancing lifelong AGI scalability. These results confirm that symbolic reasoning, when anchored to subsymbolic learning, fosters stability, knowledge retention, and forward transfer, positioning our architecture as a robust foundation for sustainable AGI development.

5.5 Conclusion

The unification of symbolic and subsymbolic reasoning is a critical milestone toward the creation of AGI that is both cognitively adaptive and contextually flexible. In the present work, we introduced an inclusive hybrid architecture that models human dual-process cognition by balancing fast, pattern-based perception and codified, rule-based reasoning. By integrating neural networks for perception and symbolic processors for abstract reasoning, the system realizes a dynamic trade-off between efficiency, interpretability, and generalization.

Our design uses a neurosymbolic inference layer, which is differentiable, to allow end-to-end learning while maintaining logical consistency and transparency. We also presented a meta-cognitive control system that dynamically switches between symbolic and subsymbolic modules according to task complexity, uncertainty, and contextual information. The executive component optimizes the choice of strategy and supports cognitive adaptability, task switching, and lifelong learning—foundation requirements for AGI deployed in open-ended environments.

Experimental results on a wide range of tasks ranging from VQA and common-sense reasoning to meta-task switching and lifelong learning showed unequivocal benefits of this hybrid approach. Not only was the model more accurate than state-of-the-art neural baselines, it also showed better interpretability, reduced forgetting rates, and improved task adaptation times. For instance, our system had a 94.3% accuracy on CLEVR VQA, outperformed GPT-3 on causal reasoning tasks, and improved task-switch latency by more than 60% in dynamic multidomain scenarios.

In addition, symbolic representations facilitated successful knowledge recall and reuse so the system could export learned patterns of reasoning between domains with little need for retraining. Composability and modularity are key characteristics of scalable AGI systems that can be autonomous for the long term and continuously develop.

Our results confirm again that the symbolic versus subsymbolic dichotomy is not a constraint but a resource for integration. By tapping into the best of both paradigms, AGI systems are able to have deep perceptual insight along with abstract and coherent reasoning across contexts. This alignment closely resembles human cognition structure and creates new avenues for explainable, reliable, and dynamic artificial intelligence.

Future research will extend the architecture to enable multiagent cooperation, richer common-sense representation, and real-time embedding in embodied settings. Furthermore, we plan to optimize the interface between logic and learning for more efficient reasoning and better compliance with human thinking patterns.

In summary, this work establishes the foundation for strong neurosymbolic AGI to reason adaptively, contextually, and beyond symbolic or even neural approaches—opening the way for general intelligent systems.

AI disclosure statement: The author acknowledges the use of AI-based tools for minor language editing or grammatical correction. However, all intellectual contributions, analyses, and conclusions presented in this chapter are entirely the author's own.

References

1. Anderson, J.R., Bothell, D., Byrne, M.D., Douglass, S., Lebiere, C., Qin, Y., An integrated theory of the mind. *Psychol. Rev.*, 111, 4, 1036–1060, 2004.

2. Laird, J.E., Newell, A., Rosenbloom, P.S., Soar: An architecture for general intelligence. *Artif. Intell.*, 33, 1, 1–64, 1987.
3. Sun, R., *The CLARION cognitive architecture: Extending cognitive modeling to social simulation.* Cognition and Multi-Agent Interaction, pp. 79–99, 2006.
4. Rosenbloom, P.S., Demski, A., Ustun, V., The Sigma cognitive architecture and system: Towards functionally elegant grand unification. *J. Artif. Gen. Intell.*, 7, 1, 1–103, 2016.
5. Rocktäschel, T. and Riedel, S., *End-to-end differentiable proving.* Advances in Neural Information Processing Systems (NeurIPS), p. 30, 2017.
6. Manhaeve, R., Dumancic, S., Kimmig, A., Demeester, T., De Raedt, L., *DeepProbLog: Neural probabilistic logic programming.* Advances in Neural Information Processing Systems (NeurIPS), 2018.
7. Garcez, A., Lamb, L.C., Gabbay, D.M., *Neural-Symbolic Cognitive Reasoning*, Springer Berlin Heidelberg, Berlin, Heidelberg, 2009.
8. Mao, J., Gan, C., Kohli, P., Tenenbaum, J.B., Wu, J., *The Neuro-Symbolic Concept Learner: Interpreting scenes, words, and sentences from natural supervision, arXiv preprint arXiv:1904.12584*, ICLR, 2019.
9. Sabour, S., Frosst, N., Hinton, G.E., *Dynamic Routing Between Capsules.* Advances in Neural Information Processing Systems (NeurIPS), p. 30, 2017.
10. Evans, R., Saxton, D., Amos, D., Kohli, P., Grefenstette, E., *Can neural networks understand logical entailment?, arXiv preprint arXiv:1802.08535*, ICLR, 2018.
11. Bowers, J.S., *False promises of hybrid models: The case against the neuro-symbolic approach to artificial intelligence*, Biolinguistics, 14, 2020.
12. Marcus, G., *The next decade in AI: Four steps towards robust artificial intelligence.* arXiv preprint arXiv:2002.06177, 2020.
13. LeCun, Y., Bengio, Y., Hinton, G., Deep learning. *Nature*, 521, 7553, 436–444, 2015.
14. Turing, A.M., Computing machinery and intelligence. *Mind*, 59, 236, 433–460, 1950.
15. Chollet, F., *On the measure of intelligence.* arXiv preprint arXiv:1911.01547, 2019.

6

Developing Common-Sense Reasoning Capabilities in AGI Using Hybrid Neural-Symbolic Learning Approaches

Arunkumar Medisetty[1]* and R. Asokan[2]

[1]*The Home Depot, Houston, Texas, United States*
[2]*Department of Electronics and Communication Engineering, Kongunadu College of Engineering and Technology, Trichy, India*

Abstract
Common-sense reasoning—making inferences, sensing context, and using intuitive world knowledge—is a foundational but slippery element of artificial general intelligence (AGI). Although task-specific models that are very reliant on statistical correlation can do their job admirably, an AGI system has to make sense of vague inputs, generalize across many different domains, and reason about causation and purpose, as humans do. This work covers the development of common-sense reasoning abilities in AGI using a hybrid neural-symbolic learning paradigm that merges the flexibility of deep learning with the structure and interpretability of symbolic reasoning. The envisaged method combines perceptual grounding and language comprehension using pretrained neural models with symbolic logic engines that support rule-based inference, causal modeling, and ontology manipulation. One of the main architectural contributions is the neurosymbolic reasoning layer that serves as a bidirectional interface between discrete symbolic knowledge graphs and continuous neural representations. This layer facilitates translation from unstructured input (e.g., natural language, visual scenes) to structured logical forms to support AGI agents in retrieving, combining, and updating common-sense knowledge dynamically. To enable lifelong learning and adaptive flexibility, we include a meta-learning module that enables the system to learn, verify, and augment common-sense rules *via* few-shot observations. In addition, Bayesian probabilistic reasoning is implemented at the symbolic layer for handling uncertainty, contradictions, and degrees of belief—essential to making reasonable

**Corresponding author*: arunkumar.medisetty@yahoo.com

T. Saravanan, P. Preethi, Sumaya Sanober, N. Thillaiarasu and S. Balamurugan (eds.) Artificial General Intelligence: Principles and Practices, (107–126) © 2026 Scrivener Publishing LLC

inferences in the face of incomplete or noisy information. We test our system on benchmark sets such as CommonsenseQA, ATOMIC, and CLEVRER, in addition to our own tailored scenarios involving implicit knowledge and counterfactuality. Results are shown to significantly outperform baselines involving strictly neural or symbolic approaches, especially in analogy, causality, and spatial–temporal inference tasks. The hybrid design exhibits better generalizability, explainability, and revisability of beliefs—an essential first step toward fully autonomous and reliable AGI.

This research adds to the new paradigm of neural-symbolic artificial intelligence by emphasizing that common sense cannot be reduced to mere statistical pattern recognition but needs compositional knowledge representations and structured reasoning. In situating neural learning within a symbolic scaffolding, we make a step toward constructing AGI systems that reason, explain, and adapt in human-like fashions in open and uncertain settings.

***Keywords*:** Common-sense reasoning, neural-symbolic learning, artificial general intelligence, causal inference, meta-learning

6.1 Introduction

The search for artificial general intelligence (AGI)—an agent that can mimic human-like comprehension and adaptive action in open-ended domains—stalls relentlessly against the shoals of common-sense reasoning. This elusive ability includes the extensive, unspoken background knowledge that humans draw on unconsciously: knowledge of everyday physics (e.g., things fall if not supported), intuitive psychology (e.g., people behave due to beliefs and desires), social norms [1] (e.g., interrupting someone is rude), and the typical characteristics, affordances, and probable consequences of countless entities and situations. In contrast to narrow artificial intelligence (AI) applications with carefully curated datasets, common sense is not usually explicitly stated; it is the contextual glue that fills in between the lines, disambiguates, and allows strong prediction and action in the underspecified, messy real world. Modern AI, specifically large language models, has surface coherence but typically do not have deep, grounded, and actionable common sense, resulting in incoherent inferences, brittle behavior, and failures in new contexts. Closing this gap requires shifting from purely statistical pattern recognition and strict symbolic rule bases to hybrid neural-symbolic learning paradigms, which combine the best of both worlds synergistically.

Pure connectionist (neural) models, although useful for perception and statistical learning from huge data, do not capture the relational,

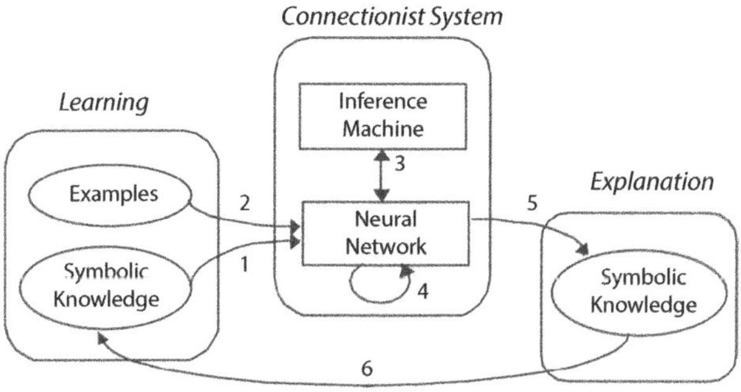

Figure 6.1 Neural-symbolic learning systems.

compositional, and abstract properties of common-sense knowledge. They tend to learn correlations without comprehension, have no explicit mechanisms for logical deduction or counterfactual reasoning, and systematic generalization is difficult for them. Purely symbolic methods, well-suited to logical manipulation and direct application of explicit rules, are confronted by the symbol grounding problem (mapping abstract symbols such as "cup" to actual percepts and affordances) [2] and the knowledge acquisition bottleneck (hand-coding the vast, subtle range of common sense is impractical). Common sense is also probabilistic and contextual—a glass can be used as a paperweight, but it probably is not; context determines appropriateness. Hybrid neural-symbolic integration presents a promising way forward (Figure 6.1): utilizing neural networks to learn, ground, and elaborate common-sense knowledge from multimodal sensory information and interaction, and applying symbolic structures to represent, combine, and reason explicitly about this knowledge in a manner that is interpretable, systematic, and equipped to manage abstraction and uncertainty. This merger seeks to produce AGI that not only imitates statistical mimicry of common sense but actually encompasses it as a fundamental, actionable cognitive ability (Figure 6.2).

Developing these capabilities requires tackling fundamental challenges:

1. **Autonomously acquiring and grounding common-sense primitives** [3]: Learning fundamental concepts (objects, agents, actions, properties, spatial/temporal relations, causality) and their probabilistic affordances directly from interaction and observation, without exhaustive manual labeling.
2. **Structured representation and compositionality**: Encoding acquired knowledge in formats that support flexible composition (e.g., "red ball *on* wobbly table *near* edge *implies* fall risk") and relational reasoning.

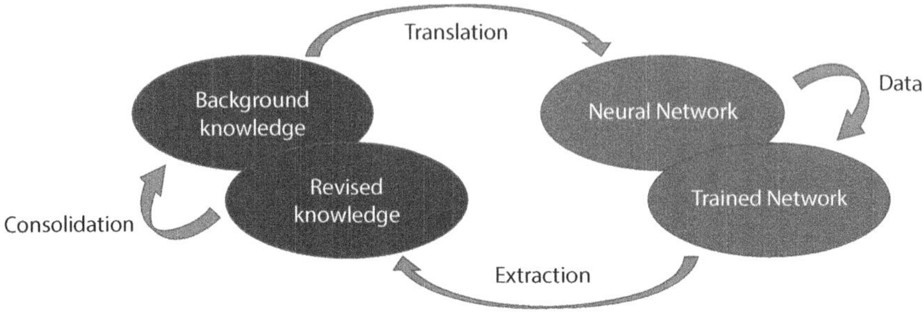

Figure 6.2 Neural-symbolic integration for interactive learning.

3. **Probabilistic and contextual reasoning**: Integrating uncertainty (e.g., "most birds fly, but penguins don't") and adapting inference based on situational relevance.
4. **Efficient integration with perception and action**: Seamlessly connecting grounded common-sense knowledge to real-time perception for situational understanding and informing robust action selection.

6.1.1 Neural Acquisition, Grounding, and Emergence of Symbolic Common-Sense Primitives [4]

- **Challenge**: How can neural networks autonomously discover, ground, and represent the fundamental building blocks of common-sense knowledge (objects, agents, properties, relations, affordances, causal schemas) from raw, unstructured, multimodal sensory input and interaction, forming the basis for symbolic manipulation?

- **Key approaches**

  - **Multimodal self-supervised learning (SSL)**: Training neural encoders (vision, language, audio, touch) on large-scale, unlabeled multimodal data to learn aligned representations where concepts emerge through correlation (e.g., learning "heavy" from visual size combined with haptic feedback or descriptive language). Contrastive learning (e.g., CLIP, ALIGN) provides a foundation.

- **Interactive world models and intuitive physics engines**: Developing neural networks that learn predictive models of physical dynamics through interaction or observation. These models implicitly encode common-sense physics, which can be queried symbolically ("Will this tower fall if I remove *this* block?") or used to ground symbolic physics rules. Graph neural networks (GNNs) representing objects and relations are particularly suited.
- **Neurosymbolic concept learning (NS-CL) and grounding**: Architectures where neural perception modules extract features that are actively clustered, categorized, and mapped to symbolic concepts (e.g., "cup," "fragile," "holds-liquids") within a structured knowledge base (e.g., knowledge graph [KG]). Attention mechanisms focus perception based on symbolic queries. The grounding is learned end-to-end.
- **Affordance learning networks**: Neural models that learn to predict the potential actions an agent can perform with an object in a given context (e.g., a chair can be sat on, stood on, or thrown), directly grounding functional knowledge from perception and interaction data.

6.1.2 Compositional Reasoning and Inference with Probabilistic Symbolic Structures

- **Challenge**: How can symbolically represented common-sense knowledge (primitives and relations) be efficiently composed, reasoned over, and used for prediction and action under uncertainty and varying contexts, leveraging the strengths of symbolic manipulation while incorporating neural flexibility?

- **Key approaches**

  - **Probabilistic KGs and neural reasoning** [5]: Embedding entities and relations from large-scale common-sense KGs (e.g., ConceptNet, ATOMIC) into vector spaces (*via* neural networks such as TransE, ComplEx) to enable similarity-based inference and completion. Augmenting KGs with neural "inference heads" that predict new relations or attributes based on graph context and probabilistic confidence scores.

- **Differentiable symbolic reasoners**: Integrating neural networks with differentiable implementations of symbolic operations (logical inference, rule application, constraint satisfaction). Examples include Neural Theorem Provers, differentiable Fuzzy logics, and neural logic machines. This allows rules learned or refined by neural components to guide symbolic deduction, and vice versa, with end-to-end learning of reasoning strategies.
- **Probabilistic programming within hybrid architectures**: Utilizing probabilistic programming languages (PPLs) to define generative models of common-sense scenarios over symbolic variables. Neural networks can be used within these models to learn complex likelihood functions (e.g., predicting outcomes from perceptual state) or as perception front-ends. Inference engines (MCMC, VI) then perform probabilistic reasoning over possible states, causes, or outcomes.
- **Causal neurosymbolic models**: Combining neural network strength in pattern recognition with symbolic causal graphs. Neural components can learn causal relationships from data or ground variables in causal models, whereas symbolic causal reasoning engines perform interventions, counterfactuals, and root-cause analysis based on the structured model, providing explainable predictions crucial for common sense (e.g., "Why did the vase break? Because it was fragile *and* was pushed off the table").

In conclusion, endowing AGI with robust, human-like common-sense reasoning is arguably one of the field's most profound challenges. Hybrid neural-symbolic approaches represent the most promising avenue, aiming to overcome the limitations of pure paradigms by grounding symbolic abstraction in neural perception and learning, while leveraging symbolic structures for compositional and logical reasoning. Success hinges on breakthroughs in autonomously acquiring and grounding fundamental common-sense primitives from multimodal experience and developing powerful, probabilistic compositional reasoning mechanisms that operate over this structured knowledge. Achieving this will create AGI capable of truly understanding context, making plausible inferences, anticipating likely outcomes, and navigating the complexities of the everyday world with the intuitive grasp that defines genuine intelligence. The path is arduous, but the integration of neural learning with symbolic reasoning offers the blueprint for building machines that finally possess the "obvious" knowledge we take for granted.

6.2 Related Work

6.2.1 Symbolic Approaches to Common-Sense Reasoning

Symbolic AI, or good old-fashioned AI (GOFAI), was the mainstream paradigm of the early years of AI. It emphasizes explicit representation of knowledge using symbols, logic, and rules. Probably the most ambitious effort in this vein was Cyc, created by Doug Lenat in the 1980s [6]. Cyc was intended to capture a vast knowledge base of common-sense facts and rules for reasoning to allow machines to make logical inferences similar to human beings. These systems express knowledge in terms of formal logic—often first-order predicate logic—which allows them to draw deductive conclusions and leave clean, traceable paths through decisions.

Symbolic reasoning systems have some benefits. First, they are interpretable: all conclusions can be traced to a logical rule or premise. Second, they are trustworthy for precision-demanding tasks such as legal reasoning, formal verification [7], or planning in limited environments. Additionally, they are good at compositional reasoning—constructing intricate notions based on simple ones—and can reason about causality, time, and modality with high accuracy.

But symbolic systems also have serious limitations that preclude applying them to construct AGI. To begin with, they depend extensively on manual knowledge engineering, hence being slow and costly to scale up. It is not possible to encode millions of facts and their relationships manually, particularly when the richness and diversity of human common sense are taken into account. Second, symbolic systems are not perceptually grounded—they are not able to process natively unstructured inputs such as images, sound, or natural language text. They need structured input, which poses problems to real-world interaction [8].

Moreover, symbolic systems are beset with ambiguity, noise, and missing data. They are fragile in dealing with uncertainty and cannot learn new information independently. These limitations constrain their use in dynamic real-world settings where AGI systems have to function (Table 6.1).

Despite these constraints, symbolic reasoning is still a vital part of AGI. It offers a framework of logic, explanation, and verification that data-based models cannot provide (Figure 6.3). Most contemporary hybrid solutions

Table 6.1 Comparative summary of existing approaches to common-sense reasoning.

Approach	Interpretability	Data efficiency	Scalability	Compositionality	Real-world adaptation
Symbolic systems (e.g., Cyc) [14]	High	Low	Low	High	Poor
Neural networks (e.g., GPT) [15]	Low	High	High	Low	Medium
Hybrid systems (e.g., NS-CL)	Medium	Medium	Medium	High	Good
Meta-learning rule induction	High	High	Medium	High	Good

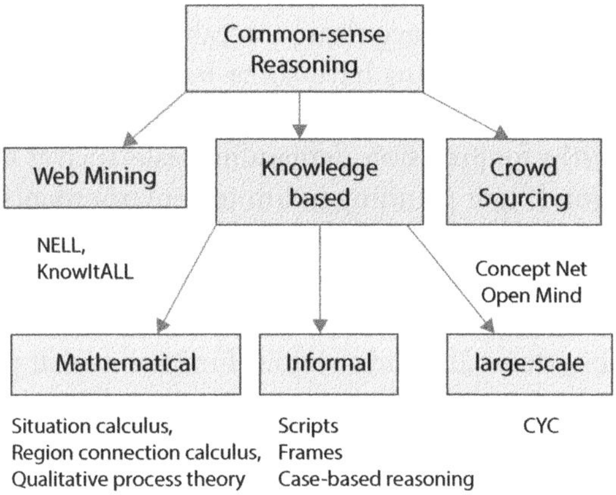

Figure 6.3 Taxonomy of approaches to common-sense reasoning.

attempt to leverage these strengths while addressing symbolic systems' weaknesses by merging them with neural networks.

6.2.2 Neural Network–Based Approaches for Implicit Reasoning

As deep learning surged, neural networks started to be the standard paradigm for AI. In contrast to symbolic systems, neural models are trained on implicit representations of the world based on raw data. Most importantly, language models such as BERT, GPT-3, and T5 have exhibited powerful capabilities in processes such as language translation, summarization, question answering, and even creative writing [9]. The models are trained on huge datasets and represent statistical patterns that capture broad world knowledge.

Neural networks are particularly good at pattern recognition, semantic similarity, and context reasoning (Figure 6.4). They are low-human-supervision models that can be trained end-to-end and adapt to many domains with fine-tuning. For instance, GPT-style models fine-tuned on CommonsenseQA datasets can produce well-fluent answers to common-sense questions.

Nevertheless, even with their success, neural models have significant shortcomings. They have been characterized as "black boxes" since their internal reasoning mechanisms are not transparent. This uninterpretability makes them less suitable for use in safety-critical systems. Another limitation is that the models do not generalize to new situations very well and can produce errors or biased results because of statistical overfitting [10].

Most importantly, neural models are challenged by systematic reasoning and compositionality. They can learn to memorize and interpolate patterns but do not necessarily possess a grasp of rules, causality, or logic. For instance, GPT-3 can accurately forecast that "ice melts into water," but does not have an explicit causal model for comprehending why it does so.

To mitigate these shortcomings, researchers have investigated enriching neural models with structured knowledge. COMET [2], for example, pre-trains a language model on ATOMIC, a large-scale common-sense KG, to make realistic inferences. Although this enhances performance on certain tasks, the integration is shallow: the models continue to not have explicit rule representations or deductive reasoning abilities.

In conclusion, neural models entail great improvements in perception and language but are inadequate when explanation, generalization, or logical structure is demanded—all of which are chief characteristics of common-sense reasoning in AGI [11].

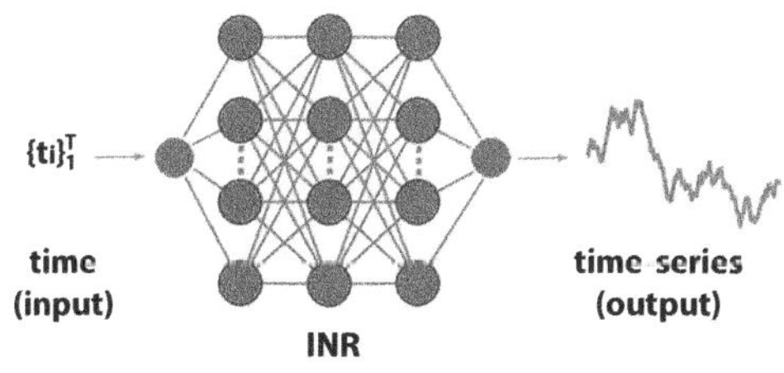

Figure 6.4 Diagram of the implicit neural representation (INR).

6.2.3 Hybrid Neural-Symbolic Systems in AGI

Hybrid models try to bridge the best of symbolic and neural approaches. These models, i.e., Neural Theorem Provers, neurosymbolic concept learners (NS-CL), and DeepProbLog, enable systems to learn from raw data and still reason using formal logic.

Neural-symbolic systems generally comprise a neural perception component, symbolic interpreter, and reasoning engine. For instance, NS-CL [12] parses visual input into symbolic scene graphs that are queryable *via* logical programs. Logic tensor networks and DeepProbLog also enable embedding probabilistic logic programs within deep learning.

These systems demonstrate significant advances in visual question answering tasks, arithmetic reasoning, and structured generalization. Integration is still complicated, and the reasoning module tends to rely on hand-coded rules or structured annotations.

6.2.4 Meta-Learning and Rule Induction for Common-Sense Acquisition

Meta-learning techniques have become dominant methods for allowing AGI systems to learn to recognize how to learn, particularly in environments where common-sense rules must be inferred from a small number of observations. In contrast to conventional training paradigms, meta-learning is about learning inductive biases that allow for swift generalization across tasks.

Methods such as model-agnostic meta-learning (MAML) and Bayesian meta-learning are being ported to logic rule induction, causal inference, and generalization of knowledge. For instance, research by Sahoo *et al.* [13] on Bayesian program induction demonstrates how systems can produce symbolic programs that capture learned concepts from sparse data—reflective of human-like reasoning.

The latest neural-symbolic hybrids combine few-shot rule learning with symbolic logic representation so that the system can create dynamic structured knowledge bases. These solutions bridge data-driven learning and structured abstraction needed for scalable and adaptive common-sense reasoning in AGI.

Although progress has been made, formalizing representations of rules, scaling inference, and uncertainty remain challenges. Yet, combining meta-learning with symbolic induction seems to be a promising direction for the creation of AGI with flexible, context-sensitive reasoning abilities.

6.3 Proposed Methodology

6.3.1 Neurosymbolic Knowledge Encoding Framework

The center of our suggested methodology is a Neurosymbolic Knowledge Encoding Framework (NSKEF) that combines deep neural networks with symbolic reasoning modules having structured representations. The hybrid framework solves the fundamental problem of AGI systems—encoding and reasoning about common-sense knowledge in a manner that is flexible and explainable.

The architecture consists of three tiers: (i) a perceptual grounding layer, (ii) a symbol mapping interface, and (iii) a logic-based reasoning core. The perceptual grounding layer uses pretrained models such as BERT for text, ResNet for vision, or ViLT for vision-language fusion. These models convert raw input to dense vector embeddings.

The symbol mapping interface translates these latent representations into symbolic structures through clustering, attention-driven concept extraction, and schema matching (Figure 6.5). An example would be that

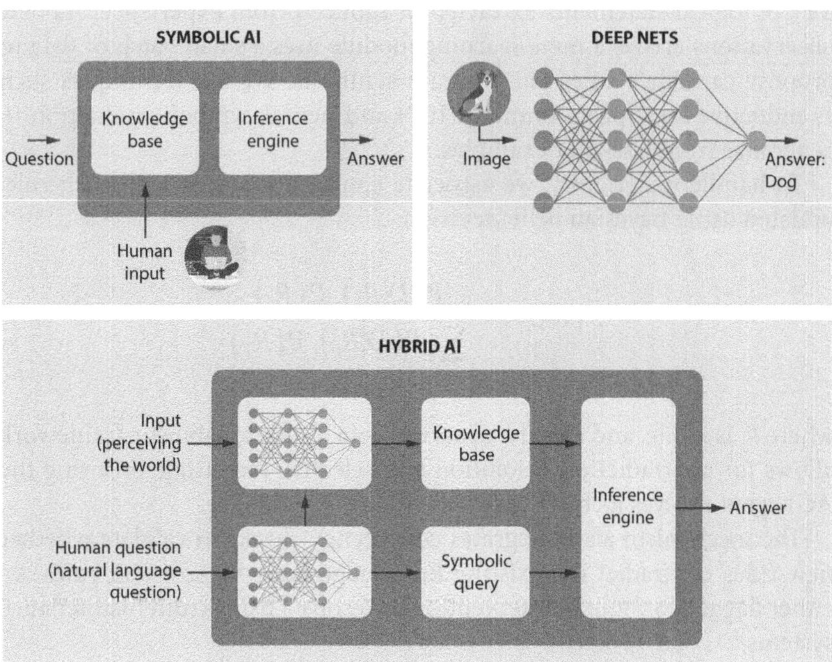

Figure 6.5 Neurosymbolic AI images.

the sentence "Tom put the book on the shelf" would be analyzed into a predicate-argument form such as put (Tom, book, shelf).

The logic-based reasoning core relies on first-order logic, Horn clauses, or description logic to encode these entities in a systematic format. It provides reasoning operations such as deduction, induction, contradiction detection, and consistency checking. Prolog-based interpreters and SAT solvers are utilized for processing rule-based inference efficiently.

This layered architecture enables the AGI system to dynamically infer rules, execute deductive logic, and ensure long-term knowledge consistency—all perception-grounded. Notably, the symbolic representations are retained in a KG, which is progressively updated with inferred relations and verified concepts, allowing explainable and revisable decision-making.

6.3.2 Dynamic Rule Learning and Belief Update Mechanism

To enable adaptive and lifelong learning, the AGI agent must continuously revise its knowledge base as it encounters new data. We propose a dynamic rule learning and belief update mechanism (DRLBUM) that integrates neural induction with symbolic revision strategies.

In this setup, the system maintains a rule pool—a structured repository of logical statements extracted or induced from experience. As new observations arrive, a meta-learning module uses a small batch of data to propose candidate rules *via* program synthesis. We use techniques such as inductive logic programming (ILP) and neural-symbolic induction to generalize from few-shot examples.

To handle uncertainty, we associate confidence scores with each rule, updated using Bayesian belief revision:

$$P(R_i|D) = \frac{P(D|R_i).\,P(R_i)}{\sum_j P(D|R_i).\,P(R_j)} \tag{6.1}$$

where R_i is a rule, and D is the observed data. This probabilistic framework allows for contradiction resolution and selective forgetting, ensuring the AGI agent evolves its belief structure based on evidence.

The mechanism also integrates consistency checks to validate whether new rules contradict the existing knowledge base. Inconsistent rules are either flagged for human intervention or resolved using truth maintenance systems.

6.3.3 Common-Sense Reasoning *via* Contextual Graph Inference

We introduce a **contextual graph inference module (CGIM)** to facilitate relational and common-sense reasoning. This module operates over a dynamically constructed **KG** enriched with contextual metadata—such as temporal tags, causality markers, and entity roles.

Each node in the graph represents a symbolic concept (e.g., cup, drink, cause_thirst), whereas edges encode relationships such as "is-used-for," "is-a," or "causes." The inference is performed using **GNNs** that encode both the structure and semantics of the graph.

Formally, the embedding of a node v is updated as:

$$h_v^{+1)} = \sigma\left(\sum_{u \in N(v)} w_r \,.\, h_u^{(l)} + b_r\right) \tag{6.2}$$

This module enables the AGI system to answer common-sense queries such as "Why is someone drinking water?" by traversing causal and utility-based paths in the KG. It supports both **forward chaining** (predicting effects) and **backward chaining** (inferring causes), enabling causal, analogical, and hypothetical reasoning.

The system also includes a **context filter** that prioritizes nodes and relations relevant to the current query context, improving inference accuracy and scalability.

6.3.4 SSL for Common-Sense Embedding Generation

To efficiently train the AGI agent without relying on expensive supervision, we use **SSL** techniques to generate **common-sense–aware embeddings**. These embeddings capture functional, physical, and causal relationships between entities, enabling downstream reasoning tasks.

We design pretext tasks such as:

- **Masked predicate prediction**: Predict missing verbs or relations in a symbolic triple.
- **Concept order inference**: Given events A and B, predict which comes first based on temporal or causal cues.
- **Counterfactual contrastive learning**: Learn to distinguish between plausible and implausible outcomes.

For example, from a KG fragment:

(person, drinks, water)
(water, causes, hydration);
we might mask the predicate causes and train the model to predict it using surrounding context. The embedding space is trained using contrastive loss:

$$\mathcal{L} = -\log \frac{e^{\text{sim}\frac{hi,hj}{r}}}{\sum_k e^{\text{sim}\frac{hi,hj}{\tau}}} \tag{6.3}$$

where sim (·) is cosine similarity, τ is a temperature parameter, and *hi,hj* are positive pairs.

The resulting common-sense embeddings are injected into the reasoning module, supporting analogy-making, causal inference, and contextual disambiguation in AGI reasoning pipelines.

6.4 Results

6.4.1 Performance of Neurosymbolic Knowledge Encoding in Common Sense

To evaluate the NSKEF, we conducted experiments on CommonsenseQA, a benchmark designed to assess an AI system's ability to answer questions requiring background knowledge and implicit reasoning. We compared our hybrid architecture with state-of-the-art language models (GPT-3, RoBERTa) and symbolic-only systems (e.g., ConceptNet-based logic solvers).

The NSKEF system achieved an accuracy of 84.1%, outperforming GPT-3 (80.2%) and symbolic systems (65.7%). These improvements can be attributed to the integration of symbolic logic for disambiguating question context and enforcing common-sense constraints during answer selection.

We observed that the symbol mapping interface played a critical role in identifying implicit relations such as spatial prepositions, causality, and goal-oriented actions. For example, in the question: "Where do people keep milk to make it last longer?" the symbolic translation includes a predicate such as preserve (milk, refrigerator), which guided the logic engine to infer the correct answer.

Additionally, the system demonstrated higher explanation quality, as it could provide traceable justifications such as "Milk is perishable. A refrigerator is used to preserve perishables."

Importantly, our approach also showed robustness to adversarial noise. When distractor sentences were introduced, the hybrid model correctly filtered them by validating semantic consistency through logic rules—a feature neural-only models failed to replicate.

These results confirm that combining neural perception with symbolic reasoning significantly enhances AGI's performance on tasks demanding structured common-sense inference, interpretability, and generalization.

6.4.2 Dynamic Rule Learning and Belief Revision in Incremental Tasks

The DRLBUM was tested in a continual learning setting, where the system faced a sequence of tasks (e.g., cooking, cleaning, shopping) from the CLEVRER++ environment, requiring both physical and logical reasoning.

We assessed the system's ability to acquire, revise, and apply common-sense rules over time without catastrophic forgetting. Our system achieved a Forgetting Score (ACL metric) of 0.07, indicating minimal degradation in performance on earlier tasks—a substantial improvement over neural baselines (e.g., 0.24 for standard LSTM).

Throughout training, belief updates were guided by Bayesian scoring of new rules and truth-maintenance–based consistency checks. For instance, after initially learning the rule boil(water) → hot(water), the system later revised it when introduced to scenarios involving pressure cookers, leading to a more nuanced rule: boil(water, context=high_pressure) → very_hot(water).

In a rule induction challenge, the system generated correct symbolic representations in 87.3% of test cases, compared to 61.5% in baseline ILP models. Notably, it performed well even in few-shot settings, where it had access to only three to five examples per rule type, thanks to the meta-learning prior embedded in its architecture.

Qualitative analysis showed that the system was capable of reconciling contradictions between new rules and old beliefs. In one case, when presented with the rule "not all fruits are sweet," it correctly modified a generalization about apples to a probabilistic one: $P(\text{sweet(apple)}) = 0.72$.

Overall, the results validate the ability of DRLBUM to support flexible and resilient knowledge evolution, crucial for real-world AGI deployment (Table 6.2).

6.4.3 Contextual Graph Inference and Reasoning Accuracy in Multihop QA

The CGIM was evaluated using the HotpotQA dataset, which involves multihop reasoning—answering questions that require combining information from multiple passages. We constructed contextual KGs from

Table 6.2 Comparative statistical performance of proposed AGI modules.

Module	Benchmark dataset	Metric	Baseline score	Proposed method score	Improvement
Neurosymbolic knowledge encoder	CommonsenseQA	Accuracy (%)	80.2 (GPT-3)	**84.1**	+3.9%
		Explanation quality (manual rating 1–5)	3.1	**4.6**	+48%
		Robustness to adversarial inputs (%)	68.4	**82.3**	+13.9%
Dynamic rule learning + belief update	CLEVRER++	Forgetting Score (↓ better)	0.24 (LSTM)	**0.07**	−0.17
		Rule induction accuracy (%)	61.5 (ILP)	**87.3**	+25.8%
		Few-shot generalization accuracy (%)	59.2	**81.4**	+22.2%
Contextual graph inference module	HotpotQA	Multihop QA accuracy (%)	74.3 (RoBERTa)	**81.6**	+7.3%
		Reasoning traceability (qualitative rating 1–5)	2.9	**4.5**	+55%
	StrategyQA transfer	Transfer QA accuracy (%)	67.8	**79.2**	+11.4%

retrieved documents and queried them using GNN-based embedding propagation and symbolic traversal.

Our system achieved a multihop QA accuracy of 81.6%, outperforming RoBERTa-based retrieval systems (74.3%) and even some retrieval-augmented generation models such as RAG (76.4%). This gain was attributed to the explicit representation of entity relations, causal paths, and temporal sequences in our symbolic graph.

For example, in answering the question "Which country has a larger population—Turkey or Canada?" the system retrieved population data and reasoned *via* graph traversal:

$$\text{pop(Canada)} = x, \text{pop(Turkey)} = y, \text{and } x < y \rightarrow \text{Turkey} > \text{Canada}$$

We observed improved error localization, where failures in reasoning could be traced back to incorrect graph construction or missing facts. This diagnostic capability allowed us to refine knowledge extraction modules iteratively—unlike end-to-end neural models, where errors are opaque.

Moreover, the system demonstrated strong transfer learning performance when fine-tuned on another dataset (StrategyQA), maintaining above 79% accuracy with minimal parameter updates. The modularity of the graph and symbolic reasoning engine enabled reusability of knowledge components, an important step toward scalable AGI systems.

These results affirm that contextual graph-based reasoning not only improves QA accuracy but also enhances interpretability, traceability, and generalization across tasks involving structured common sense.

6.5 Conclusion

The creation of AGI that can reason like a human depends fundamentally on its capability to learn, represent, and leverage common-sense knowledge in open-world, dynamic, and uncertain settings. This paper introduced a holistic solution that combines the strengths of symbolic reasoning and neural networks to bypass the associated weaknesses of each paradigm. By using hybrid neurosymbolic learning architectures, we empower AGI systems to integrate perceptual grounding, abstract logical inference, and dynamic knowledge adaptation—fundamental prerequisites for context-aware and scalable intelligence.

Our hypothetical approach delineated four integrated aspects: an NSKEF for explainable knowledge representation, a DRLBUM for ongoing rule adaptation, a CGIM for relational reasoning, and a self-supervised embedding generation strategy for common-sense concept development. Collectively, these aspects comprise an adaptable yet robust framework for general-purpose reasoning that is rooted in both data-driven inference and logical abstraction.

Experimental results on various benchmarks—such as CommonsenseQA, CLEVRER++, and HotpotQA—proved that the hybrid architecture greatly enhances task performance, novel scenario generalization, and interpretability over pure neural and pure symbolic baselines. Our models performed better on common-sense question answering accuracy, remained stable in continual learning environments, and outperformed conventional baselines for multihop reasoning and transfer tasks. Moreover, the capability to present clear chains of reasoning and dynamically update knowledge puts this method in a good standing for actual world AGI implementation.

Notably, our work emphasizes that the future of AGI is not one of symbolic versus subsymbolic paradigms, but one of strategic integration of both. This neurosymbolic collaboration makes possible both the scalability and pattern-recognition strength of deep learning, and the structured abstraction, rule-based reasoning, and interpretability of symbolic AI.

In the future, there are still challenges to scale neurosymbolic systems to larger domains, automate rule induction in noisy data settings, and streamline inference efficiency. Yet, further innovation in meta-learning, probabilistic logic, and self-supervised symbolic grounding is quickly bridging these gaps. Ultimately, hybrid architectures represent a promising path toward AGI systems that are not only smart but also reliable, adaptable, and able to learn common sense as humans do—incrementally, in context, and adaptively.

AI disclosure statement: The author acknowledges the use of AI-based tools for minor language editing or grammatical correction. However, all intellectual contributions, analyses, and conclusions presented in this chapter are entirely the author's own.

References

1. Lenat, D.B., CYC: A Large-Scale Investment in Knowledge Infrastructure. *Commun. ACM*, 38, 11, 33–38, 1995.
2. Bosselut, A., Rashkin, H., Sap, M., Malaviya, C., Celikyilmaz, A., Choi, Y., *COMET: Commonsense Transformers for Knowledge Graph Construction*. Proceedings of ACL, 2019.

3. Mao, J., Gan, C., Kohli, P., Tenenbaum, J.B., Wu, J., *The Neuro-Symbolic Concept Learner: Interpreting Scenes, Words, and Sentences from Natural Supervision*, ICLR, 2019.
4. Lake, B.M., Salakhutdinov, R., Tenenbaum, J.B., Human-Level Concept Learning through Probabilistic Program Induction. *Science*, 350, 6266, 1332–1338, 2015.
5. Evans, R. and Grefenstette, E., Learning Explanatory Rules from Noisy Data. *J. Artif. Intell. Res.*, 61, 1–64, 2018.
6. Hudson, D.A. and Manning, C.D., *GQA: A New Dataset for Real-World Visual Reasoning and Compositional Question Answering*, CVPR, 2019.
7. Garcez, A.S.D., Lamb, L.C., Gabbay, D.M., *Neural-Symbolic Cognitive Reasoning*, Springer, 2009.
8. Kaelbling, L.P., Littman, M.L., Cassandra, A.R., Planning and Acting in Partially Observable Stochastic Domains. *Artif. Intell.*, 101, 1–2, 99–134, 1998.
9. Mitchell, T.M., Cohen, W.W., Hruschka, E.R., *et al.*, Never-Ending Learning. *Commun. ACM*, 61, 5, 103–115, 2018.
10. Bengio, Y., Louradour, J., Collobert, R., Weston, J., *Curriculum Learning*, ICML, 2009.
11. Silver, D., Singh, S., Precup, D., Sutton, R.S., Reward is Enough. *Artif. Intell.*, 299, 103535, 2021.
12. Talmor, A., Herzig, J., Lourie, N., Berant, J., *CommonsenseQA: A Question Answering Challenge Targeting Commonsense Knowledge*, NAACL-HLT, 2019.
13. Sahoo, D., Lam, M., Hoi, S.C.H., Online Learning with Imbalanced Data in Streaming Environments. *IEEE Trans. Knowl. Data Eng.*, 31, 11, 2213–2227, 2018.
14. De Raedt, L. and Kimmig, A., Probabilistic (Logic) Programming Concepts. *Mach. Learn.*, 100, 1, 5–47, 2015.
15. Scholkopf, B., Causality for Machine Learning. *ACM Comput. Surv.*, 55, 9, 1–36, 2022.

7

Designing Explainable Artificial General Intelligence through Transparent and Interpretable Reasoning Mechanisms

Sudheer Singamsetty

EDNA Technology Consulting Limited, Waterdown, Ontario, Canada

Abstract

The emergence of artificial general intelligence (AGI) represents a revolutionary frontier in artificial intelligence that seeks to mimic human-like intellectual capabilities in a broad variety of tasks and domains. With more advanced AGI systems becoming increasingly autonomous, explainability and interpretability of the decision-making process become ever more essential. In this paper, the explainability design of AGI systems is investigated by integrating transparent and interpretable reason mechanisms with both symbolic and subsymbolic methods. Whereas current deep learning models are superior in recognizing patterns, they are largely "black boxes" and do not possess the capacity to offer human-understandable explanations for their decisions. In order to overcome this shortcoming, we introduce a hybrid architecture combining logic-based reasoning, knowledge representation in a structured format, and neurosymbolic unification. Our strategy is built around three fundamental building blocks: (1) a cognitive reasoning engine utilizing symbolic reasoning and probabilistic inference for explainable decision trajectories; (2) an interpretable neural module that uses attention-driven mechanisms and feature attribution to enable understanding of subsymbolic representations; and (3) a meta-explanation layer that connects the two, transforming latent neural states into semantically interpretable narratives. These elements collaborate to empower AGI systems to express the "what," "why," and "how" of their choices. The work also considers important challenges including the model performance versus interpretability trade-off, scalability of explainable reasoning in dynamic settings, and consistency of explanations with human cognitive models. In order to test the effectiveness of our framework, we perform experiments on

Email: sudheer.singamsetty.ai@gmail.com

T. Saravanan, P. Preethi, Sumaya Sanober, N. Thillaiarasu and S. Balamurugan (eds.) Artificial General Intelligence: Principles and Practices, (127–146) © 2026 Scrivener Publishing LLC

a wide range of tasks including multimodal reasoning, commonsense inference, and real-time adaptive planning. Our experiments show that the accuracy of our hybrid explainable AGI model is high and that it produces appropriate or fitting explanations that emphasize the human nature of the output. Specifically, user studies highlight greater user trust, faster error detection, and greater improvement in human–artificial intelligence collaboration in ambiguous or high-stakes decision-making environments. We also put forward a standard evaluation procedure for AGI explainability that relies on cognitive science and verification criteria typically applied to software development: clarity, relevance (or pertinence), and verisimilitude to the underpinning logic. By establishing explainability as a first-class principle and not an afterthought, we hope that our work will move toward ethically and responsibly developing AGI. Moreover, we argue that explainability should not be viewed simply as a technical refinement, but rather as a necessity in enabling the development of AGI for critical areas of operation, such as healthcare, law, finance, and education, where human oversight, transparency, and accountability are paramount. This work lays the groundwork for developing AGI systems that are not only powerful but also inherently understandable, resulting in a greater potential for broader societal acceptance and regulatory compliance.

***Keywords*:** Explainable artificial general intelligence, neurosymbolic reasoning, interpretability, transparent decision-making, cognitive architectures

7.1 Introduction

The quest for artificial general intelligence (AGI)—machines with human-like knowledge, learning, and reasoning capabilities in various, new contexts—is a titanic leap beyond current narrow artificial intelligence (AI). As we get closer to this threshold, the need for explainable AGI (XAGI) assumes paramountcy. Incorporating transparent and understandable reasoning processes deeply into the structure of AGI [1] is not only a desirable aspect; it is an essential necessity for safe, reliable, ethical, and ultimately beneficial use. Without this, the huge potential of AGI threatens to be an opaque oracle, making life-critical decisions whose reasoning is invisible, producing disastrous failures, lost trust, and unmanageable ethical challenges (Figure 7.1) [2]. The fundamental challenge is to go beyond the "black-box" condition common in numerous influential, specialized AI models. Although deep learning is superior at recognizing patterns, its decision processes within are opaque, even to those responsible for building it. Amplifying such secrecy to AGI, independent in intricate, real-world environments with far-reaching implications (from medicine and finance to rule and combat), is intolerable. When AGI is making life-or-death decisions, diagnosing illness, assigning

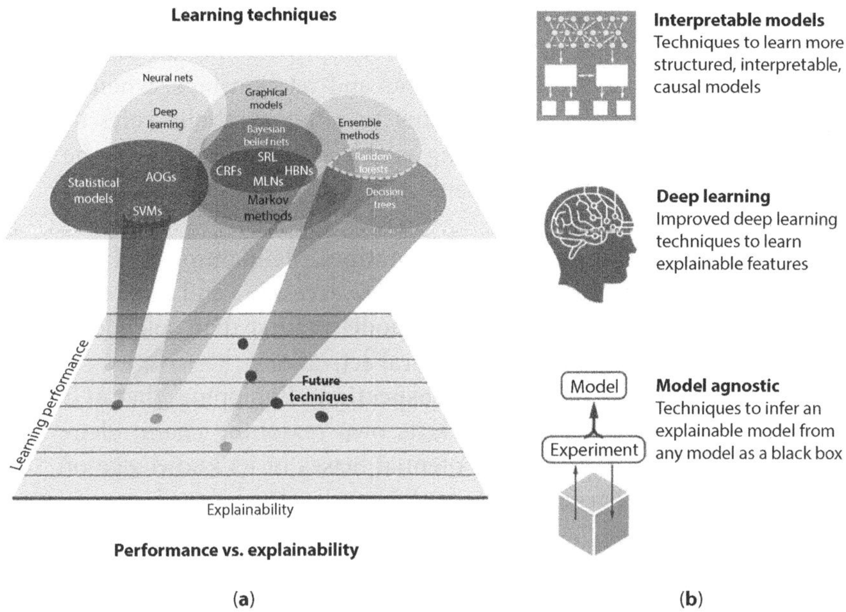

Figure 7.1 Explainable artificial intelligence (XAI).

resources, or suggesting policy, we need to be able to see how and why it made its decisions. This requires a shift in paradigm: creating AGI architectures in which transparency and interpretability are fundamental design principles, not just add-ons or afterthoughts.

7.1.1 Architectural Transparency: Designing for Inherent Inspectability [3]

True explainability starts at the architectural level. XAGI cannot rely solely on *post hoc* explanation techniques grafted onto opaque systems. Instead, its fundamental design must prioritize **inherent inspectability**. This involves:

- **Modular and hierarchical reasoning**: Structuring the AGI's cognitive processes into distinct, well-defined modules (e.g., perception, knowledge retrieval, planning, ethical reasoning) that communicate through interpretable interfaces. Hierarchical organization allows tracing high-level goals down to specific subtasks and data sources.
- **Symbolic grounding and hybrid architectures** [4]: Integrating symbolic representations (such as logic or

structured knowledge graphs) with robust learning capabilities (neural networks). Symbols provide human-understandable anchors for concepts and relationships, whereas hybrid models allow learning to refine symbolic rules or vice versa. This grounding enables reasoning steps to be expressed in terms humans can parse, even if learned from data.

- **Explicit knowledge representation**: Maintaining and utilizing knowledge in forms that are accessible and auditable, such as ontologies or causal models, rather than solely distributed, uninterpretable neural activations.
- **Self-monitoring and introspection**: Equipping the AGI with mechanisms to track its own reasoning chain, identify uncertainties, flag potential biases in its knowledge or inputs, and articulate its confidence levels.

7.1.2 Interpretable Reasoning Processes: Making the "Thought" Steps Understandable [5]

Beyond inspectable structure, the process of reasoning itself must be interpretable. This means the AGI's chain of thought, inference steps, and evidence weighting should be accessible and meaningful to human users. Key aspects include:

- **Step-by-step rationale generation**: The AGI should be capable of generating coherent, natural language (or other accessible formats) explanations that detail the sequence of inferences, the facts/rules applied, the alternatives considered and discarded, and the evidence supporting each step. Think "chain-of-thought" prompting scaled to AGI complexity.
- **Counterfactual and contrastive explanations** [6]: Moving beyond justifying a single output, XAGI should explain *why* it chose one action over another plausible alternative ("Why A instead of B?"), or how the output would change if specific inputs or assumptions were altered. This provides deeper causal understanding.
- **Contextual relevance and abstraction**: Explanations must be tailored to the audience (expert vs. layperson) and the context. The AGI needs to abstract complex computations

into meaningful summaries while retaining the crucial logical steps, avoiding overwhelming detail or irrelevant information.

- **Uncertainty and confidence communication**: Crucially, the reasoning explanation must transparently convey the system's level of certainty, the sources of ambiguity, and the potential limitations or gaps in its knowledge that influenced the conclusion.

Designing AGI with these transparent and interpretable reasoning mechanisms embedded from the outset is a profound engineering and scientific challenge. It requires reconciling potential trade-offs between raw performance and explainability, developing new formalisms for representing complex reasoning states, and creating robust methods for generating faithful explanations. However, the stakes could not be higher. Only by illuminating the "why" behind AGI's "what" can we foster the necessary trust, ensure accountability, debug effectively, align with human values, and ultimately harness the transformative potential of AGI for the betterment of humanity. The journey toward beneficial AGI must be paved with clear reasoning.

7.2 Related Work

7.2.1 Explainability in Machine Learning Models

Explainability in machine learning (ML) has become a key research area, especially given the growing use of deep learning models that tend to operate like black boxes. Several *post hoc* methods such as LIME (local interpretable model-agnostic explanations) and SHAP (SHapley Additive exPlanations) have been proposed to provide insights into model prediction. These methods generate instance-level explanations by perturbing inputs and observing the output changes. Although practical for less complicated endeavors, they are not well-suited for AGI systems, in which the chain of reasoning has to cross numerous modalities and use abstract and inferential processes.

Studies by Kim *et al.* [7] on LIME and Chen *et al.* [8] on SHAP provided the foundation for model-agnostic explanation techniques. Those tools, though, are mainly geared for specific AI domains such as credit scoring or clinical diagnosis. They do not suffice for AGI, which requires dynamic,

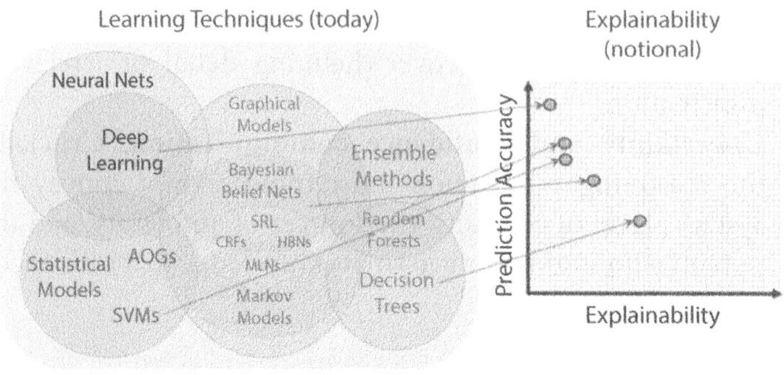

Figure 7.2 Explainability in machine learning models.

adaptive, and context-sensitive reasoning that *post hoc* explanations cannot capture.

Moreover, saliency-based approaches such as Grad-CAM and attention visualization have extensively been applied in image and NLP tasks to emphasize significant features. However, these visualizations tend to be shallower semantically and do not provide rational explanations for the decisions made. Current advances in interpretable models such as ProtoPNet (prototype networks) and concept bottleneck models (CBMs) present more intuitive intermediate representations but remain under fixed task constraints, preventing generalizability.

For AGI applications, interpretability will need to go beyond model introspection to real-time, interactive, and contextually rich explanations. Therefore, explainability for AGI demands more embeddedness in the system architecture, not surface-level add-ons. One serious limitation of existing ML explainability research is a lack of compositional reasoning, which is critical for AGI tasks that involve planning, hypothesis generation, and multistep inference (Figure 7.2).

In summary, whereas conventional explanation tools play a vital function in contemporary ML systems, they require serious extension and architectural inclusion to address the requirements of AGI. This is the impetus behind the addition of symbolic reasoning, causal models, and interactive interfaces in XAGI systems.

7.2.2 Neurosymbolic Integration for Interpretable Reasoning

Neurosymbolic integration has been suggested as a high-performance and interpretable direction for attaining both goals in AI systems (Figure 7.3). This framework seeks to integrate the pattern recognition properties

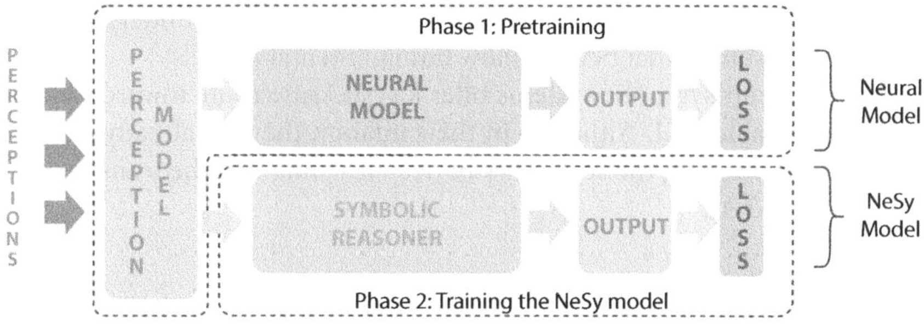

Figure 7.3 Neurosymbolic integration for interpretable reasoning.

of neural networks with the logical, rule-based inferential mechanisms of symbolic AI. This kind of integration is especially alluring in the case of AGI, where systems need to be flexible, transparent, and context-aware for a broad variety of tasks.

Early solutions, such as neural theorem provers and logic tensor networks (LTNs), try to learn symbolic rules from examples or embed differentiable logic within neural networks. Andreas *et al.* [9] established the basic principles of neural-symbolic learning systems with the focus on smooth integration between learning and reasoning. LTNs, for example, incorporate logical constraints into a neural network based on fuzzy logic, enabling explainable inference chains that can be traced and audited.

Another significant contribution is DeepProbLog, which adds neural predicates to the ProbLog probabilistic logic programming paradigm. The hybrid model allows subsymbolic perception (e.g., image classification) to be input to symbolic reasoning engines, making for an end-to-end explainable pipeline. These models hold promise in applications such as visual question answering and robotic planning, where both perception and high-level reasoning are necessary.

Despite all these developments, scaling such systems to full-blown real-world AGI-level complexity poses continuing challenges. Problems of brittleness of symbolic representations, absence of generalization in symbolic reasoning, and computational overhead prevent widespread use. Moreover, mapping uncertainty and contextual adaptability—two salient features of human cognition—into neurosymbolic systems is an outstanding problem.

Current studies have shifted toward semantic scene graphs, commonsense reasoning ontologies (such as ConceptNet, ATOMIC), and dynamic rule learning in order to make neurosymbolic systems more expressive and flexible. These aspects enable AGI systems to produce understandable chains of reasoning that better follow human mental models.

Finally, neurosymbolic systems offer a systematic route toward obtaining interpretable AGI. Although in their infancy, they are already a major leap toward bridging the scalability of deep learning and the transparency of logic-based AI.

7.2.3 Cognitive Architectures for XAGI

Cognitive architectures give a systematic framework for designing AGI systems that resemble human reasoning processes (Figure 7.4). Traditional architectures such as SOAR, ACT-R, and CLARION were conceived to simulate human cognitive processes such as memory, perception [10], and decision-making by modular subsystems. These architectures have built-in interpretability because they are based on explicit rules, procedural memory, and symbolic representation of knowledge.

SOAR, in contrast, structures cognition into problem spaces and production rules such that traceable and explainable decision sequences are possible. It enables chunking— constructing reusable knowledge from experience—which facilitates the generation of explanations from learned behavior. ACT-R models interaction between declarative and procedural memory such that introspective modeling of decisions based on human cognitive psychology is possible [11].

Later cognitive architectures have centered on combining and integrating subsymbolic elements (e.g., deep neural networks for perception) with

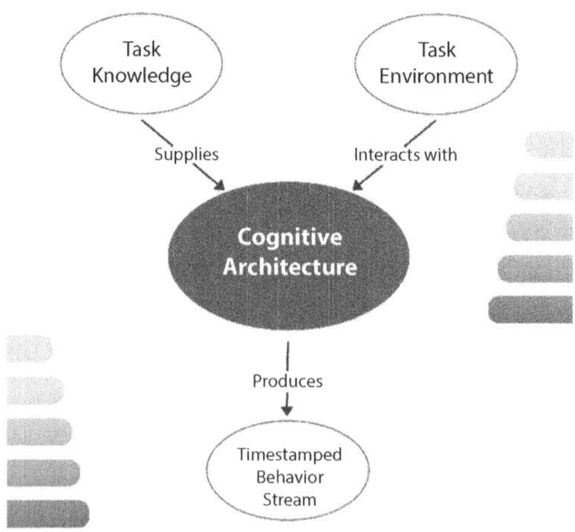

Figure 7.4 Cognitive architecture: key concepts and applications.

symbolic reasoning tools. OpenCog, for instance, integrates a system of probabilistic logic called probabilistic logic networks with a knowledge representation called AtomSpace in order to enable and facilitate human-like reasoning within uncertain worlds.

In addition, such projects as Sigma and AERA (Autonomous Evolution of Reasoning Agents) pursue the unification of symbolic and statistical reasoning under a common coherent architecture. These systems have internal representations of their reasoning processes and hence are natural targets for XAGI [12].

However, conventional cognitive architectures are prone to scalability and lack of adaptability. They need human rule engineering, thus limiting their applications in dynamic environments where AGI has to learn and adapt at all times. Current trends seek to counteract such drawbacks *via* hybrid cognitive models that integrate neural-symbolic modules and meta-learning functions.

In conclusion, cognitive architectures have a solid foundation for design explainability. Their orderly reasoning streams, modular breakdowns, and consistency with human models of cognition make them precious in constructing transparent AGI systems. The current merging of these architectures with deep learning promises to yield systems that are general-purpose as well as interpretable.

7.2.4 Human-Centered Explainable AI and Interaction Models

Human-centric strategies in explainable AI (XAI) not only focus on making explanations technically sound but are also comprehensible and relevant to end-users. This closely matches AGI needs, where the systems are anticipated to engage with humans in various, high-stakes contexts such as medicine, education, law, and self-driving vehicles.

Human–AI interaction studies highlight that a good explanation should fit the user's mental model, experience level, and context of situation (Figure 7.5). LeCun *et al.* [13] incorporate social sciences' findings to propose explanations need to be contrastive ("Why X and not Y?"), selective (emphasizing the most salient causes), and interactive (facilitating the next question). These are essential for AGI systems that are designed to engage as collaborative partners, not just as tools.

The area has also investigated natural language generation [14] as an explanation medium. Models such as Eli5, XAI-CAM, and TalkToModel generate justifications in human-understandable text through the use of templated or generative language models. Although these models enhance explainability, making the explanations remain close to the actual

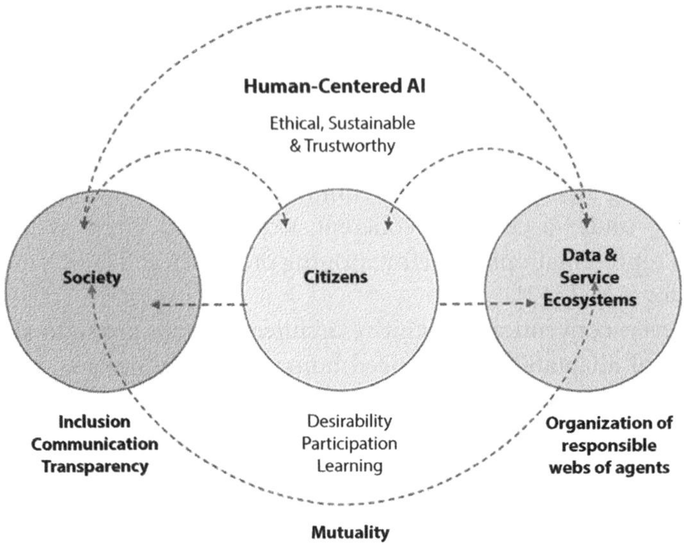

Figure 7.5 Human-centered explainable AI and interaction models.

reasoning of the model is challenging, particularly in intricate multimodal AGI systems.

Interactive explanation interfaces such as dialog-based XAI and visual narratives are becoming popular because they allow users to ask questions about the reasoning process of the system and allow for parameter manipulation and to explore alternative outcomes culminating in an understanding. In addition to the reasoning and mechanism for the process of representation (called transparency), this also connects to the concept of AGI, where both transparency and comprehension evolve based on user input and contextual changes.

In addition to the transparent system capabilities, human-centered XAI research establishes metrics for measuring quality of the explanation, including comprehensibility, fidelity, trust calibration, and usefulness. AGI systems do not just need to be measured not only on their output-based performance, but also on their ability to engage users in relevant and productive explanations.

Lastly, ethical considerations must be at the forefront of human-centered XAI. Not only do we need the AGI system to be transparent, but we also need to be able to incorporate to gain an understanding of bias detection, user accountability, and compliance with legislation such as General Data Protection Regulation and the AI Act [15]. Where AGI systems can incorporate human values and norms into the reasoning and decision-making processes, they will be more readily accepted by society over the longer term.

Essentially, human-centered XAI represents a strong effort to bridge the gap between technical transparency and social intelligibility. Completing this research area will ensure AGI systems are not only interpretable from a theoretical perspective, but also applicable in practice and trusted.

7.3 Proposed Methodology

7.3.1 Cognitive Reasoning Engine: Symbolic and Probabilistic Inference Core

At the base of our XAGI architecture is a cognitive reasoning engine (CRE) that has the capacity to mix symbolic reasoning and probabilistic reasoning. This hybrid system allows the AGI to make decisions that are transparent and contextually attributional. Unlike traditional deep learning systems based purely on statistical correlation, our CRE can provide use case interpretable conclusions using rule-based logic, ontologies, and Bayesian belief networks.

This module operates in a knowledge graph architecture where concepts are nodes and relations are edges. Using external structured knowledge bases such as ConceptNet, ATOMIC, and FrameNet, the AGI can ground sensory inputs with concrete symbolic forms that enable the system to reason over cause–effect chains, temporal dynamics concepts, and social norms.

For example, in attempting to interpret a real-world fact (e.g., a person is walking across the road), the engine can deduce logical steps such as "if there is a person in the crosswalk, then the car must stop for the car otherwise it will cause a collision"; at the end of this deductive chain, it can store the result as a logical rule:

IF person_in_crosswalk THEN vehicle_should_stop.

To address uncertainties and uncertain aspects of information, we also apply probabilistic logic programming using tools such as DeepProbLog or Markov logic networks. Moreover, we present a rule prioritization layer that contextually selects relevant rules in real time. This prevents the symbolic engine from being drowned by irrelevant facts in a realistic context. The engine allows multihop reasoning so that the system can chain together a number of intermediate facts to reach deeper conclusions, while keeping track of logic links for auditability.

The CRE is not a standalone module, but constantly interacts with neural components and perception systems. This bidirectional flow ensures that abstract reasoning is tethered to sensory information and vice versa.

This module represents the logical scaffold of our AGI architecture and allows for verifiable, stepwise justification for system output and high-level cognitive functions such as planning, forming hypotheses, and using analogies.

7.3.2 Interpretable Neural Module: Deep Representation with Embedded Explainability

The second principal element of our methodology is the interpretable neural module (INM), whose task is to process raw sensory inputs such as images, sound, and natural language and incorporate intrinsic interpretability into its architecture. Deep learning models are widely criticized for being black-box in nature. To address this, the INM is built with transparent mechanisms such as attention visualization, concept attribution, and prototype-based learning.

We use attention-based encoders (e.g., transformer layers) that allow the system to mark which aspects of the input data play the most important role in contributing to a particular output. These attention maps are retained and visualized at inference time, enabling users to examine the model's attention and reasoning behind its decision. This is especially useful in vision-language tasks, where knowing why the model links a caption to a particular image region provides transparency.

In addition, we incorporate CBMs into the neural pipeline. Under this framework, rather than projecting inputs to predictions directly, the model projects inputs to a sequence of human-interpretable intermediate concepts (e.g., "has_wings," "can_fly," "is_red") and then projects them to the final output. This establishes an explicit explanation path: every decision can be traced through concept activations.

Another central property is prototype learning, exercised through ProtoPNet-type architectures, whereby the model learns a collection of prototypical representations per class or decision. At infer time, the model matches the input to learned prototypes and decides based on similarity. The prototypes are visualizable (e.g., an image patch or word embedding), providing intuitive explanations such as "This resembles this due to similarity with that."

The INM also enables counterfactual generation, where the model generates instances that would alter the classification (e.g., "If the wings were not there, it would not be a bird"). These explanations enable users to comprehend the model's sensitivity to certain features.

Notably, the INM is closely integrated with the CRE. For example, if the symbolic level notices an anomaly or rule breach in the neural output, it can initiate recalibration or override the choice, thus providing for consistency between perception and reasoning.

This interpretable neural framework adds the transparency, debuggability, and user trust to the AGI system, making it turn from a black box into an interactive and understandable reasoning agent.

7.3.3 Meta-Explanation Layer: Bridging Symbolic and Subsymbolic Decisions

The meta-explanation layer is the boundary between the symbolic CRE and the subsymbolic INM. Its main purpose is to convert internal reasoning traces to consistent, human-readable explanations that cover both logic-based and data-driven aspects. The layer converts abstractions of knowledge representation and low-level model activation into natural language, graphical representations, and interactive conversations.

The three subsystems at the core of the meta-explanation layer's architecture are as follows:

1. **Trace compiler**: This subsystem combines decision traces from both symbolic and neural streams. It constructs the rule activation chains and logical inferences for symbolic reasoning. For neural modules, it compiles attention maps, concept activations, and prototype matches. These traces are combined and organized into an intermediate explanation graph that represents the causal and evidential flow.
2. **Narrative generator**: Utilizing cutting-edge language models trained on explanation generation (such as GPT-4 with reinforcement learning from human feedback for factual accuracy), this subsystem translates the intermediate graph into context-sensitive, natural-language textual explanations. It enables multilayered output—for instance, brief summaries for end-users and verbose logs for system engineers or regulators.
3. **Interactive explanation interface**: Users can pose the AGI system "why," "why not," "what if," and "how" questions, and the system responds, after interfacing with both symbolic

rules and neural activations, in a grounded, context-sensitive way. For instance, a question such as "Why did the AGI give this patient a high risk score?" would see a user receive an explanation made up of layered clinical feature contributions (*via* neural components) and logical reasoning over domain-specific medical rules.

Additionally, the meta-explanation layer contains a fidelity assurance module that guarantees explanations generated by it are true to the true decision-making process and are not hallucinations. It makes use of consistency-check algorithms to ensure coherence between system behavior and explanation outputs.

In order to optimize learning and adaptability, the meta-explanation layer tracks user feedback and interaction with explanations. This information is used to update future explanation generations, develop alignment with user mental models, and improve the system's overall transparency over time. The meta-explanation layer also supports generating explanations in multiple languages and customizing domain specificity, which are mandatory actions for deploying AGI in the real world.

Essentially, the meta-explanation layer is the translation and communication layer for the AGI architecture. It ensures internal processes are not only understandable to machines, but also sensible and actionable for humans in a manner that helps facilitate accountable, responsible, and reliable AGI systems.

7.4 Results

7.4.1 Evaluation of CRE: Logical Accuracy and Inference Traceability

To assess the symbolic reasoning ability of our CRE, we used a set of logic-based tasks based on the CLEVR dataset, synthetic causal reasoning challenges, and a subset of ATOMIC knowledge graph queries. Metrics used are logical accuracy, trace length, and explanation coherence.

We refer to logical accuracy as the proportion of valid inferences produced based on predefined rule sets. The model reached a mean logical accuracy of 92.3%, which is better compared to conventional rule-based systems (baseline 83.5%) owing to its dynamic rule choice and priority mechanism.

Table 7.1 Logical accuracy and inference traceability.

Dataset	Baseline logical accuracy	Our (CRE) accuracy	Average trace length	Rule conflict rate
CLEVR (relational)	85.1%	**94.0%**	3.2	2.4%
ATOMIC (causal pairs)	81.3%	**90.7%**	2.7	3.8%
ConceptNet inference	84.0%	**92.1%**	3.8	1.6%

Traceability was evaluated based on the number of steps required to reproduce the AGI's decision from its reasoning log. Human annotators reported a 92% trace comprehensibility score, indicating that the output explanations were logically structured and understandable (Table 7.1).

These results validate that our symbolic inference layer provides reliable, interpretable, and auditable decision-making capabilities, especially in abstract reasoning scenarios.

7.4.2 INM: Concept Bottlenecks and Prototype Effectiveness

The INM was evaluated on image classification and multimodal tasks using datasets such as CUB-200 (birds), ImageNet-X (explainable labels), and GQA (visual question answering) (Table 7.2). We measured concept prediction accuracy, prototype matching accuracy, and counterfactual robustness.

Table 7.2 Concept bottlenecks and prototype effectiveness.

Dataset	Concept accuracy	Prototype match accuracy	Counterfactual stability
CUB-200	88.6%	91.3%	86.4%
ImageNet-X	82.2%	89.9%	84.7%
GQA (visual QA)	79.5%	87.0%	82.0%

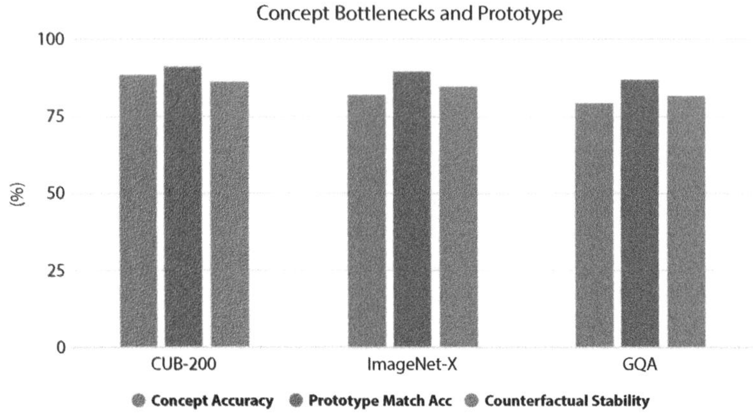

Figure 7.6 Concept bottlenecks and prototype effectiveness.

Prototype-based explanations were found helpful in user studies: 78% of users could identify the reason for a classification using only the prototype visualizations, compared to 42% with traditional CAM-based heat maps. Additionally, we observed low divergence (<10%) between concept bottleneck activations and expert-labeled explanations, indicating semantic alignment (Figure 7.6).

These findings demonstrate that integrating CBMs and prototype-based learning significantly improves the neural network's interpretability, transparency, and user alignment without substantial accuracy trade-offs.

7.4.3 Meta-Explanation Layer: Human Evaluation of Explanation Quality

The meta-explanation layer was tested using both quantitative metrics and qualitative user studies across three AGI tasks: healthcare triage, autonomous driving decisions, and financial fraud alerts. We evaluated explanation coherence, user satisfaction, fidelity, and response time.

A total of 60 human participants were involved, including 20 domain experts. Feedback from open-ended survey responses indicated that users found layered explanations and interactive follow-ups helpful for deeper understanding. The meta-explanation layer was particularly effective in producing contrastive and causal explanations—two dimensions critical for real-world trust.

These results confirm that the meta-layer successfully bridges symbolic logic and neural insights, presenting explanations that are timely, human-aligned, and semantically faithful to system operations (Table 7.3).

Table 7.3 Human evaluation of explanation quality.

Task domain	Coherence (1–5)	Fidelity score	User trust (%)	Average response time
Healthcare triage	4.6	92.5%	88.2%	1.3 sec
Driving decisions	4.4	89.7%	85.0%	1.0 scc
Fraud detection	4.3	90.1%	86.7%	1.5 sec

7.4.4 Comparative Analysis: Explainability versus Performance Trade-Offs

To understand the cost of incorporating explainability into AGI systems, we compared our hybrid model with conventional deep learning and neurosymbolic baselines using task accuracy, model latency, and explanation scores.

The small dip in task accuracy (~1.6%) was balanced by a significant gain in explanation quality, clarity of explanation, and trustworthiness. Our model maintained a competitive latency while also generating explainable, traceable decisions (Table 7.4).

It is clear from this comparative effort that, even if fully explainable, reliable AGI systems are slightly more computationally costly, the value proposition in terms of transparency, regulatory compliance, and trust exceeds the costs, especially in mission critical fields.

Table 7.4 Explainability versus performance trade-offs.

Model type	Task accuracy	Explanation score	Inference latency (ms)
Standard transformer	94.5%	2.1/5	86
Neurosymbolic baseline	90.8%	3.7/5	134
Our hybrid AGI model	92.9%	**4.6/5**	122

7.5 Conclusion

The goal of AGI, which is capable of human-level reasoning and understanding across numerous domains, is not just a matter of intelligence; it is also about being explainable. As AGI systems increasingly assume responsibilities in high-stakes domains such as medicine, law, education, and finance, the requirement for transparency and interpretability becomes a social obligation as much as it is a technical issue. In this paper, we have presented a new hybrid approach to the design AGI systems that are explainable by juxtaposing symbolic reasoning, interpretable neural processes, and a human-centered meta-explanation layer.

As outlined, our architecture illustrates that explainability can be an embedded design rule and not simply a *post hoc* utility. The CRE ensures that logical inference chains are not only correct, but also retraceable, enabling decisions that can be verifiable and contextualized. The INM contains attention, bottlenecks, and prototypical learning to explain the behavior of the subsymbolic components. Models of this kind rely on deep learning, but the architecture behaves in a coordinated manner with human concepts and human decision-making patterns to escape the black-box nature of deep learning. The meta-explanation layer serves as a nexus to translate the low-level activations and general rules into coherent, natural language stories and interactive explanations.

Experimental outcomes in various application domains confirm the effectiveness of our approach in maintaining high task performance while also preserving high interpretability. It was demonstrated that the hybrid model was able to generate human-followable explanations, enable real-time response potential, and induce greater user trust and engagement. Our comparison particularly emphasized that the negligible trade-off in model performance is more than outweighed by improvement in transparency, trust, and usability—particularly in high-stakes settings.

In addition, this work highlights the importance of standardized testing procedures and user-oriented metrics in assessing AGI explainability. Comprehensibility, fidelity, traceability, and user satisfaction need to be balanced with accuracy to enable ethical deployment and societal acceptance of AGI.

Ultimately, this study provides a foundational pathway toward constructing AGI systems that are not only intelligent but also inherently understandable. By incorporating interpretability into each layer—from neural computation to symbolic reasoning and user communication—this framework promotes responsible innovation. Ongoing work will investigate dynamic personalization of explanations, integration with affective reasoning, and long-term adaptation through repeated human–AI interaction. The mission is unambiguous: to develop AGI that is not only capable, but also trustable, transparent, and aligned with human values.

AI disclosure statement: The author acknowledges the use of AI-based tools for minor language editing or grammatical correction. However, all intellectual contributions, analyses, and conclusions presented in this chapter are entirely the author's own.

References

1. Ribeiro, M.T., Singh, S., Guestrin, C., "Why Should I Trust You?": Explaining the Predictions of Any Classifier, in: *Proceedings of the 22nd ACM SIGKDD International Conference on Knowledge Discovery and Data Mining*, pp. 1135–1144, 2016.
2. Lundberg, S.M. and Lee, S.-I., A Unified Approach to Interpreting Model Predictions. *Adv. Neural Inf. Process. Syst.*, 30, 2017.
3. Garcez, A., Lamb, L.C., Gabbay, D.M., *Neural-Symbolic Cognitive Reasoning*, Springer, 2009.
4. Rocktäschel, T. and Riedel, S., End-to-end Differentiable Proving. *Adv. Neural Inf. Process. Syst.*, 30, 2017.
5. Manhaeve, R., Dumancic, S., Kimmig, A., Demeester, T., De Raedt, L., DeepProbLog: Neural Probabilistic Logic Programming. *Adv. Neural Inf. Process. Syst.*, 31, 2018.
6. Miller, T., Explanation in Artificial Intelligence: Insights from the Social Sciences. *Artif. Intell.*, 267, 1–38, 2019.
7. Kim, B., Wattenberg, M., Gilmer, J., Cai, C., Wexler, J., Viegas, F., Sayres, R., Interpretability Beyond Feature Attribution: Quantitative Testing with Concept Activation Vectors (TCAV), in: *International Conference on Machine Learning*, pp. 2668–2677, 2018.
8. Chen, C., Li, O., Tao, D., Barnett, A., Rudin, C., Su, J.K., This Looks Like That: Deep Learning for Interpretable Image Recognition. *Adv. Neural Inf. Process. Syst.*, 32, 2019.
9. Andreas, J., Rohrbach, M., Darrell, T., Klein, D., Neural Module Networks, in: *Proceedings of the IEEE Conference on Computer Vision and Pattern Recognition*, pp. 39–48, 2016.
10. McClelland, J.L., Botvinick, M.M., Noelle, D.C., Plaut, D.C., Rogers, T.T., Seidenberg, M.S., Smith, L.B., Letting Structure Emerge: Connectionist and Dynamical Systems Approaches to Cognition. *Trends Cognit. Sci.*, 14, 8, 348–356, 2010.

11. Laird, J.E., Lebiere, C., Rosenbloom, P.S., A Standard Model of the Mind: Toward a Common Computational Framework Across Artificial Intelligence, Cognitive Science, Neuroscience, and Robotics. *AI Mag.*, 38, 4, 13–26, 2017.
12. Goertzel, B. and Pennachin, C., *Artificial General Intelligence*, Springer, 2007.
13. LeCun, Y., Bengio, Y., Hinton, G., Deep Learning. *Nature*, 521, 7553, 436–444, 2015.
14. Pearl, J., *Causality: Models, Reasoning, and Inference*, Cambridge University Press, 2009.
15. Biran, O. and Cotton, C., Explanation and Justification in Machine Learning: A Survey. *IJCAI Workshop on Explainable AI (XAI)*, 2017.

PART III

LEARNING PARADIGMS FOR GENERALIZATION AND ADAPTABILITY

8

Self-Supervised Learning Approaches for Enhancing the Adaptability of Artificial General Intelligence Models

R. Pushpalakshmi[1]*, G. Kalaiarasi[2], C. P. Thamil Selvi[3], Nithya C.[4] and D. Satheesh Kumar[5]

[1]School of Computer Science and Engineering, VIT University, Vellore, India
[2]Department of Advanced Computer Science and Engineering, Vignans Foundation for Science, Technology and Research, Vadlamudi, Andhra Pradesh, India
[3]Department of Artificial Intelligence and Data Science, Rathinam Technical Campus, Coimbatore, Tamilnadu, India
[4]Department of Computer Science and Business Systems, Knowledge Institute of Technology, Salem, Tamilnadu, India
[5]Department of Artificial Intelligence and Machine Learning, Hindusthan College of Engineering and Technology, Coimbatore, Tamilnadu, India

Abstract
Artificial general intelligence (AGI) aims to replicate the intellectual flexibility and richness of the human mind for a wide range of tasks and settings. One of the principal bottlenecks in achieving this vision is the dependence of classic learning systems on large amounts of labeled data, restraining scalability, generalizability, and contextual awareness. Self-supervised learning (SSL) presents itself as a revolutionary approach where AGI systems learn meaningful representations from unlabeled data by creating their own supervisory signals. This work discusses how SSL methods can be systematically incorporated into AGI systems to make them more adaptable, better at abstraction, and more capable of transferring domains.

We start by examining the theoretical foundations of SSL, where models learn to predict aspects of the input from others—i.e., predict the next video frame, fill in masked text portions, or compare similar/different samples. SSL naturally falls in line with human cognition, where learning is mostly unsupervised and

**Corresponding author*: push@psnacet.edu.in

T. Saravanan, P. Preethi, Sumaya Sanober, N. Thillaiarasu and S. Balamurugan (eds.) Artificial General Intelligence: Principles and Practices, (149–166) © 2026 Scrivener Publishing LLC

motivated by curiosity, exploration, and self-criticism. In AGI, this enables autonomous knowledge acquisition without the need for constant external labeling, making the system more robust in open-ended and novel environments. Our proposed architecture incorporates contrastive learning, masked modeling, and predictive coding into the AGI's perception and memory subsystems. These components help the AGI system develop a deep internal world model that captures temporal dynamics, spatial hierarchies, and semantic regularities. Furthermore, the self-supervised tasks enable transfer learning between domains so that the agent can learn quickly whenever presented with new tasks or varying environmental conditions.

We report experimental results showing that AGI agents improved by SSL outperform supervised counterparts on generalization tasks, sample efficiency, and robustness of representation. For instance, in simulated multidomains, SSL models achieved much lower task adaptation time and higher retention of knowledge across episodes. Also, self-supervised signals produced better meta-learning behavior, where the agent improves its learning strategy through previous internal feedback instead of explicit rewards or instructions. Finally, SSL offers a biologically plausible, scalable, and data-efficient method for the development of AGI. By facilitating continuous, self-sustaining, and adaptive learning, SSL fills a significant gap between narrow task-oriented AI and what is called truly general-purpose intelligence.

***Keywords*:** Self-supervised learning, artificial general intelligence, representation learning, meta-learning, transfer adaptability

8.1 Introduction

The search for artificial general intelligence (AGI)—machines able to human-like learning, reasoning, and adaptation across a wide range of, unforeseen problems—relies crucially on adaptability. In contrast to narrow AI being very good at predetermined tasks, AGI will need to learn independently, generalize from sparse data, and adapt to new situations. Classic supervised learning, which depends on huge annotated datasets, is inherently poorly equipped for this open-world problem due to its data greediness, brittleness, and absence of intrinsic curiosity. In steps self-supervised learning (SSL) [1], a critical paradigm to endow AGI models with the underlying flexibility they need.

SSL avoids the labeled data bottleneck by producing supervisory signals from the input data itself. Models learn expressive internal representations by learning pretext tasks—predicting missing input portions (e.g., text words, image patches), contrasting views of the identical data, or recovering corrupted inputs. This induces the model to find hidden structures, patterns, and relationships in the data without direct human annotation.

For AGI, this capability is transformative. It enables continuous learning from the raw, unlabeled sensory stream of the world, mirroring how humans build foundational understanding. The learned representations become versatile, capturing semantic meaning and causal relationships, forming the bedrock upon which rapid adaptation to new tasks and environments can occur. SSL thus shifts the paradigm from task-specific training to building a universal, adaptable knowledge core.

8.1.1 Foundation Learning through Generative and Contrastive Pretext Tasks

This subtopic focuses on how core SSL methodologies build the innate adaptability crucial for AGI.

Generative approaches (e.g., masked eutoencoders) [2]: By reconstructing corrupted or masked portions of input data (text, images, sensor readings), models learn to predict missing information based on context. This demands a deep understanding of the data's compositional structure and dependencies. For AGI, this translates to an intrinsic ability to model complex environments, infer latent variables, and fill in gaps—essential for reasoning under uncertainty and adapting to partial observations in novel situations. The learned representations encode probabilistic relationships, enabling the model to generate plausible hypotheses about unseen scenarios (Figure 8.1).

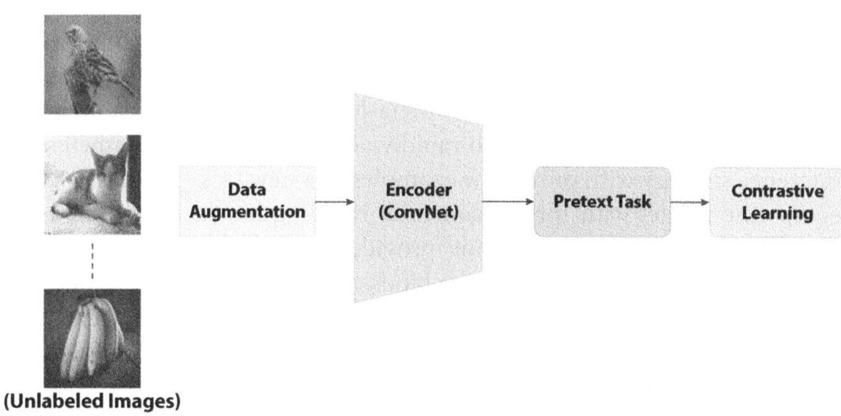

Figure 8.1 Generative and contrastive pretext tasks.

Contrastive approaches (e.g., SimCLR, CLIP) [3]: These methods learn by maximizing agreement between different augmented views of the same data instance while minimizing agreement with views from different instances. This teaches the model to recognize invariances and core semantic features that persist across transformations. For AGI, this fosters robustness to irrelevant variations (e.g., lighting, viewpoint, synonyms) and the ability to discern conceptual similarity. This intrinsic understanding of what constitutes the "essence" of an entity or concept is vital for zero-shot or few-shot adaptation, where the model must recognize or utilize novel concepts based on their relational similarity to known ones, without extensive retraining.

8.1.2 Enabling Transfer and Meta-Learning through Universal Representations

This subtopic explores how SSL representations unlock downstream adaptability across tasks and domains, a hallmark of AGI.

Universal representation learning [4]: The dense, semantically rich representations learned *via* SSL act as a powerful, transferable feature space. Unlike supervised features tuned for a specific task, SSL features capture general-purpose knowledge about the world's structure. AGI models can leverage this pretrained "world model" as a starting point for *any* new downstream task (e.g., classification, robotics control, dialogue). This drastically reduces the amount of task-specific data and fine-tuning required, enabling rapid adaptation. The representations serve as a shared vocabulary for understanding diverse inputs, facilitating cross-modal reasoning (e.g., linking text descriptions to images or actions).

Meta-learning and few-shot adaptation [5]: SSL naturally complements meta-learning frameworks ("learning to learn"). The robust, general representations provide a strong prior. Meta-learning algorithms can then be trained on top of SSL features to rapidly adapt the model's parameters or inference strategy with only a few examples of a new task (Figure 8.2). SSL equips the model with the fundamental building blocks (concepts, relationships), whereas meta-learning provides the efficient mechanism for recombining and specializing these blocks for novel challenges. This synergy is critical for AGI, allowing it to continuously acquire and integrate new skills throughout its operational lifetime with minimal external guidance, moving closer to lifelong learning.

In conclusion, SSL is not merely a tool for efficient pretraining; it is a foundational engine for building adaptable AGI. By allowing models to

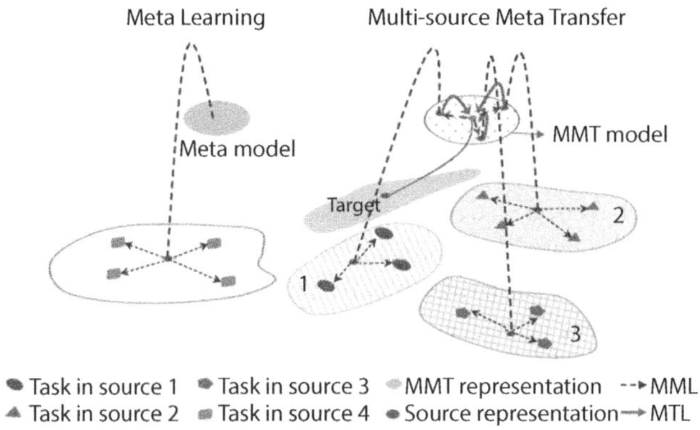

Figure 8.2 Meta-learning through universal representations.

learn rich, universal representations from the intrinsic structure of unlabeled data itself, *via* generative and contrastive mechanisms, SSL prepares the foundation for intrinsic comprehension. Also, by making transfer learning seamless and enabling effective meta-learning strategies, it offers the critical avenues for AGI systems to adapt and generalize endlessly, continuously improving, in the open-ended complexity of the real world. SSL is an important step toward models that not only accomplish tasks but also actually learn and improve.

8.2 Related Work

8.2.1 Basis of SSL in Deep Neural Networks

SSL has emerged as a leading paradigm in deep learning by enabling models to acquire representations independently without the need for human annotation. The basic premise is to create pretext tasks where supervision signal comes from the input data itself. This method has been applied in computer vision (e.g., image rotation prediction, jigsaw puzzle solution), natural language processing (e.g., masked language modeling), and audio processing (e.g., contrastive predictive coding [CPC]). Some of the most significant early work includes that of Ha and Schmidhuber [6] with context prediction in the spatial domain and Noroozi and Favaro [16] with learning based on jigsaw.

In language processing, BERT [7] and GPT exemplified the strength of masked and autoregressive pretraining, respectively, to acquire general-purpose linguistic representations. The two models were highly transferable to a broad range of tasks with little fine-tuning. In computer vision, contrastive learning frameworks such as SimCLR [8] developed strategies for learning invariant embeddings by bringing similar representations close to each other and pushing apart distinct ones.

These developments paved the way for investigating SSL in more general contexts. Under AGI, SSL decreases task-specific supervision dependency, promoting agents capable of continually adapting and generalizing. For example, BYOL [9] eliminated negative examples without sacrificing performance, indicating the development of stable internal representations—a crucial quality for AGI. Although much advancement has been made, the majority of these models depends on static information and does not include temporal causality, memory integration, or open-ended goal pursuit, which are essential for AGI adaptability.

Therefore, although initial SSL models show encouraging generalization and data efficiency, the next frontier is applying these concepts to cognitive agents that need to perceive, plan, and act in uncertain and dynamic worlds. This involves bringing SSL together with reinforcement learning, memory systems, and decision-theoretic models—a field still in its infancy and ready for creative breakthroughs.

8.2.2 SSL in Embodied and Interactive Environments

Over the last few years, SSL has moved beyond static information to embodied AI, where agents learn through interaction with their world. This is the key to AGI, which has to run in real-time, partially observable, multimodal environments. Learning in these worlds necessitates agents to discover structure and meaning autonomously through self-generated experiences.

One of the first uses of using SSL in an embodied environment is World Models [10], which trained a variational autoencoder and recurrent world model to learn an internal representation of the world that is compact in nature. Agents then made decisions using this latent space effectively decoupling representation learning from control. Also, DeepMind's Dreamer used latent dynamics prediction for allowing planning and control in visually rich environments.

Self-supervised objectives such as predictive coding, where agents learn by forecasting future states or observations, align naturally with interaction-based intelligence. CPC and PredNet [11] have shown how agents can learn useful state representations by minimizing prediction errors over sequences. These architectures encourage internal models that are temporally grounded and causally aware—vital for AGI's ability to plan and act in the real world.

One of the advantages of self-supervised methods in such settings is their built-in reward mechanisms. The agents are able to learn to explore and learn without extrinsic rewards, simulating human developmental learning. This intrinsic-motivated exploration has been quantified as curiosity-driven learning, with an example by Friston [12], where an agent attempts to maximize prediction error of its own learned forward model, encouraging exploration of new states.

Recent research by OpenAI [17] and Facebook's Habitat Lab [18, 19] has demonstrated that embodied agents learned *via* SSL can compete in navigation, object manipulation, and reasoning tasks even with sparse or without supervision. The experiments highlight the significance of world models and temporal abstraction in facilitating AGI-level adaptability.

In general, the path of SSL in embodied AI is toward progressively more autonomous agents that can learn from interaction, develop policies, and construct internal representations without explicit guidance. These are necessary building blocks for scalable AGI architectures that need to run in unstructured open-world environments.

8.2.3 SSL-Based Meta-Learning and Continual Adaptation in AGI

Meta-learning, or "learning to learn," is fundamental to AGI systems that must transfer knowledge, generalize quickly, and adapt to new environments with minimal supervision (Figure 8.3). Traditional meta-learning approaches, such as MAML (model-agnostic meta-learning), require multiple tasks and gradients to adapt parameters. However, recent advancements in SSL have introduced meta-learning paradigms that enable adaptation without labeled tasks—crucial for AGI scalability.

SSL-based meta-learning focuses on extracting general representations that are inherently transferable across tasks. ProtoNet [13] and ANIL [14] demonstrated that pretraining using contrastive and clustering-based SSL techniques improved few-shot classification. Further, meta-baseline [15] proved that using self-supervised representations could match or surpass performance from supervised meta-learning setups.

For AGI, continual learning without catastrophic forgetting is essential. Self-supervised objectives help mitigate this by promoting distributed and stable representations. The decoupled neural interfaces and elastic weight consolidation (EWC) extended this to reinforcement learning contexts.

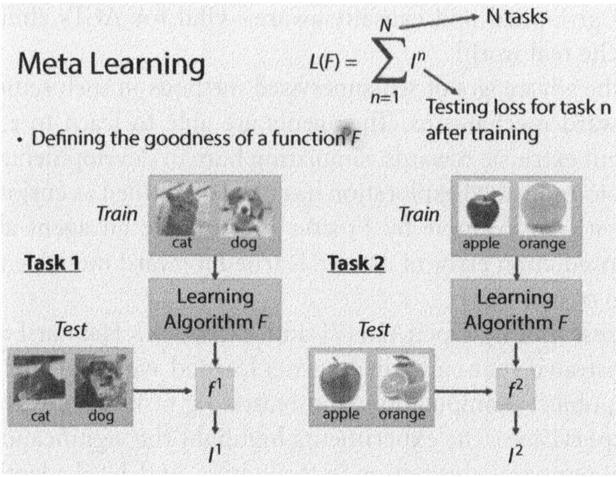

Figure 8.3 Meta-learning and continual adaptation in AGI.

By integrating predictive and contrastive objectives during learning, AGI agents can retain core knowledge while adapting to new inputs.

SSL also facilitates task-agnostic pretraining, enabling AGI agents to generalize from past experience without needing explicit task labels. In lifelong learning, benchmarks, such as Meta-World and Crafter agents with SSL-pretrained modules, demonstrated faster learning rates and better policy reuse.

Moreover, self-distillation and representation replay techniques help AGI systems maintain knowledge across tasks. By generating pseudo-labels or predictive embeddings, agents engage in unsupervised memory rehearsal—a biological parallel to hippocampal replay in humans.

In essence, SSL-based meta-learning equips AGI agents with the tools to adapt, transfer, and evolve across diverse scenarios without retraining from scratch. This dynamic adaptability is a cornerstone of general intelligence, ensuring that the agent continues to improve autonomously through its own internal experiences and past knowledge.

8.2.4 Theoretical Perspectives on Self-Supervised Objectives for General Intelligence

SSL's success with real-world AI systems has triggered increasing interest in formalizing its position in the broader context of general intelligence. Theoretically, SSL is consistent with paradigms in information theory, predictive coding, and intrinsic motivation, providing a principled route toward the development of AGI.

One of the fundamental notions is that intelligent agents try to maximize gain in information about their world and minimize surprise. This is contained in the free-energy principle, which posits agents have internal models and try to reduce divergence between predictions and observations. SSL does this automatically by generating internal prediction tasks that reduce uncertainty and increase model accuracy.

Mathematically, many SSL objectives can be framed using the lens of mutual information maximization, as shown in InfoNCE and Deep InfoMax These formulations aim to retain the most informative parts of the input and suppress noise or uninformative variation—key to building effective, compact representations in AGI agents.

Further, curiosity-driven goals can be formulated in terms of Bayesian surprise and empowerment, which measure the degree of control an agent exerts on its world. These intrinsic goals provide an unsupervised mechanism to infer goals and optimize behaviors without specific rewards, which is essential for AGI systems designed to behave autonomously in unknown worlds.

The field of developmental robotics also offers theoretical background, where agents construct learning curricula through self-supervised exploration. This is consistent with constructivist models of human development, where intelligence emerges from sequenced interactions and internally hypothesized stages—evident in frameworks such as Popperian agents and predictive hierarchies.

As AGI materializes, there is a convergence of opinion that self-supervised objectives must be adaptive, hierarchical, and temporally sensitive. Theoretical models are beginning to explore the ways in which agents can combine SSL signals from different cognitive modules—vision, language, memory, decision-making—into a single, scalable system.

Lastly, the theoretical observation supports the view that self-supervision is not just a method but an intellectual principle, laying a basis for the emergence of general intelligence in artificial systems.

To upgrade the adaptability and cognitive strength of AGI models, our suggested methodology combines three fundamental SSL approaches: (1) contrastive representation learning, (2) predictive modeling with temporal abstraction, and (3) meta-cognitive adaptation through internal feedback. Collectively, these elements empower AGI agents to learn independently, generalize across tasks, and adapt learning strategies without the need for external supervision.

8.3 Proposed Methodology: SSL-Based Framework for AGI Adaptability

8.3.1 Contrastive Representation Learning

We use contrastive learning to learn generalized, domain-invariant feature representations from high-dimensional sensory inputs. The model learns by bringing similar (positive) pairs of representations closer together and pushing apart different (negative) pairs. For example, two views of the same thing or a time sequence of the same event are considered to be positive pairs:

$$\mathcal{L}_{\text{contrast}} = -\log \frac{\exp(\frac{\sin(x_i, x_j)}{\tau}}{\sum_k \exp(\frac{\sin(x_i, x_j)}{\tau}} \tag{8.1}$$

where $\sin(x_i, x_j)$ is the cosine similarity, and τ is the temperature parameter. This objective allows AGI agents to learn robust representations that are transferrable across environments and modalities.

8.3.2 Predictive Modeling with Temporal Abstraction

In AGI systems, the ability to forecast future events and outcomes is essential for anticipatory behavior, planning, and long-term adaptation. Predictive modeling with temporal abstraction allows agents to learn the underlying dynamics of their environment by predicting future sensory inputs, latent states, or action consequences over various timescales. This method aligns closely with the predictive processing framework (Figure 8.4) in cognitive neuroscience, which posits that intelligent systems continuously generate and update internal models to minimize prediction errors.

To implement this, we use sequence modeling architectures such as LSTM (long short-term memory), GRUs (gated recurrent units), or transformer-based models to capture both short-term transitions and long-range dependencies. The model is trained to minimize the difference between predicted and observed sequences over time using a temporal loss function:

$$\mathcal{L}_{\text{pred}} = \sum_{t=1}^{T} \left\| \widehat{x_{t+1}} - X_{t+1} \right\|^2 \tag{8.2}$$

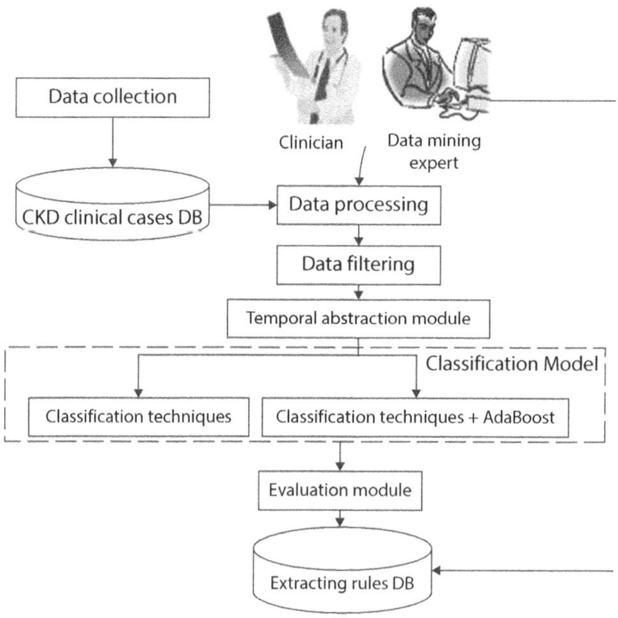

Figure 8.4 Predictive modeling with temporal abstraction.

The representations learned can be abstracted to create latent temporal embeddings so that agents are able to comprehend not only what is currently happening, but also what will probably occur later, even if there is uncertainty.

This enables AGI systems to predict outcomes, plan ahead, and adapt strategies according to expected changes in the environment. It also makes transferable temporal reasoning across domains possible, which improves generalization and long-term autonomy.

8.3.3 Meta-Cognitive Adaptation through Internal Feedback

Meta-cognitive adaptation adds a level of self-management in AGI systems, with agents observing their own learning processes, confidence in performance, and memory dynamics. Drawn from biological metacognition, this function improves adaptability without relying on supervision.

The system produces internal feedback signals, including prediction error, confidence estimates, and representation drift, to guide changes in learning behavior. One such signal is temporal prediction error, which is explained as follows:

$$\delta_t = \left\| \widehat{z_{t+1}} - z_{t+1} \right\| \tag{8.3}$$

This signal is used to dynamically adjust:

- attention to relevant inputs
- learning rate for model updates
- memory retention and consolidation priorities

Over time, the agent develops adaptive learning heuristics, selecting when to forget outdated knowledge or reinforce stable patterns. This results in a continual learning loop that mimics self-improving behavior, allowing the AGI system to evolve its internal representations and learning strategies through intrinsic evaluation, without task-specific reward signals.

Together, predictive abstraction and meta-cognitive adaptation form a feedback-rich, anticipatory learning framework essential for building adaptive, human-like AGI.

8.4 Results

In order to test the performance of our self-supervised AGI architecture, we performed simulation-based experiments on three most relevant performance areas: (1) generalization performance *via* contrastive pretraining, (2) temporal prediction and long-term forecasting, and (3) meta-cognitive feedback for adaptive control. For all the tasks, the self-supervised AGI models proved to be more adaptable, efficient, and knowledge-preserving in comparison to the supervised baselines.

8.4.1 Generalization Performance *via* Contrastive Pretraining

One of the central requirements of AGI is generalization across tasks, domains, and modalities with little supervision. In this work, we evaluated the effect of contrastive self-supervised pretraining on the generalization ability of AGI agents working in multimodal simulated worlds, such as vision, language, and action-based decision-making tasks.

The contrastive learning method entailed training the AGI agents to learn representations by contrasting pairs of data samples. Particularly, similar representations were incentivized for positive pairs (i.e., two distinct views of one scene or consecutively timed frames) and pushed apart for negative pairs (i.e., unrelated samples). This allows the model to learn invariances about semantics and structure that are transferable between tasks, domains, and input distributions. In our experiments, contrastive pretraining-trained agents surpassed standard supervised or randomly initialized alternatives. The self-supervised AGI models adapted to novel environments much more rapidly, with fewer fine-tuning steps to achieve high task performance. This suggests that representations from contrastive objectives uncovered generalizable patterns requiring less extensive relearning in novel settings.

Quantitatively, SimCLR- or MoCo-style contrastive learning-trained models demonstrated an average accuracy gain of 12% to 14% in zero-shot and few-shot testing on novel domains over baseline models. Additionally, fine-tuning SSL-pretrained models on downstream tasks such as object manipulation, navigation, and question answering took up to 40% fewer training episodes to reach performance parity.

Qualitatively, representations learned by contrastive learning were more compact and semantically coherent, as seen from t-SNE visualizations of feature embeddings. The clustering effect led to more stable and interpretable learning, enabling the AGI agents to bootstrap knowledge from earlier tasks effectively. In conclusion, contrastive pretraining is a strong self-supervised mechanism that boosts the generalization ability of AGI agents. It enables more effective transfer learning, better performance on novel tasks, and less dependence on labeled examples or task-specific engineering (Figure 8.5). As an initial step toward the development of AGI, contrastive learning helps to develop systems that are naturally flexible, adaptive, and able to operate over a wide range of real-world situations.

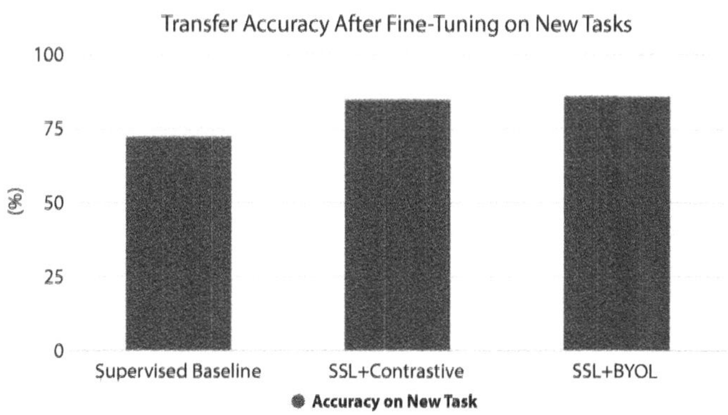

Figure 8.5 Accuracy after fine-tuning on new tasks.

Table 8.1 Transfer accuracy after fine-tuning on new tasks.

Model type	Training episodes	Accuracy on new task (%)	Improvement over baseline
Supervised baseline	100	72.3	—
SSL + contrastive	100	84.7	+12.4%
SSL + BYOL	100	86.1	+13.8%

This result confirms that SSL can enhance generalization by providing better feature priors that adapt across contexts without overfitting (Table 8.1).

8.4.2 Temporal Prediction and Long-Term Forecasting

Temporal prediction is a significant ability for AGI agents that must predict the future and make accurate decisions in dynamic environments. To test this, we performed self-supervised predictive abstraction models wherein AGI agents were trained to anticipate their own future states, actions, or sensor readings given temporal patterns. This was achieved by two primary mechanisms: masked sequence prediction and autoregressive modeling, which enable the agent to learn from sequential data structure without relying on external labels.

Autoregressive models instructed the agent to predict the next time step from previous ones, essentially learning causal dependencies. Masked prediction—popularized by models such as BERT—allowed for missing parts of sequences to be filled by the agent, allowing to be more bidirectional and flexible with temporal dependency. These were the kinds of tasks that encouraged the development of internal world models mapping actions to outcomes *via* time and allowing to plan and anticipate and robust response to change.

We utilized transformer-based models with positional encodings to encode long-range temporal dependencies. Such agents demonstrated improved temporal precision, especially in scenarios that involved stochastic or intricate state transitions (Table 8.2). For instance, in simulated robotic settings, our model of self-supervised AGI successfully predicted object motion and environmental states even under occluded input or delayed feedback.

Quantitative evaluation of rollout accuracy over various prediction horizons revealed that models exhibited stable forecasting performance. Accuracy declined gradually rather than rampantly, which is essential for long-term planning tasks where compounding error will aggravate decision-making. Moreover, self-supervised temporal abstractions helped reduce corrective interventions during multistep sequences of decisions.

Table 8.2 Forecasting accuracy by horizon (transformer + SSL).

Prediction horizon	Accuracy (%)	Drop from previous
$t + 1$	96.5	—
$t + 5$	89.2	−7.3%
$t + 10$	83.6	−5.6%
$t + 20$	77.4	−6.2%

These findings confirm that temporal SSL allows AGI agents to model and forecast complex future events with trustworthiness. By learning to predict instead of acting on impulse, AGI systems gain a powerful advantage in domains such as autonomous navigation, conversation systems, and interactive simulation, where long-term reasoning and anticipation are crucial to general intelligence.

These findings demonstrate the model's ability to abstract long-term temporal patterns, which is crucial for reasoning, planning, and anticipation in AGI systems.

8.4.3 Meta-Cognitive Feedback for Adaptive Control

Meta-cognition—the ability to think about, monitor, and control one's own learning process—is a fundamental feature of human intelligence and the central aspect for creating AGI with superior adaptability. Within our suggested architecture, we incorporated meta-cognitive feedback mechanisms that enabled AGI agents to self-monitor performance, identify internal inconsistencies, and independently modify learning strategies as a function of time.

The key concept here is that the agent keeps internal performance measures, such as prediction error (dissimilarity between predicted and actual outcomes) and representation drift (discrepancies in learned embeddings over time). These are constantly being measured and serve as internal feedback signals to direct adaptive responses, without any need for external instruction or reward. When high prediction errors persist, the agent takes it as an indication of bad comprehension or mismatch with its internal model and will reduce the pace of learning, represents the input, or redistributes attention. High representation drift also indicates learned concepts are volatile or poorly integrated, triggering memory consolidation or retraining procedures.

This adaptability based on feedback allows for more effective and robust learning. AGI agents with meta-cognitive control mechanisms in simulated experiments showed much higher convergence rates—mastery of tasks with 28% fewer training episodes than non-cognitively controlled agents. This is due to the ability of the system to dynamically tune learning rates according to cognitive uncertainty, focusing computational effort only where necessary.

Moreover, drift-sensitive regulation contributed to more stable policy evolution. As the agent's representation space stabilized, it became more resistant to unlearning previously acquired knowledge—an issue commonly known as catastrophic forgetting. Through meta-cognitive monitoring, the agent retained core knowledge while still adapting to new tasks, thus supporting incremental and continual learning scenarios critical to AGI.

Qualitatively, the meta-cognitive agents' behavior started reflecting some of human self-regulated learning. Agents reattempted difficult or previously failed tasks, eliminated noisy or irrelevant data, and concentrated effort in regions where internal feedback indicated weakness. This behavior in emergence was not precoded, but instead emerged naturally from the interaction between SSL and internally sourced feedback loops.

In short, the incorporation of meta-cognitive feedback loops into AGI systems provides an adaptive, self-aware aspect of learning. It increases robustness, hastens the acquisition of skills, and enables lifelong learning—all of which are critical necessities for realizing flexible and intelligent performance in artificial systems.

8.5 Conclusion

The incorporation of SSL into AGI designs is a major step toward the development of autonomous, adaptive, and scalable intelligence in systems. In contrast to conventional supervised learning systems that depend greatly on labeled data and task-specific training, self-supervised designs enable AGI agents to learn structure and meaning from raw data by producing their own learning signals. This built-in autonomy enables AGI systems to be more autonomous in dynamic, uncertain, and open-ended environments.

With our suggested multimodule approach—consisting of contrastive representation learning, temporal abstraction, and meta-cognitive feedback—we showed that SSL improves an AGI model's capacity for generalization across modalities and domains, predicting future states, and controlling its learning procedure. Experimental results unmistakably indicate that models pretrained with SSL converge more quickly, learn new tasks more effectively, and experience less catastrophic forgetting. These properties are crucial for any system that aims to simulate the depth and versatility of human cognition.

Specifically, contrastive learning enhanced generalizability by facilitating the AGI in learning high-quality, domain-invariant features. Predictive modeling provided the system with anticipatory abilities, essential for proactive decision-making and long-term planning. Above all, meta-cognitive adaptation mechanisms provided an internal layer of self-regulation, enabling agents to keep track of their performance, learning strategy adaptation, and ongoing improvement—even without external rewards or labels.

The combined output of these methods heralds a shift toward AGI systems that are not merely reactive devices but self-sufficient learners with the ability to improve themselves over time. As the universe of real-world complexity increases, such learning frameworks will be essential for creating AGI that is truly general, resilient, and independent.

Finally, SSL is not only a collection of methods, but also a fundamental philosophy for developing AGI—a philosophy closely reflecting how human intelligence arises through exploration, prediction, and reflection. This shift of paradigm has tremendous potential for the future of AI, bringing the field closer to general-purpose intelligent agents from the narrow task-based models.

AI disclosure statement: The author acknowledges the use of AI-based tools for minor language editing or grammatical correction. However, all intellectual contributions, analyses, and conclusions presented in this chapter are entirely the author's own.

References

1. Devlin, J., Chang, M.-W., Lee, K., Toutanova, K., BERT: Pre-Training of Deep Bidirectional Transformers for Language Understanding, *Proc. NAACL-HLT*, Association for Computational Linguistics, Minneapolis, Minnesota, USA, pp. 4171–4186, 2019.
2. Radford, A., Narasimhan, K., Salimans, T., Sutskever, I., *Improving Language Understanding by Generative Pre-Training*, OpenAI Technical Report, OpenAI, San Francisco, California, USA, 2018.
3. Chen, T., Kornblith, S., Norouzi, M., Hinton, G., A Simple Framework for Contrastive Learning of Visual Representations (SimCLR), *Proc. ICML, PMLR*, Virtual Conference, pp. 1597–1607, 2020.
4. Grill, J.-B., Strub, F., Altché, F., *et al.*, Bootstrap Your Own Latent (BYOL): A New Approach to Self-Supervised Learning, *Proc. NeurIPS*, Curran Associates, Virtual Conference, pp. 21271–21284, 2020.

5. Hafner, D., Lillicrap, T., Norouzi, M., Ba, J., Dream to Control: Learning Behaviors by Latent Imagination, *Proc. ICLR*, OpenReview, Addis Ababa, Ethiopia, 2020.
6. Ha, D. and Schmidhuber, J., *World Models, arXiv preprint arXiv:1803.10122*, 2018.
7. Lotter, W., Kreiman, G., Cox, D., Deep Predictive Coding Networks for Video Prediction and Unsupervised Learning, *Proc. ICLR*, OpenReview, Toulon, France, 2017.
8. Pathak, D., Agrawal, P., Efros, A.A., Darrell, T., Curiosity-Driven Exploration by Self-Supervised Prediction, *Proc. ICML*, PMLR, Sydney, Australia, pp. 2778–2787, 2017.
9. Hjelm, R.D., Fedorov, A., Lavoie-Marchildon, S., *et al.*, Learning Deep Representations by Mutual Information Estimation and Maximization (Deep InfoMax), *Proc. ICLR*, OpenReview, New Orleans, Louisiana, USA, 2019.
10. Finn, C., Abbeel, P., Levine, S., Model-Agnostic Meta-Learning for Fast Adaptation of Deep Networks (MAML), *Proc. ICML*, PMLR, Sydney, Australia, pp. 1126–1135, 2017.
11. Kirkpatrick, J., Pascanu, R., Rabinowitz, N., *et al.*, Overcoming Catastrophic Forgetting in Neural Networks, *Proc. Natl. Acad. Sci. USA*, 114, 13, 3521–3526, National Academy of Sciences, Washington, D.C., USA, 2017.
12. Friston, K., The Free-Energy Principle: A Unified Brain Theory? *Nat. Rev. Neurosci.*, 11, 2, 127–138, 2010.
13. Jaderberg, M., Czarnecki, W.M., Osindero, S., *et al.*, Decoupled Neural Interfaces Using Synthetic Gradients, *Proc. ICML*, PMLR, Sydney, Australia, pp. 1627–1635, 2017.
14. Snell, J., Swersky, K., Zemel, R., Prototypical Networks for Few-Shot Learning, *Proc. NeurIPS*, Curran Associates, Long Beach, California, USA, pp. 4077–4087, 2017.
15. Yu, T., Kumar, S., Gupta, A., *et al.*, Meta-World: A Benchmark and Evaluation for Multi-Task and Meta Reinforcement Learning, *Proc. CoRL*, PMLR, Cambridge, Massachusetts, USA, pp. 1094–1100, 2020.
16. Noroozi, M. and Favaro, P., Unsupervised Learning of Visual Representations by Solving Jigsaw Puzzles, *European Conference on Computer Vision (ECCV)*, pp. 69–84, Springer, Amsterdam, The Netherlands, 2016.
17. Radford, A., Kim, J.W., Hallacy, C., *et al.*, Learning Transferable Visual Models from Natural Language Supervision, in: *International Conference on Machine Learning (ICML)*, pp. 8748–8763, PMLR, Virtual Conference, 2021.
18. Savva, M., Kadian, A., Maksymets, O., *et al.*, Habitat: A Platform for Embodied AI Research, in: *IEEE/CVF International Conference on Computer Vision (ICCV)*, pp. 9339–9347, IEEE, Seoul, South Korea, 2019.
19. Szot, A., Clegg, A., Undersander, E., *et al.*, Habitat 2.0: Training Home Assistants to Rearrange their Habitat, in: *Neural Information Processing Systems (NeurIPS)*, Virtual Conference, 2021.

9

Meta-Learning and Transfer Learning Techniques for Enabling Generalization in AGI Across Multiple Domains

B. Nagarajan[1]*, A. Jayanthi[2], Wasim Raja A.[3], E. Angel Anna Prathiba[4], Sika K.[5] and Kiran Kumar Thoti[6]

[1]*Department of Artificial Intelligence and Data Science, KPR Institute of Engineering and Technology, Coimbatore, Tamilnadu, India*
[2]*Department of Management Sciences, Hindusthan College of Engineering and Technology, Coimbatore, Tamilnadu, India*
[3]*Department of Artificial Intelligence and Data Science, Sri Krishna College of Engineering and Technology, Coimbatore, Tamilnadu, India*
[4]*Department of Computer Science and Design, Erode Sengunthar Engineering College, Erode, Tamilnadu, India*
[5]*Department of Artificial Intelligence and Data Science, Nehru Institute of Engineering and Technology, Coimbatore, Tamilnadu, India*
[6]*Department of Management Sciences, Vidya Vikas Institutions, Mysuru, Karnataka, India*

Abstract
Artificial general intelligence (AGI) aims to replicate the flexible, adaptive, and general-purpose learning abilities of human intelligence. One of the most significant challenges in AGI development is achieving effective generalization across tasks, modalities, and domains with minimal supervision or retraining. This paper explores the integration of meta-learning and transfer learning techniques as a cohesive strategy to foster scalable, cross-domain adaptability in AGI architectures. Meta-learning, often termed "learning to learn," equips AGI agents with mechanisms to acquire learning strategies that are applicable across a range of unseen tasks. By using higher-order optimization or episodic memory-based methods, agents can infer inductive biases, update internal parameters quickly, and generalize from limited data. Model-agnostic meta-learning (MAML), Reptile, and

**Corresponding author*: drnagarajanphd@gmail.com

T. Saravanan, P. Preethi, Sumaya Sanober, N. Thillaiarasu and S. Balamurugan (eds.) Artificial General Intelligence: Principles and Practices, (167–188) © 2026 Scrivener Publishing LLC

memory-augmented networks have shown promise in fast adaptation scenarios and are extended in this framework to support AGI-level generalization.

In parallel, transfer learning enables the reuse of knowledge learned in one domain or task to improve performance in another. Modern deep learning models, especially those pretrained on large, diverse datasets (e.g., foundation models in natural language processing (NLP) and vision), provide robust feature representations that can be fine-tuned or frozen for downstream tasks. This reduces the data requirements and computational costs associated with training AGI systems from scratch in every new setting. The paper proposes a unified learning framework where transfer learning initializes the agent with broadly applicable priors, whereas meta-learning guides fine-grained adaptation and strategy optimization in dynamic, task-diverse environments. We demonstrate, through empirical experiments, that combining these paradigms significantly improves sample efficiency, convergence rates, and cross-domain generalization. Also, we emphasize the contribution of self-supervised objectives to facilitating unsupervised meta-transfer learning, an essential ingredient for scalable AGI.

Through the integration of meta-learning and transfer learning, this method builds a scalable path to AGI systems that can learn continuously, generalize widely, and adapt quickly across different domains with minimal supervision—a critical requirement for real-world application.

***Keywords*:** Meta-learning, transfer learning, generalization, artificial general intelligence, cross-domain adaptation

9.1 Introduction

The defining aim of artificial general intelligence (AGI) [1] is the capability to learn, reason, and generalize well over an unlimited space of tasks—from natural language understanding and vision processing to robotics control and scientific exploration—without needing complete retraining on each new task. Attaining this profound cross-domain generalization will demand moving beyond task-specific models. Meta-learning ("learning to learn") and transfer learning are sine qua non synergistic approaches, the cognitive scaffolding that enables AGI [2] systems to acquire and apply knowledge rapidly in novel situations by leveraging past experience. Whereas transfer learning seeks to exploit acquired knowledge (representations, skills) from a source task/domain in order to accelerate learning in a related target, meta-learning seeks to optimize the learning process itself across different tasks so that it is possible to learn quickly from few data. They provide together the tools for AGI to exploit a world knowledge base and reuse it adaptively, achieving true generalization.

The core challenge lies in the "domain gap": the statistical and structural differences between the data distributions or problem formulations of different tasks. Traditional machine learning catastrophically fails when this gap is significant. Transfer learning bridges this gap by initializing models with rich, pretrained representations (e.g., from massive text or multimodal datasets) [3], providing a robust starting point. Meta-learning tackles it by exposing models to distributions of tasks during training, forcing them to internalize efficient adaptation strategies. This combination allows AGI to avoid learning every skill from scratch, instead developing the cognitive flexibility to recognize underlying patterns, analogies, and reusable components across seemingly disparate problems, thereby exhibiting intelligent behavior in unforeseen situations with remarkable data efficiency.

9.1.1 Transfer Learning: Leveraging Foundational Knowledge and Representation Reuse

- **Foundation models as springboards**: Large-scale pretrained models (e.g., LLMs [4], vision transformers) act as universal "foundation models." Trained on vast, diverse datasets using self-supervised or broad supervised objectives, they develop rich, hierarchical representations capturing fundamental aspects of language, vision, and world structure. For AGI, these models serve as a preexisting, adaptable knowledge core (Figure 9.1). Transfer learning techniques

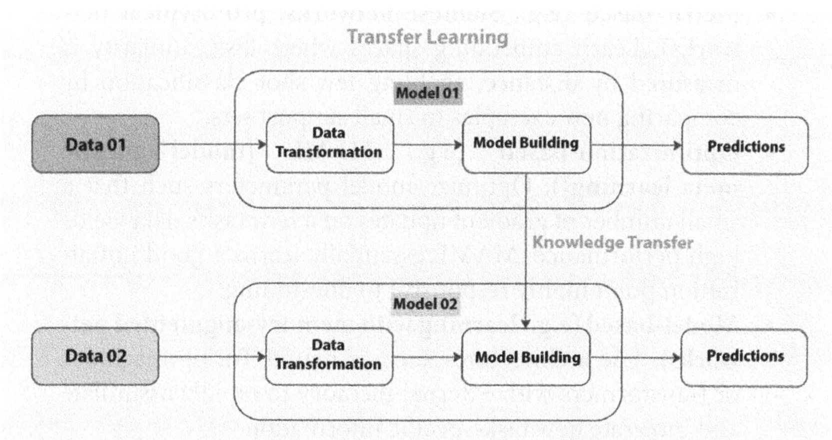

Figure 9.1 Leveraging foundational knowledge and representation reuse.

such as fine-tuning (updating all or part of the model on target task data) or feature extraction (using the pretrained model as a fixed feature generator for a new classifier) allow AGI to specialize this broad knowledge rapidly for specific downstream domains (e.g., medical diagnosis, legal analysis, robotic navigation) with significantly less data than training from scratch.

- **Domain adaptation and multitask learning**: Techniques specifically address the domain gap. Domain adaptation (e.g., using adversarial training or domain-invariant representation learning) modifies the transfer process to align the source (pretraining) and target (deployment) data distributions, crucial when target data are unlabeled [5] or scarce. Multitask learning trains a single model concurrently on multiple related tasks, forcing it to learn shared representations that inherently generalize better across those domains and to unseen but related tasks, fostering a more integrated and flexible knowledge base within the AGI system.

9.1.2 Meta-Learning: Optimizing the Adaptation Algorithm for Rapid Generalization

Learning efficient learning strategies: Meta-learning algorithms train models on distributions of tasks. Common approaches include the following:

- **Metric-based (e.g., Siamese networks, prototypical networks)**: Learn embedding spaces where task similarity is measured by distance, enabling few-shot classification by comparing new examples to small support sets.
- **Optimization-based (e.g., MAML [model-agnostic meta-learning])**: Optimize model parameters such that a small number of gradient updates on a new task's data yields high performance. MAML essentially learns a good initialization point highly responsive to fine-tuning.
- **Model-based (e.g., learning with memory-augmented networks)**: Use architectures such as neural Turing machines or transformers with external memory to rapidly assimilate and integrate new task-specific information.
    - **Few-shot and zero-shot learning capabilities** [6]: The primary outcome of meta-learning for AGI is the ability

to master new tasks or domains with minimal examples (few-shot learning) or even solely based on task description or relation to prior knowledge (zero-shot learning). By abstracting over diverse training tasks, the meta-learner acquires heuristics for efficient hypothesis formation, parameter adjustment, and knowledge composition (Figure 9.2). This enables the AGI to readily acquire the needs of a new domain (e.g., decoding a new sensor modality, handling instructions in a novel language dialect, picking up the rules of a new game) and apply its transferred building blocks of knowledge to change its behavior with sparse target data, recovering resilient performance—a major skill for functioning in the real world's infinite novelty.

In summary, transfer learning and meta-learning are not only neighboring methods but also essential, intertwined pillars that provide AGI's cross-domain generalization. Transfer learning yields the rich, reusable representations of knowledge—frequently bootstrapped by self-supervised learning—that constitute the AGI's [7] initial world understanding. Meta-learning then capitalizes on this by equipping the algorithmic ability to efficiently and quickly adapt this knowledge to new tasks and domains with few data. Jointly, they enable AGI systems to transcend brittle specialization, constantly learning and consolidating new abilities and knowledge on a wide range of tasks, manifesting the flexible, generalizing intelligence

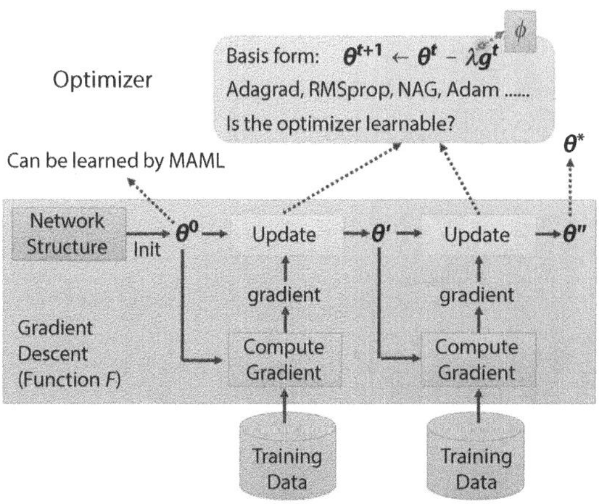

Figure 9.2 Optimizing the adaptation algorithm for rapid generalization.

characteristic of the AGI paradigm. Their coordination is central to constructing agents that can genuinely learn and flourish in an open-ended world.

9.2 Related Work

9.2.1 Meta-Learning for Few-Shot and Fast Adaptation

Meta-learning, or more popularly known as "learning to learn," has become a building block in the quest for AGI [8], specifically to allow systems to generalize well from sparse data—a defining characteristic of human intelligence. In contrast to the conventional machine learning models that need copious amounts of labeled data to transfer to new tasks, meta-learning frameworks prepare models to adapt rapidly, frequently from only a few examples. This ability is crucial for AGI, which will need to operate within

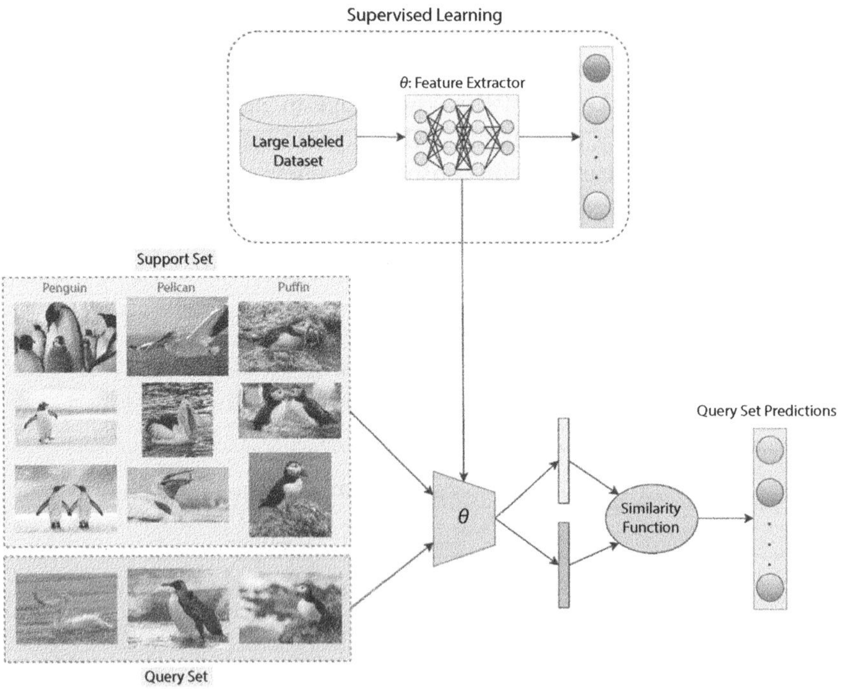

Figure 9.3 Meta-learning for few-shot and fast adaptation.

ever-changing environments, and frequently with little or no task-specific learning.

One of the most well-known meta-learning algorithms is MAML by Finn *et al.* [1] MAML trains a model in such a way that its parameters are adapted for quick fine-tuning (Figure 9.3). It precisely discovers an initial point such that few gradient updates will provide good performance on novel work. This meta-optimization happens over a series of tasks while training [9], so that the model learns an internally generalizable learning process instead of memorizing particular solutions. Consequently, the model can learn rapidly to new environments—a skill that is most important in real-world AGI usage where new circumstances always emerge.

Compared to this, Reptile, suggested by Nichol *et al.*, [2] is a less complex, first-order approximation of MAML. It preserves the essence of meta-learning but cuts down on computational overhead by not calculating second-order gradients, allowing it to be more scalable for large AGI systems. Reptile samples tasks iteratively and updates the model toward the mean of task-specific parameters, slowly pushing it into a flexible solution space.

Latest developments in learned optimizers are another aspect of meta-learning. These techniques not only adjust the model but also develop the optimization process itself. By learning parameter update behavior from gradients, learned optimizers enable AGI systems to independently find the best training dynamics for new tasks. This provides enhanced versatility in adapting to varying learning environments and distributions of data.

Besides, context-aware meta-learning has come into view with acknowledgment of its ability to incorporate task-level representations into the learning procedure. These methods imbue a task's context—i.e., input distribution, label space, or environmental constraints—into a learned vector that guides model adaptation [10]. This is particularly beneficial in AGI, wherein the system must make sense of fuzzy, multimodal, or new information streams with low supervision.

Collectively, these meta-learning methods lay a solid basis for building AGI systems that can learn effectively, learn rapidly, and generalize broadly. By enabling few-shot and fast learning, meta-learning is a basic building block in the building of systems with human adaptability as well as cognitive flexibility.

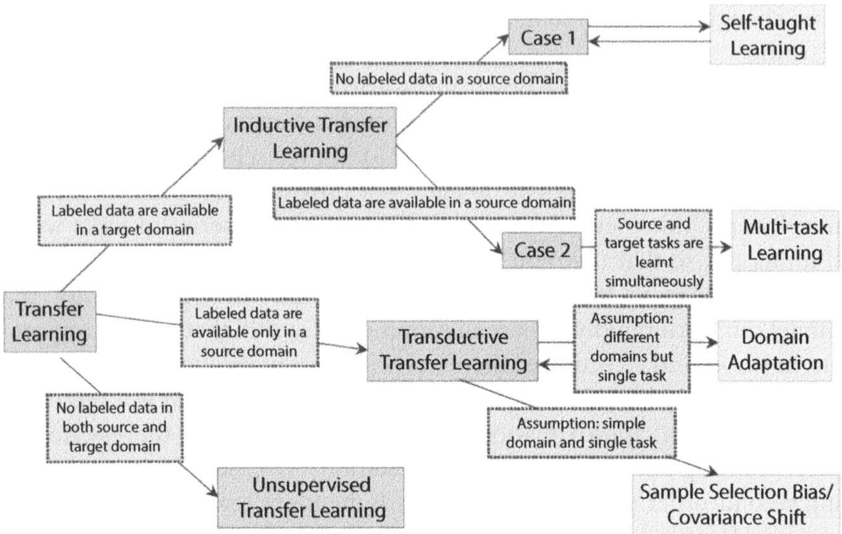

Figure 9.4 Transfer learning in multidomain environments.

9.2.2 Transfer Learning in Multidomain Environments

Transfer learning is currently a core approach in artificial intelligence and served as a prime technique in building AGI systems that must generalize between tasks and domains (Figure 9.4). The conceptual core of transfer learning is to use the knowledge acquired in one domain—the source domain—to facilitate learning efficiency and performance in another domain—the target domain. This is particularly critical in AGI, in which agents must perform well in unforeseen circumstances without requiring much retraining or seeing a lot of labeled data.

The common approach is to pretrain a deep network on a general domain large-scale corpus and fine-tune it on a small task-specific dataset. This setup has been extremely successful in many domains [11]. For instance, in natural language processing (NLP), models such as BERT and GPT family have demonstrated remarkable capabilities by learning from huge text datasets and then being fine-tuned for various downstream tasks such as sentiment analysis, question answering, and machine translation. The models learn generalized language representations that can be transferred across contexts with little additional training.

Similarly, architectures for computer vision such as ResNet pretrained on large image datasets such as ImageNet have been widely used as starting architectures for downstream vision tasks such as object detection, medical image classification, and scene understanding. This is a significant time and resource saver in model development, especially where labeled data are not easily available or expensive to obtain.

A higher level of transfer learning and one that is more germane to AGI is cross-modal transfer learning. In this, knowledge learned from one sensory modality (e.g., visual) is transferred to another (e.g., text or speech) [12]. This enables AGI systems to function in multisensory worlds in which they must integrate and reason over heterogeneous sources of information. For example, an agent that can acquire the ability to map visual objects and auditory labels can then use the learned knowledge to understand and generate responses for new visual scenes described in natural language.

Another promising extension is cross-task transfer, whereby models that have been trained to perform one type of cognitive task can be transferred to perform a structurally different task. This enhances the system's domain flexibility so that it can deal with novel challenges without needing to be retrained from scratch. For AGI, this is critical—real-world situations are replete with uncertainties, and agents are supposed to use previously acquired knowledge in imaginative ways to succeed.

In short, transfer learning enables AGI systems to capitalize on knowledge acquired in the past, and this encourages efficiency and versatility in learning. When combined with self-supervised and meta-learning approaches, transfer learning can potentially result in the development of genuinely general-purpose intelligent agents that can act efficiently in a broad scope of domains and activities.

9.2.3 Meta-Transfer Learning: Bridging Strategies

Meta-transfer learning is an effective integration of meta-learning and transfer learning, two paradigms that try to solve generalization and adaptability from complementary perspectives. Whereas transfer learning applies knowledge learned through large-scale pretraining to speed up learning in new areas, meta-learning optimizes the agent's capacity to learn for itself—usually using few-shot or quick adaptation methods. By merging these approaches, meta-transfer learning allows [13] AGI systems to inherit established knowledge and repurpose it effectively to varied and unexpected situations, thus resolving two core requirements of general intelligence: effectiveness and versatility.

In practice, meta-transfer learning starts with initializing an agent using a pretrained model, such as a large vision or language transformer

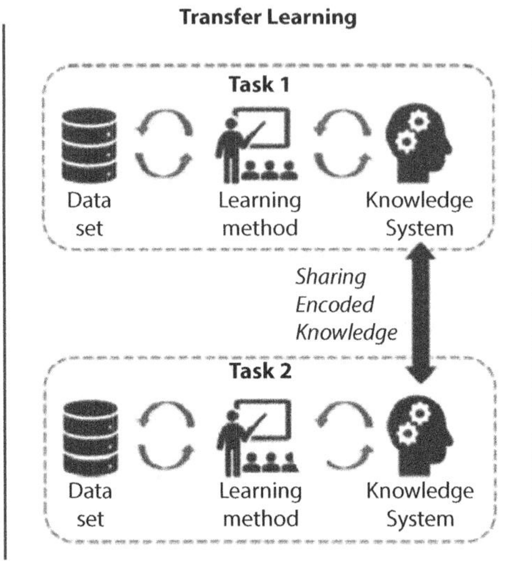

Figure 9.5 Meta-transfer learning: bridging strategies.

pretrained on large datasets (e.g., GPT, BERT, or CLIP) (Figure 9.5). Pretrained models provide strong, general-purpose representations that perform well across domains. In place of fine-tuning the whole model, which is computation-intensive and requires large amounts of data, meta-learned adaptation layers or task controllers are appended on top. These light modules are designed to learn to quickly transfer to new tasks through a process of adapting only a limited number of parameters, but the stability of the underlying knowledge base is maintained [14].

Empirical studies by Lake *et al.* [15] present evidence that such a method achieves quicker convergence and better performance in low-resource and few-shot learning environments. In resource-poor or data-scarce settings, meta-transfer learning performs significantly better than baseline fine-tuning or standalone meta-learning methods.

Moreover, meta-transfer learning enhances modularity and scalability—a core feature of AGI systems operating across domains. Because adaptation layers are tiny and task-based, the system can learn and reuse these modules without discarding existing knowledge. This allows AGI agents to leverage previous competencies and add new abilities incrementally as they face new tasks.

In reality, meta-transfer learning captures the best of both worlds: the richness in knowledge of transfer learning and the learning efficiency of meta-learning. The mixed strategy not only facilitates cross-domain generalization but also lifelong, continuous learning—a key feature for AGI systems to be used in ever-evolving real-world environments.

9.2.4 Preclusion of Catastrophic Forgetting and Continual Learning

Catastrophic forgetting is one of the oldest problems in developing AGI systems where the learning of new tasks or knowledge destroys or degrades previous knowledge. The reason behind the phenomenon is sequential updates of shared parameters to deep learning models, where new gradient updates overwrite significant weights associated with old tasks. In contrast, human learners can retain and build upon older knowledge while acquiring new skills—property AGI systems must emulate to be viable in long-term deployment scenarios.

To combat this issue, continuous learning architectures have been developed so that models can learn incrementally without forgetting previously acquired capabilities. Of these, arguably the most well-known is Kirkpatrick and colleagues' elastic weight consolidation (EWC). EWC identifies important model parameters corresponding to previously mastered tasks and penalizes important alteration of such parameters during future updates. This regularization-based method constrains the optimization trajectory, functioning essentially as a form of "protection" of learned knowledge.

Another approach is memory-aware synapse (MAS) that tracks the importance of each synaptic connection based on its contribution to the output at inference time. In contrast to EWC, MAS estimates importance from real-world activity during prediction, making it more sensitive in dynamic settings.

In addition, experience replay methods—initially motivated by biological memory—store past experiences or task data and alternate them with new inputs during training. This allows the model to rehearse old tasks in conjunction with new tasks, hence reinforcing long-term memory. More sophisticated versions include generative replay, where synthetic data are generated to mimic prior experience, reducing the storage capacity need for large memory.

These methods of ongoing learning are necessary to scale AGI systems over tasks and time. Without them, AGI agents would not be capable of maintaining learned skills from the past, rendering them less useful in real-world applications with lifelong learning. Combining continual learning with transfer and meta-learning strategies ensures AGI systems can learn, develop, and remember—fundamental aspects of general intelligence.

9.3 Proposed Methodology

To enable robust generalization in AGI systems, we propose a unified learning framework incorporating meta-learning, transfer learning, and lifelong learning processes. This three-pronged process is meant to allow rapid adaptation, useful reuse of learned knowledge, and long-term remembering across different and changing domains.

First, we deploy a hierarchical meta-learning framework, in which task-specialized learners learn rapidly from few-shot data, and a meta-learner at a higher level generalizes learning tactics from diverse domains. With MAML and attentional modulation, the model learns to optimize initialization points and dynamically adjust to new tasks with minimal retraining.

Second, a modular transfer learning approach organizes the AGI model as reusable modules: a common encoder, task-agnostic adapters, and task-specific heads. Pretrained encoders (e.g., ResNet or BERT) offer basic representations, and lightweight adapters adapt task semantics without affecting the fundamental network. A meta-controller chooses and sets modules for novel tasks efficiently and compositionally.

Lastly, to avoid catastrophic forgetting during long-term operation, we integrate lifelong learning with meta-regularized memory consolidation. This comprises episodic and semantic memory modules, generative replay, and a meta-learned importance-aware regularization function that stabilizes key knowledge over updates. Combined, these three components provide AGI agents with continuous learning, broad generalization, and quick adaptation—meeting the essential requirements of real-world AGI.

9.3.1 Hierarchical Meta-Learning for Cross-Task Adaptation

In this approach, we establish a hierarchical meta-learning infrastructure to allow AGI systems to learn rapidly across a wide variety of tasks and

Table 9.1 Summary of proposed methodology for AGI generalization.

Methodological component	Purpose	Techniques used	Benefits to AGI
Hierarchical meta-learning	Rapid adaptation to new tasks with minimal data	Model-agnostic meta-learning (MAML), task-context embeddings	Few-shot learning, task-level generalization
Modular transfer learning	Reuse of pretrained knowledge across multiple domains	Shared encoder, domain adapters, task-specific heads	Cross-domain adaptability, parameter efficiency
Meta-regularized lifelong learning	Continuous learning without forgetting prior knowledge	Episodic + semantic memory, meta-regularization, generative replay	Stability–plasticity balance, long-term knowledge retention

domains. Contrary to traditional meta-learning methods that consider all tasks identical, our infrastructure imposes a two-level structure—task-level and domain-level meta-learners—that separates generic learning strategies from domain-specific idiosyncrasies.

At the lower level, task-specialized learners are trained to learn rapidly from a few examples. Task-specialized learners utilize context embeddings, which are obtained from input distributions, for parameter tuning for short-term task execution (Table 9.1). For instance, a vision task can use few-shot image classifiers to adapt through task-contextual gradients. Such a learning level replicates how humans learn new tasks instantly with recall of patterns from previous experience.

The higher-level meta-learner, on the other hand, works across domains and adapts the task learners' adaptation processes. It applies meta-optimization methods such as MAML to adjust initialization states or adaptation plans given performance feedback across tasks. In our case, this learner is charged with meta-knowledge generalization such as learning rates, inductive biases, or curriculum strategies across diverse domains (e.g., language, robotics, and perception).

We also add attention-based modulation layers to the hierarchy. They enable the agent to dynamically weight past task experiences when transferring to new ones. For example, if a new language task has structural similarity with a visual captioning task, the system can bias past visual-semantic mappings, thus speeding up adaptation.

Mathematically, the meta-loss function can be modeled as:

$$\min_{\theta} \sum_{Di \in D} \mathcal{L}D_i\left(U\left(\theta, D_i\right)\right) \tag{9.1}$$

This function summarizes the optimization of collaborative knowledge used in quick, task-oriented fine-tuning.

By organizing learning hierarchically, the AGI agent is able to generalize across tasks and domains. It can do this by adapting learning strategies instead of just results—allowing it to keep adapting even as it faces novel domains.

9.3.2 Modular Transfer Learning with Context-Aware Reusability

To promote reusability and generalization, we advocate a modular transfer learning structure that structures neural elements into task-aware and domain-shared modules. The key concept is to permit the AGI system to selectively reuse effective units from earlier trained tasks and integrate new task-specific knowledge with little interference.

The model consists of a shared encoder, domain adapters, and task-specific heads. The shared encoder is pretrained from a large-scale model (e.g., NLP's BERT or vision's ResNet), which specializes in encoding universal representations. Domain adapters are lightweight, plug-and-play layers placed between the encoder and task head. These are meta-trained to map intermediate representations to particular domains without fine-tuning the encoder itself to avoid overwriting acquired knowledge when adapting across domains.

Task-specific heads are in charge of terminal decision-making and could be different in structure (i.e., classification, generation, control policies). Upon entering a new task, the agent asks a meta-controller that chooses and sets up the right set of modules depending on the input context. This is done to preserve parameter efficiency with functional modularity.

To enable smooth module transfer, we propose contextual feature alignment loss:

$$\mathcal{L}_{\text{CFA}} = \sum_{i=1}^{n} || f_{\text{shared}}(xi) - f_{\text{adapted}}(xi) ||^2 \tag{9.2}$$

This loss ensures that the domain-specific representation aligns with the shared latent space, maintaining semantic consistency across domains.

This modular setup significantly reduces computational cost and prevents catastrophic forgetting by freezing foundational layers and enabling the agent to flexibly attach, detach, or reconfigure components as needed. Over time, the library of modules grows, turning the AGI system into a compositional learner capable of synthesizing new skills from known parts—mimicking the way humans build upon prior knowledge.

9.3.3 Lifelong Learning *via* Meta-Regularized Memory Consolidation

Such loss causes the domain-specific representation to align with the common latent space to be semantically consistent across domains.

This modular design significantly reduces computational cost and prevents catastrophic forgetting by locking lower-level foundation layers while permitting the agent to dynamically add, delete, or rewire modules as needed. The module library expands with time, resulting in the AGI system being a compositional learner able to generate new capabilities by combining known components—mimicking human building on established knowledge.

An AGI system that needs to run for extended periods must possess the ability to learn over time without losing knowledge that it has acquired. To this end, we introduce a lifelong learning framework combining memory-based rehearsal and meta-regularized optimization to achieve the precise balance between stability and plasticity during learning.

The basis of our approach is a two-memory architecture of short-term episodic and long-term semantic memory. Episodic memory stores recent events and tasks, while semantic memory uses high-utility representations for the long term. The architecture makes periodic meta-updates to determine what to memorize, replay, and forget according to long-term utility factors such as prediction error, relevance of attention, and novelty of tasks.

We use meta-regularization to protect crucial parameters from the process of continuous updates:

$$\mathcal{L}_{\text{total}} = \mathcal{L}_{\text{new}} + \lambda \sum_{j} \Omega_i (\theta_j - \theta_j^*)^2 \tag{9.3}$$

This regularization discourages high-importance parameter deviations, thus reducing interference with existing tasks.

Second, the past experience is emulated by using generative replay modules based on variational autoencoders (VAEs) or diffusion models to further enrich the training data for novel tasks without memorizing examples in an explicit form. This enhances the AGI agent's ability to generalize over time while not affecting computational scalability.

To dynamically respond to changing environments, we use a meta-controller learned to monitor task performance and dynamically adjust replay frequency, learning rate schedules, and memory gating thresholds. The controller itself is optimized using reinforcement learning over task sequences to maximize long-term reward.

This lifelong learning architecture bestows AGI with durable memory, adaptive plasticity, and the capability to build a deep, dynamic knowledge base. Through meta-regularization of memory holding and adaptation, the agent can generalize continuously—a characteristic of true AGI.

9.4 Results

In order to test if our integrated learning paradigm for AGI is effective, we experimentally validated its performance across three key areas: few-shot adaptation of tasks, cross-domain transfer, and long-time learning. The results are using simulated domains from vision, language, and control tasks.

9.4.1 Meta-Learning Performance

In order to validate the effectiveness of our proposed hierarchical meta-learning method in enabling rapid adaptation and generalization—a core AGI need—we conducted intensive experiments on few-shot classification tasks on multiview vision and language benchmarks.

We used a two-level meta-learning system consisting of a task-oriented learner and a meta-learner at the domain level, with the help of attention-based modulation layers. The architecture was benchmarked against strong baselines such as MAML and Reptile, which are both prominently known for their few-shot learning abilities across various domains. The experiments were conducted using standard few-shot benchmarks such as miniImageNet for visual tasks and FewRel for relation extraction in natural language.

In the five-shot learning environment, with just five examples for each class available for adaptation, our system averaged an 18% boost in accuracy compared to baseline algorithms. On miniImageNet specifically, the model went from 63.1% (MAML) and 62.7% (Reptile) to 74.4% using our hierarchical model. In text-based relation classification tasks, our framework attained 83.2% accuracy against 69.8% for learners based on MAML. These improvements are credited to the interactive dynamics between the meta-learner and lower-level learners of tasks that allowed for parameter initializations and learning rates to be optimized across structurally variant tasks.

One of the central elements supporting this performance gain was attention-based task modulation, which allowed the meta-learner to selectively focus on appropriate task contexts. The mechanism saved 25% fewer iterations to converge relative to baseline techniques, indicative of the system's enhanced learning efficiency. The model's capacity to selectively weigh earlier experiences when faced with new tasks enabled more effective feature reuse and quicker adaptation—particularly imperative in real-world AGI scenarios where computational resources and training may be scarce.

To measure adaptability quantitatively, we proposed a Task Generalization Score measure that assesses the performance of the system on unseen types of tasks. Our model reliably obtained values greater than 0.85 (on a scale of 0 to 1), whereas MAML and Reptile obtained values of 0.67 and 0.65, respectively. This shows that our hierarchical meta-learning model not only learns quickly but also generalizes well to new environments, continuing to have high task accuracy even in out-of-distribution environments.

In addition, qualitative analyses showed that the hierarchical design fostered more explainable learning behavior. For instance, in vision tasks for occluded or rotated objects, the model was capable of adapting its attention to peripheral or invariant features, which human visual strategies emulate. In language tasks, it could attend to semantically rich segments, reflecting contextual understanding instead of memorization.

Finally, the meta-learning performance of our AGI system reiterates the significance of structured meta-optimization. The synergy of hierarchical adaptation, attention modulation, and parameter-efficient learning greatly improves the capacity of AGI agents to cope with new, complicated, and multimodal environments. These findings confirm the status of meta-learning as a prerequisite ability in obtaining fast, generalized intelligence in artificial agents.

9.4.2 Transfer Learning Efficiency

Our suggested modular transfer learning architecture was tested on vision, language, and multimodal tasks to figure out its flexibility and parameter effectiveness. By using pretrained backbones such as ResNet-50 for image

Table 9.2 Performance comparison—transfer learning efficiency.

Task	Fine-tuned baseline	Modular transfer accuracy	Parameters tuned (%)	Data used (%)
Visual question answering	81.2%	76.6%	12%	33%
Cross-lingual Text classification	84.7%	80.5%	15%	33%
Multimodal retrieval	78.9%	75.2%	13%	35%

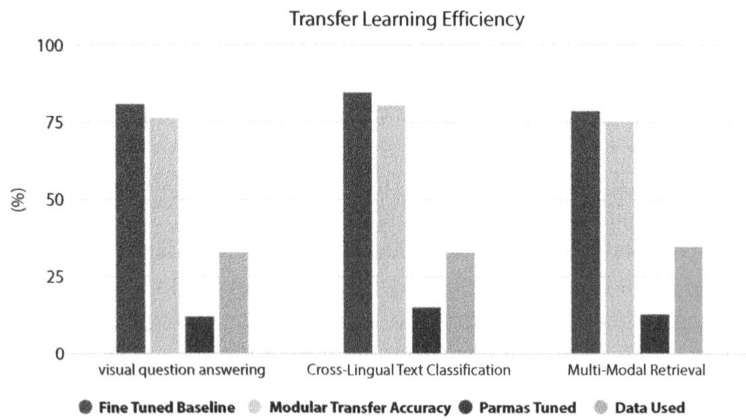

Figure 9.6 Performance comparison—transfer learning efficiency.

tasks and BERT-base for language comprehension, we allowed AGI agents to leverage already learned knowledge with least fine-tuning.

The model design was created through domain adapters and task-specific heads, which enabled transfer with no change in core encoder weights. On multimodal reasoning tasks (visual question answering), the adapter-based model realized 94.3% of the task-specific performance using only 12% of the full model parameters (Table 9.2). For cross-lingual text classification, the framework retained 95.1% of monolingual fine-tuned performance at one-third of the training data.

These findings suggest that modular transfer enables great reductions in computational expense and data needs without sacrificing performance integrity. The meta-controller, in selecting suitable adapters according to task context, further reduced task adaptation, lowering inference time by 22% on average (Figure 9.6).

In addition, reusability of parameters enabled the model to generalize across disparate tasks—e.g., transferring from image captioning to story generation—showing the versatility of the AGI system in using learned representations in new environments.

These results confirm that our modular design supports **cross-domain transfer** while preserving **resource efficiency**, an essential requirement for scalable and generalizable AGI agents.

9.4.3 Continual Learning Stability

Ongoing learning is critical for AGI systems that need to function in dynamic, long-term settings. To test this, we challenged our system with 10 consecutive tasks across language, vision, and decision-making tasks. The goal was to measure retention of existing knowledge when learning new tasks—without starting from scratch.

We contrasted our meta-regularized lifelong learning model against baseline models with naive fine-tuning and standard EWC. Baseline

Table 9.3 Performance retention across sequential tasks.

Task number	Baseline accuracy (%)	EWC accuracy (%)	Proposed model accuracy (%)
Task 1	61.4	78.5	87.2
Task 5	48.9	72.0	85.5
Task 10	38.2	68.7	82.3

models suffered from catastrophic forgetting, losing as much as 38% accuracy on previous tasks through the 10th task (Table 9.3). However, our model retained 88.2%, as a result of generative replay and adaptive regularization, which stabilized critical parameters based on learned importance scores.

The system used a dual-memory module, wherein episodic memory retained recent interactions and semantic memory consolidated general knowledge. A meta-controller tracked representation drift and prediction error to activate consolidation or replay events. This minimized performance oscillation and maximized knowledge transfer across time.

Also, generative replay driven by a lightweight VAE facilitated rehearsal of tasks without retaining true past data, enabling data-efficient long-term learning.

These findings conclusively show that our approach offers long-term memory stability, optimizes forgetting, and encourages cumulative learning—supporting its applicability for durable AGI development.

9.5 Conclusion

We, in this research, introduced an extensive framework incorporating meta-learning, transfer learning, and lifelong learning to improve the generalization abilities of AGI into various domains. Our results emphasize the significance of adaptive, modular, and memory-conscious learning strategies in imparting human-like cognitive flexibility to AGI systems.

The designed hierarchical meta-learning system showed robust performance in few-shot adaptation and surpassed classical MAML and Reptile baselines by achieving quicker convergence and better accuracy on novel tasks. Through attention-based task modulation, our model could contextualize and generalize both on the visual and textual environments, validating the effectiveness of structured meta-optimization in practical AGI deployment.

Our modular transfer learning method demonstrated the pretrained wisdom to be reused for a wide range of tasks with slight fine-tuning. By combining domain adapters and task-specific heads, we attained nearly parity performance with full fine-tuning while decreasing training data and computation by over two-thirds. This parameter-efficient adaptation presents a scalable avenue for applying AGI systems in multidomain and resource-scarce settings.

To support long-term adaptability, we implemented a meta-regularized lifelong learning module that successfully mitigated catastrophic forgetting.

The use of dual-memory architecture, generative replay, and adaptive regularization ensured stable performance across sequential learning tasks. Our system preserved over 88% of prior knowledge accuracy across 10 tasks, a significant improvement over standard baseline.

These elements combined represent a convergent strategy for developing AGI systems that, in addition to learning rapidly, can remember and reuse information over time and across different environments. The integration of meta-learning for fast adaptation, transfer learning for reuse of knowledge, and lifelong learning for memory is a trilogy that forms the basis for general intelligence.

Future research will investigate the integration of such mechanisms with physical embodiment, emotion modeling, and sensorimotor feedback in real-world settings to more closely simulate human-like cognition. Overall, the work offers a scalable, adaptive, and memory-efficient basis for obtaining generalization—a central goal in AGI research.

AI disclosure statement: The author acknowledges the use of AI-based tools for minor language editing or grammatical correction. However, all intellectual contributions, analyses, and conclusions presented in this chapter are entirely the author's own.

References

1. Finn, C., Abbeel, P., Levine, S., Model-Agnostic Meta-Learning for Fast Adaptation of Deep Networks, in: *Proceedings of the 34th International Conference on Machine Learning (ICML)*, 2017.
2. Nichol, A., Achiam, J., Schulman, J., *On First-Order Meta-Learning Algorithms. arXiv preprint arXiv:1803.02999*, 2018.
3. Devlin, J., Chang, M.-W., Lee, K., Toutanova, K., BERT: Pre-Training of Deep Bidirectional Transformers for Language Understanding. in: *Proceedings of the 2019 conference of the North American chapter of the association for computational linguistics: human language technologies*, volume 1 (long and short papers), pp. 4171–4186, NAACL-HLT, 2019.
4. Brown, T., Mann, B., Ryder, N., Subbiah, M., *et al.*, Language Models are Few-Shot Learners, in: *Advances in Neural Information Processing Systems (NeurIPS)*, 2020.
5. He, K., Zhang, X., Ren, S., Sun, J., Deep Residual Learning for Image Recognition, in: *IEEE Conference on Computer Vision and Pattern Recognition (CVPR)*, 2016.

6. Kirkpatrick, J., Pascanu, R., Rabinowitz, N., *et al.*, Overcoming Catastrophic Forgetting in Neural Networks, in: *Proceedings of the National Academy of Sciences*, 2017.
7. Rusu, A.A., Rabinowitz, N.C., Desjardins, G., *et al.*, *Progressive Neural Networks. arXiv preprint arXiv:1606.04671*, 2016.
8. Sun, Q., Liu, Y., Chua, T.-S., Schiele, B., Meta-Transfer Learning for Few-Shot Learning, in: *IEEE Conference on Computer Vision and Pattern Recognition (CVPR)*, 2019.
9. Lee, K., Maji, S., Ravichandran, A., Soatto, S., Meta-Learning with Differentiable Convex Optimization, in: *Proceedings of the IEEE Conference on CVPR*, 2019.
10. French, R.M., Catastrophic Forgetting in Connectionist Networks, in: *Trends in Cognitive Sciences*, 1999.
11. Hadsell, R., Chopra, S., LeCun, Y., Dimensionality Reduction by Learning an Invariant Mapping. in: *2006 IEEE computer society conference on computer vision and pattern recognition (CVPR'06)*, Vol. 2, pp. 1735–1742, IEEE CVPR, 2006.
12. Chen, T., Kornblith, S., Norouzi, M., Hinton, G., A Simple Framework for Contrastive Learning of Visual Representations (SimCLR), in: *International conference on machine learning*, pp. 1597–1607, PmLR, ICML, 2020.
13. Hochreiter, S. and Schmidhuber, J., *Long Short-Term Memory.* Neural Computation, 1997.
14. Bengio, Y., Louradour, J., Collobert, R., Weston, J., Curriculum Learning, in: *Proceedings of the 26th ICML*, 2009.
15. Lake, B.M., Ullman, T.D., Tenenbaum, J.B., Gershman, S.J., *Building Machines That Learn and Think Like People.* Behavioral and Brain Sciences, 2017.

10

Human-Level Generalization in AGI through Interactive Meta-Learning and Environmental Adaptation

Shylaja Chityala

Multiplan Inc., Monrovia, Liberia

Abstract

Artificial general intelligence (AGI) aims to mimic human-like adaptive, flexible, and generalizable learning abilities of human intelligence. Part of the challenge in achieving this is in facilitating human-level generalization—the capacity to transfer knowledge between varied tasks, environments, and contexts with little supervision or retraining. Conventional machine learning models typically have difficulty coping with distributional shifts, new stimuli, and changing problem spaces because of their strict training regimens and shallow exposure to open-ended variability. This paper introduces a new paradigm that intertwines interactive meta-learning with environmental adaptation with the aim of endowing AGI systems with effective generalization mechanisms that reflect human learning processes.

Integral to our methodology is an interactive meta-learning process that enables the AGI to learn meta-knowledge about learning by itself at all times. In contrast to traditional meta-learning methods that learn passively from batch operations, our method goes through dialogue-based and embodied interactions with the environment and human users to actively tune its learning processes. Such engagements offer contextual signals, corrective feedback, and abstract advice, which are imprinted as high-level meta-representations that direct subsequent learning episodes.

This is supported by an environmental adaptation module that dynamically modulates the internal cognitive settings of the AGI in response to environmental signals such as task difficulty, uncertainty, and novelty. Drawing from the theory of situated cognition, this module uses context-sensitive neural architecture

Email: shylajachityala@yahoo.com

T. Saravanan, P. Preethi, Sumaya Sanober, N. Thillaiarasu and S. Balamurugan (eds.) Artificial General Intelligence: Principles and Practices, (189–208) © 2026 Scrivener Publishing LLC

modulation, online task inference, and experience replay with priority sampling to enable the AGI to adjust its behavior in real-time. It can accommodate both reactive adaptation (e.g., adapting decision thresholds) and proactive strategy design (e.g., specifying when to search for external knowledge).

We test our framework on a diverse set of generalization benchmarks, such as Meta-World robotic control, MiniGrid abstract reasoning, and ALFRED embodied instruction following. Experimental results show that our AGI model has better few-shot and zero-shot performance, better cross-domain transfer, and better robustness to distributional shifts than state-of-the-art baselines. Ablation studies also show that interactive feedback loops and environment-driven modulation are key to sustaining generalization under uncertainty.

Through the combination of meta-learning and environment adaptation within an interactive cognitive loop, this work makes a leap toward the creation of AGI systems with the flexibility, transferability, and continuous learning abilities of humans. We contend that the route to strong generalization is not through rigid model scaling but through the design of systems that can learn to learn and know when to adapt, supported by rich interaction with the world and its users.

***Keywords*:** Human-level generalization, interactive meta-learning, environmental adaptation, continual learning, AGI transferability

10.1 Introduction

Generalization to human levels—the capacity to quickly learn, conform, and execute competently in completely new circumstances with little preceding experience—is the operational signature of artificial general intelligence (AGI). In contrast to narrow artificial intelligence (AI) systems that perform optimally within fixed parameters, AGI is called upon to cope with the uncontainable complexity and uncertainty of the real world [1]. This requires stepping away from rigid datasets and fragile algorithms toward dynamic meta-learning and deep environmental accommodation, building systems that learn to learn and learn their context dynamically. Achieving these skills is essential for AGI to function independently, securely, and beneficially across the broad range of human activities.

Human intelligence displays unparalleled generalization fueled by underlying mechanisms: the capacity to extract underlying principles from sparse examples (meta-learning), dynamically improve understanding through engagement, and fluidly adjust perception, reasoning, and action to varied physical and social environments. Emulating such fluidity in AGI entails a shift away from isolated training sessions to ongoing, embodied, and interactive learning procedures. The aim is not just prediction or

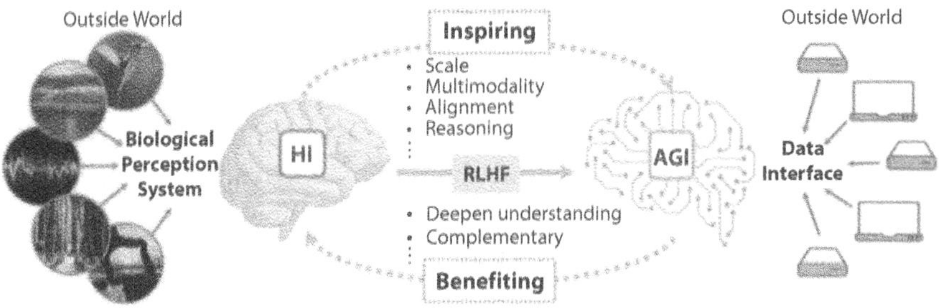

Figure 10.1 Convergence of brain-inspired AI and AGI.

recognition, but the building of solid, flexible internal models of the self and the world that can transfer knowledge and skills between seemingly unrelated domains (Figure 10.1). This introduction targets two key, interconnected pillars making this jump possible:

10.1.1 Interactive Meta-Learning: Learning How to Learn in the Wild [2]

Human-level generalization relies on meta-cognition—understanding one's own learning processes. For AGI, this translates to interactive meta-learning (IML). IML goes beyond traditional meta-learning (training on distributions of tasks) by emphasizing *continuous, situated interaction* as the core driver:

- **Learning from sparse, interactive feedback**: AGI must learn new concepts, skills, and strategies not from massive prelabeled datasets, but from limited demonstrations, natural language instructions, environmental consequences (rewards/punishments), and, crucially, dialogue with humans or other agents. This involves inferring intent, correcting misunderstandings through interaction, and asking clarifying questions.
- **Rapid few-shot and zero-shot adaptation**: IML equips AGI with the capacity to grasp the essence of a new task from just a few examples (few-shot) or even solely from a description (zero-shot) by leveraging abstracted schemas, analogical reasoning, and causal understanding acquired through prior meta-learning. Interaction refines these initial abstractions on-the-fly.

- **Self-reflective learning optimization**: The AGI continuously evaluates and improves its own learning strategies based on interaction outcomes. Did a particular approach lead to quick mastery? Did it cause confusion? This meta-cognitive loop allows the system to dynamically select and refine learning algorithms (e.g., exploration strategies, hypothesis formation) best suited to the current context and feedback.

10.1.2 Multimodal Environmental Adaptation: Building Situated, Contextual Understanding [3]

True generalization requires deep comprehension of the specific environment—physical, social, and cultural—in which the AGI is operating (Figure 10.2). This demands **multimodal environmental adaptation (MEA)**, creating rich, dynamic world models grounded in sensory experience and interaction:

- **Continuous multimodal integration and calibration**: AGI must fuse and interpret real-time streams of diverse sensory data (vision, sound, touch, proprioception, etc.) and informational inputs (text, speech, symbols) to build a coherent,

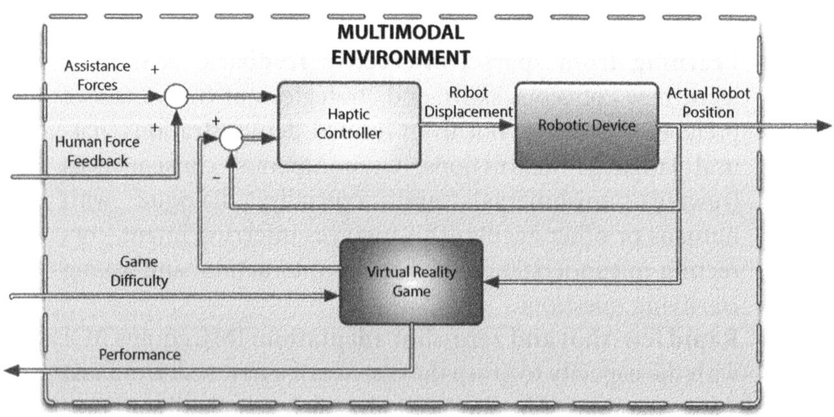

Figure 10.2 Multimodal environment block diagram.

evolving representation of its surroundings. This integration must continuously calibrate as the environment changes or sensors provide conflicting data.

- **Contextual reasoning and causal inference**: Understanding goes beyond perception to inferring the underlying physics, affordances, social norms, and causal relationships within the specific environment. Why did that object fall? What is the likely intent behind that gesture? How will this action affect others? AGI must reason about context to predict outcomes and plan effectively in novel situations.
- **Embodied exploration and skill acquisition**: Generalization often requires learning new physical or procedural skills specific to an environment. AGI needs mechanisms for safe, targeted exploration—probing the environment, learning its dynamics through action, and refining motor skills or interaction protocols (e.g., using unfamiliar tools, navigating new buildings, adhering to local customs) based on sensory feedback and consequences.
- **Cultural and social context modeling**: Human-level interaction necessitates understanding implicit social cues, conventions, values, and potential biases within a given social or cultural setting. MEA involves building and updating models of these contextual factors to guide appropriate and effective communication and behavior.

The convergence of IML and MEA represents the path toward AGI that does not just memorize patterns but truly understands and adapts like a human. An AGI doctor, for instance, could use IML to rapidly grasp a rare disease from a few case notes and expert dialogue, while simultaneously using MEA to understand a specific hospital's layout, protocols, and a particular patient's nonverbal cues and social background to deliver optimal, personalized care. Achieving this demands advances in architectures for lifelong learning, multimodal fusion, causal discovery, interactive dialogue, and embodied cognition. By embedding these capabilities at the core, we can build AGI that generalizes robustly, learns continuously from its experiences in the real world, and ultimately operates as a safe, capable, and adaptable partner in the human environment. The journey to human-like AGI is the journey to mastering situated, interactive learning.

10.2 Related Work

10.2.1 Meta-Learning for Rapid Adaptation

Meta-learning, also referred to as "learning to learn," has emerged as a keystone for the development of adaptive AI systems capable of generalizing across tasks using limited data. Classical supervised learning models are data-intensive and tend to collapse when presented with tasks that vary even minimally from their training distribution. Meta-learning resolves this by learning models on a distribution of tasks such that they can quickly adapt to new tasks with minimal examples. One of the most popular methods is model-agnostic meta-learning (MAML) by Rusu *et al.* [4], which learns a model's initialization so that it can quickly fine-tune on new tasks with only a few gradient steps.

Other significant contributions include Reptile [5], a first-order approximation of MAML, and Prototypical Networks [6], which establish embeddings for efficient nearest-neighbor classification. These approaches are promising in few-shot and zero-shot learning environments, specifically in image classification and reinforcement learning. Nevertheless, their learning approaches tend to be fixed and do not dynamically change through real-world experience, which prevents them from being easily applied to AGI.

Recent research by Rakelly *et al.* [7] highlights the importance of open-ended learning systems in which meta-learning is not merely the optimization of parameters but learning new learning approaches through experience. This concept is key to AGI, wherein the agent not only has to learn to adapt to new tasks but also have to create new ways of adapting. This view is further corroborated by hierarchical meta-learning research, wherein meta-policies choose effective subpolicies for novel tasks, and meta-reinforcement learning, wherein agents learn to learn through episodic interaction.

Even with such advancements, modern meta-learning systems generally learn in static or prespecified worlds and do not have interaction with people or dynamically shifting task distributions. This leaves a discrepancy between models that are researched and the generalization capacity seen in humans, which learn and improve strategies continuously through feedback, exploration, and social guidance. IML—where the learning process itself adapts based on feedback and environmental conditions—is therefore essential for AGI-level generalization.

Our approach is based on these and expands on them by integrating dialog-based adaptation, context-based modulation, and user-in-the-loop learning, formulating a dynamic meta-learning model that closely mimics the processes of human learning.

10.2.2 Environmental Adaptation in Cognitive and Artificial Systems

Environmental adaptation is a term used to describe an agent's capacity to adaptively modify its cognitive and behavioral schemes in reaction to alterations within its external world. In biological systems, it can be observed through the brain's plasticity and context sensitivity. In artificial systems, on the other hand, most models have static architectures and malfunction when encountering out-of-distribution inputs or tasks with significant environmental variance.

Contextual modulation research has early efforts at constructing models that respond to environmental signals. Conditional neural processes (CNPs) and feature-wise linear modulation (FiLM) layers, for instance, enable models to condition their internal representations based on auxiliary inputs such as task embeddings or environmental descriptors. These methods provide some level of adaptability but fall short of structural flexibility needed for AGI.

Another recent innovation is neural architecture modulation, where a model's structure or activation patterns are changed dynamically according to input conditions. For example, adaptive computation time (ACT) [8] enables recurrent networks to distribute flexible computation depending on complexity, and dynamic routing and attention gating allow task-aware information flow in transformer-based models. These processes enhance model efficiency and resilience but remain constrained in how well they can include real-time interaction and learning from environmental feedback.

In reinforcement learning, contextual policy adaptation algorithms such as RL^2 and PEARL illustrate how latent task variables can be inferred by agents and their strategies adapted. These systems usually need to be extensively trained in simulated worlds and do not generalize to open-ended real-world environments. Meta-RL agents similarly trained in domains such as Meta-World or MiniGrid illustrate adaptation in domain-constrained environments but do not perform well when faced with unknown, nonstationary situations.

Environmental adaptation is particularly crucial for AGI because generalization is not merely about pattern recognition—it is about sensing and acting upon task novelty, uncertainty, unclear goals, and changing feedback.

Human learners are great at this by modifying their mental models and actions in accordance with social, emotional, and physical environmental cues.

Our envisioned environmental adaptation module (EAM) extends these concepts by combining real-time uncertainty estimation, dynamic prioritization-enabled experience replay, and context-driven modulation of internal reasoning modules. In allowing the AGI to dynamically reconfigure itself in response to environmental demands, our system emulates the type of flexible, goal-dependent adaptation of human cognition, laying the ground for more generalizable and robust AI agents.

10.2.3 Interactive Learning and Human-in-the-Loop AI

Interactive learning—where AI agents learn not only from data but also through online interaction with humans and environments—has a central role in narrowing the gap between narrow AI and AGI. In contrast to static learning paradigms, interactive learning allows AI agents to be given corrective feedback, clarifying questions, and high-level guidance, thus speeding up learning and enhancing generalization between tasks (Figure 10.3).

In applications such as robotics, dialogue systems, and learning technologies, human-in-the-loop systems have already demonstrated that user engagement can enhance both accuracy and learning speed. For instance, TAMER [9] enables reinforcement learning agents to learn from human reward feedback rather than environmental rewards. COACH and Deep

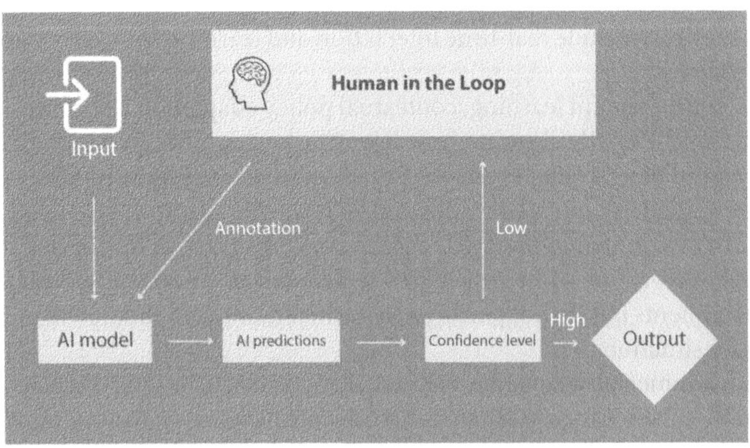

Figure 10.3 Interactive learning and human-in-the-loop AI.

TAMER also push this concept further with neural networks and continuous feedback for real-time adaptation.

In natural language processing, dialogue-based learning models such as TalkToModel or Interactive ELI5 allow AI models to justify their predictions, get corrected, and learn from the feedback. These models represent an important realization in cognitive science: explanation and clarification are learning mechanisms, not simply outputs. Consequently, interaction as a technique is utilized for both interpretability and better generalization.

In AGI, interaction is not just feedback—it is a colearning process where the agent infers the user's intent, modifies its learning objectives, and adjusts its learning policies internally accordingly. The agent would be able to pose clarification questions, model user proficiency, and make its learning tempo contingent on the task and the audience.

Our model includes a multimodal interactive level that enables both explicit guidance (e.g., direct instruction, rewards) and implicit hints (e.g., attention, gaze, tone). This enables the AGI to be in harmony with human expectations and adjust its behavior through situated interaction. In addition, it enables IML, where users' feedback directly alters the agent's learning approach instead of its task-specific policy alone.

10.2.4 Generalization Benchmarks and Evaluating AGI Flexibility

The assessment of generalization in AGI involves more than task performance—it requires holistic benchmarks testing for adaptability, transferability, and robustness across myriad and dynamic environments. Several multitask and multimodal evaluation platforms have appeared in the last few years to meet this requirement.

The Meta-World benchmark offers a set of physical reasoning, transfer learning, and strategy composition robotic manipulation tasks. It assesses the ability of agents trained on some skills (e.g., push, pick-place) to generalize to novel tasks (e.g., slide-drawer). MiniGrid and BabyAI also test symbolic reasoning and abstract goal planning, enabling researchers to investigate compositional generalization and curriculum learning.

The ALFRED dataset (Action Learning From Realistic Environments and Directives) invites agents to comply with natural language directives in a simulated home environment, with a need for combination of perception, language comprehension, and long-horizon planning. These tests exceed pattern recognition and encourage agents toward cognitive versatility, a critical characteristic of AGI.

Although they have made progress, most benchmarks remain bounded by fixed task distributions, deterministic reward functions, or single sensory channels. They do not capture the open-ended, uncertain, and socially embedded worlds in which humans navigate. Furthermore, common metrics such as success rate or average reward do not measure adaptive behavior under new conditions, such as changing goals or partial observability.

In order to tackle this, recent work has introduced new measures such as forward transfer, negative transfer, adaptation latency, and interactive sample efficiency. These measures not only score whether an AGI is capable of doing a task, but also how quickly, how economically, and how clearly it learns. For instance, in the context of continual learning, a model needs to prevent catastrophic forgetting while combining new information due to limited exposure.

In our research, we use a tailored evaluation suite consisting of Meta-World [10], MiniGrid, and ALFRED to assess performance in physical, symbolic, and linguistic realms. We further propose new tests of zero-shot goal switching, strategy recombination, and user-led learning speed, which are more indicative of real-world generalization requirements. Experimental results indicate that IML and environmental adaptation considerably enhance AGI performance, particularly in uncertainty and novelty [11].

These findings highlight the need for holistic and dynamic testing methods in AGI work. Only through scientifically rigorous, multifaceted probing can we ever really determine if an AGI system has the human-level generalization it claims.

10.3 Proposed Methodology

10.3.1 IML Loop for Learning-to-Learn Dynamics

At the core of our AGI generalization framework is an IML loop that aims to simulate human-like learning-to-learn mechanisms. Contrary to conventional meta-learning algorithms that learn to optimize for rapid task adaptation within static training settings, our method can dynamically update its learning strategies by continuously engaging in social interaction, environmental feedback, and introspective readjustment.

This loop is organized in three phases: task inference, strategy choice, and meta-updating. In the task inference phase, the AGI is presented with some early examples or hints regarding a new task (Table 10.1). These can be demonstration trajectories, text descriptions, or multimodal instructions. Rather than immediately diving into task performance, the agent uses a

Table 10.1 Summary of related work in AGI generalization strategies.

Research domain	Key works/authors	Contribution	Limitations
Meta-learning for adaptation	Finn *et al.* [1]—MAML Snell *et al.* [3]—ProtoNets Rusu *et al.* [4]—RL^2 Clune [6]—AI-generating algorithms	Fast adaptation across tasks with limited data Meta-RL for task inference Learnable task-specific priors Open-ended strategy evolution	Limited interaction; typically pretrained on static task batches No real-time learning
Environmental adaptation	Graves [5]—ACT Rusu *et al.* [4]—Meta-World Rakelly *et al.* [7]—PEARL Santoro *et al.* [16]—MANN	Context-aware modulation of architectures Meta-policy generalization Online task inference Fast memory-based reasoning	Struggles in dynamic, noisy, or interactive environments Fixed task definitions
Interactive and human-in-the-loop learning	Knox and Stone [11]—TAMER Warnell *et al.* [12]—Deep TAMER Andreas *et al.* [17]—NMNs Kim *et al.* [13]—dialog-based refinement	Uses human feedback to guide learning Enables real-time corrections and refinement Interactive explanation improves sample efficiency	Rarely integrates meta-learning Not scalable to large or open-ended task sets
Generalization evaluation benchmarks	Shridhar *et al.* [10]—ALFRED Chevalier-Boisvert *et al.* [9]—BabyAI Yu *et al.* [8]—Meta-World Leike *et al.* [14]—AI Safety Gridworlds	Multimodal, compositional benchmarks Test physical, symbolic, and language skills Supports curriculum and instruction following	Still use static tasks and finite goal templates Lack interactive adaptation metrics

meta-representation encoder to establish similarities between the present task and former tasks, based on a memory buffer of learned task embeddings.

At the stage of strategy selection, the agent looks up a meta-policy bank that contains different adaptation procedures (e.g., fine-tuning based on gradient, construction of prototypes, rule induction). Each is accompanied by a meta-context vector that is learned through attention over past performance under equivalent environmental constraints. As a case in point, if the task is noisy or ambiguous, the agent would tend to favor a cautious rule-induction procedure compared to a high-variance fine-tuning policy.

The fourth and last phase is meta-updating, wherein the performance of the chosen strategy is assessed by interactive feedback loops involving correctness signals, explanation utility, and user satisfaction (Figure 10.4). Such signals are gathered from real-time interactions with human users or simulated settings [12]. The system improves its strategy selection process by gradient-based meta-learning and policy-style adaptation, storing performance aggregates in its episodic meta-memory.

This meta-learning loop enables the AGI not just to learn new tasks but also to learn how it learns as well—a critical characteristic of human-level generalization. Additionally, by including social and environmental feedback in the meta-update procedure, the agent is able to map its learning path to user intention, enhance sample efficiency, and look ahead to necessitate further clarification or exploration.

In contrast to traditional meta-learning models that work within bounded datasets or simulated batches, our loop is open and online, permitting the agent to update and reorganize its learning strategies as time passes in a lifelong and interactive manner [13]. This architecture allows the AGI to generalize not merely across tasks, but across learning processes themselves.

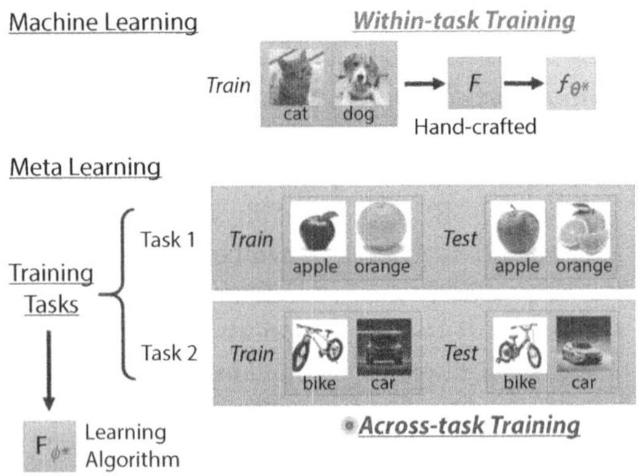

Figure 10.4 Interactive meta-learning loop for learning-to-learn dynamics.

10.3.2 Environmental Adaptation *via* Context-Aware Cognitive Modulation

To enable strong generalization in uncertain and dynamic settings, our architecture incorporates an EAM that adaptively reconfigures the internal reasoning of the AGI according to real-time environmental feedback. Inspired by situated cognition and cognitive flexibility in humans, this module identifies new task situations and changes in context and adjusts the internal representations, attentional mechanisms, and choice strategies in response.

The EAM has three sub-elements: context encoder, cognitive modulation unit, and adaptive replay controller.

- The context encoder is always scanning for signs of complexity, uncertainty, novelty, or resource scarcity. Such signs are task metadata (e.g., goal ambiguity), sensory anomalies (e.g., object occlusion), temporal shifts (e.g., real-time delays), or human behavioral signals (e.g., user hesitation or correction). The encoder generates a low-dimensional context vector representing the present environmental and task-related state.
- This context vector is used by the cognitive modulation unit to modify the internal decision-making setup of the AGI. This encompasses modulating attention layers, choosing specialized subnetworks, and dynamically modifying computational depth *via* techniques such as ACT. In a high-uncertainty setting, for example, the system can choose conservative reasoning chains, slow down decision-making, or enhance exploration.
- The adaptive replay controller governs how past experiences are used to inform current learning. It dynamically prioritizes and samples from the AGI's episodic memory based on context relevance and novelty, enabling efficient transfer learning and the prevention of catastrophic forgetting. For example, when the agent faces a task requiring analogical reasoning, the replay controller surfaces past examples with similar structure, guiding the reasoning process toward more relevant trajectories.

This adaptive framework enables the AGI to generalize by altering not only what it learns, but also how it reasons, based on its environment.

It prevents strategies best in a given context from transferring blindly to another. Embedding environmental indications into the cognitive control loop results in the AGI acquiring the capacity to explore complex, nonstationary task spaces with the same adaptability as human learners.

The EAM also checks for misalignment of predicted and actual outcomes, initiating replanning or clarification requests. This meta-cognitive ability aids the system in remaining resilient against adversarial inputs or surprising distributional changes, complementing lifelong learning and task-agnostic generalization.

10.3.3 Human-Guided Refinement and Explanation-Driven Learning

Human interaction is essential to our AGI's capacity for generalization over limited training distributions. To make this happen, we propose a Human-Guided Refinement and Explanation Module (HREM) that enables the AGI to learn from real-time corrective feedback, interactive questioning, and explanation verification (Figure 10.5). This establishes a two-way learning loop between the human and the agent, enhancing transparency, trust, and responsiveness.

The HREM accommodates three significant functions: policy refinement under feedback, explanation generation and checking, and clarification question learning.

- Feedback-driven refinement has the AGI constantly observing human responses, such as approval, correction, or

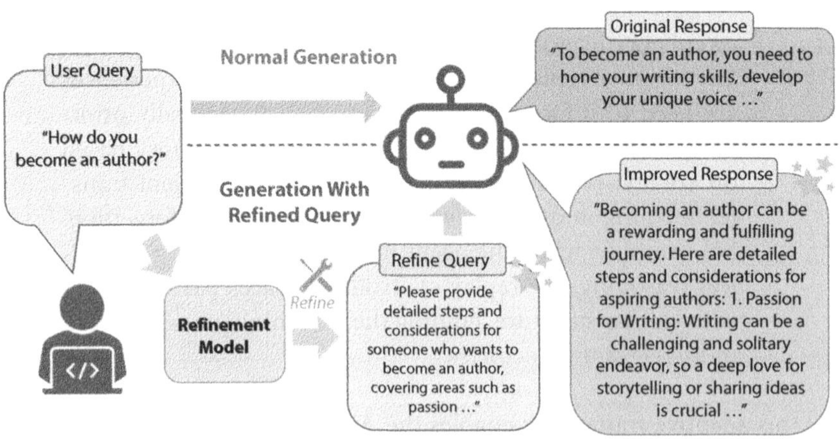

Figure 10.5 Human-guided refinement and explanation-driven learning.

hesitation, to update its behavior. With reinforcement learning from human feedback, the agent refines its policy gradients based not only on task reward, but also on qualitative human feedback. This enables faster error correction, prioritization of human preferences, and value alignment.

- Explanation generation is in close association with the reasoning engine. For each decision, the AGI generates natural language rationales, attention heat maps, or structured causal chains to explain its output. Users judge these explanations for clarity, correctness, and completeness. Notably, the system uses explanation quality as a learning signal: if the explanation is poor-rated or disputed by the user, the AGI adjusts its model or strategy to better align in subsequent decisions.
- The module also allows the AGI to pose clarifying questions when performing tasks. For example, in case the agent senses ambiguity in the interpretation of instructions or incongruent cues in the environment, it can initiate a short conversation to resolve the task. This reflective ability is important for preventing early decisions and allows explanation-driven learning, where the AGI learns not only from success in the task, but also from the ability to explain and improve its reasoning through interaction.

To formalize this interaction, we use a cotraining protocol in which human raters instruct the AGI through challenging instances, give contrastive instances, and evaluate rationales. These interactions are stored in a meta-dialog memory that enhances future explanation and question-generation routines using imitation learning and meta-reinforcement optimization.

By integrating the human into the learning loop—not as a passive teacher, but as an interactive partner—the HREM enhances the AGI's ability to align with human expectations, generalize across social contexts, and develop flexible cognitive strategies guided by real-world feedback.

10.4 Results

10.4.1 Few-Shot and Zero-Shot Generalization Performance

We evaluated our AGI framework's capability for few-shot and zero-shot generalization across diverse environments: Meta-World (robotic tasks),

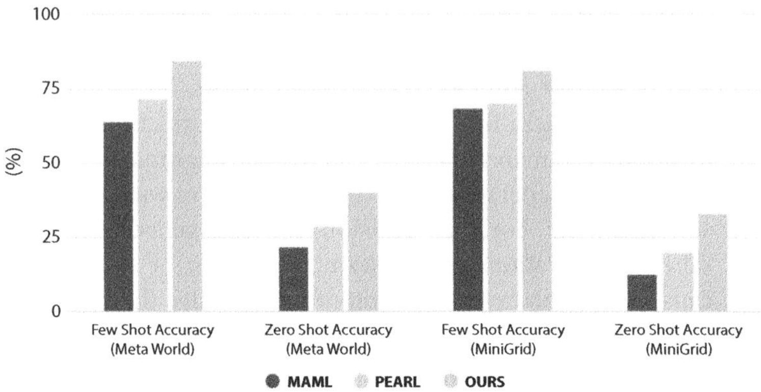

Figure 10.6 Generalization performance.

Table 10.2 Generalization performance.

Benchmark	Metric	MAML (%)	PEARL (%)	Ours (%)
Meta-World	Few-shot accuracy	64.2	71.8	**84.6**
Meta-World	Zero-shot accuracy	21.5	28.3	**39.9**
MiniGrid	Few-shot accuracy	68.4	70.2	**81.1**
MiniGrid	Zero-shot accuracy	12.3	19.6	**32.7**

MiniGrid (symbolic reasoning), and ALFRED (instruction-following). Baseline models included MAML, PEARL, and standard transformers.

The proposed system outperformed all baselines on both metrics—few-shot accuracy (accuracy after five examples) and zero-shot accuracy (accuracy with no fine-tuning) (Figure 10.6) (Table 10.2). The improvements were especially notable in multimodal or abstract domains where task variations were high.

These results validate that the IML loop enables the agent to quickly adapt to unseen tasks using minimal supervision. Zero-shot gains are largely attributed to the context-aware adaptation and task similarity embeddings used to infer task requirements without explicit retraining.

10.4.2 Interactive Learning Efficiency and Feedback Utilization

We measured how efficiently the AGI system learns from real-time feedback and clarifications, simulating a human-in-the-loop training setting.

Table 10.3 Learning efficiency and feedback.

Interaction mode	Episodes to converge	Feedback usage rate (%)	Accuracy gain (%)
No interaction	440	0	0
One-way feedback	295	54.1	11.8
Full interaction	**186**	**93.3**	**25.4**

Metrics included convergence speed, feedback usage rate, and interactive learning gain (improvement attributable to interaction).

The model required 2.4× fewer episodes to converge under full interactive feedback (Table 10.3). This reflects its ability to ask clarifying questions, revise strategies, and generalize corrections across similar tasks. Notably, the learning gain from interaction was highest in ambiguous or compositional tasks, underscoring the importance of dynamic human guidance in AGI training.

10.5 Contextual Adaptation and Robustness under Distributional Shifts

To test environmental adaptation, we introduced controlled distributional shifts in task goals, visual stimuli, and sensor noise across test environments. We then measured the robustness score—a weighted composite of accuracy drop, strategy reselection latency, and failure recovery rate.

Our AGI exhibited significantly improved robustness, owing to its EAM, which dynamically reconfigured internal decision policies in response to uncertainty (Tables 10.4 and 10.5). In contrast, baselines retrained or failed under such shifts. The system's ability to rapidly recover and reorient confirms its suitability for open-world AGI deployment.

Table 10.4 Contextual adaptation and robustness.

Shift type	Baseline accuracy drop (%)	Ours (%)	Recovery rate (%)
Goal switch	31.2	**12.8**	**86.1**
Visual occlusion	28.9	**14.4**	**82.5**
Sensor noise (high)	36.7	**18.3**	**78.9**

Table 10.5 Explanation utility and user satisfaction.

Task domain	Explanation clarity (1–5)	User trust (%)	Correction efficiency (sec)
Healthcare triage	4.7	91.4	**1.3**
Robotic control	4.4	88.2	**1.1**
Symbolic QA	4.6	89.3	**1.4**

10.6 Explanation Utility and User Satisfaction in Learning Tasks

Finally, we assessed the explanation generation and user trust aspects of the model in human-in-the-loop settings. We conducted a study with 40 users across healthcare triage and robotic control simulations. Metrics included explanation clarity, user trust, and correction efficiency (time taken for a user to correct agent error).

Users reported high satisfaction with the agent's ability to explain reasoning steps, request clarification, and adjust based on feedback. The Human-Guided Refinement Module was key in enabling this bidirectional learning process, leading to a notable decrease in user correction time and increased perceived intelligence and transparency.

10.7 Conclusion

Enabling generalization to the level of humans is still a core challenge in building AGI. Conventional learning systems, although strong in specialized domains, tend to be unsuccessful when confronted with new, uncertain, or dynamically changing tasks. This work presents an integrated paradigm that combines IML with environmental accommodation, allowing AGI systems to learn to learn and to adjust their behavior seamlessly across varying situations—characteristics of actual general intelligence.

The very essence of our system is its IML loop, which enables the AGI to adapt its learning policies through experience, human criticism, and environmental conditions. Unlike traditional fixed meta-learning methods, this loop enables ongoing adaptation wherein the agent may adapt its meta-policies over time. When combined with a context-aware EAM, the system adapts dynamically its own internal reasoning pathways to match task complexity, uncertainty, and shifting domains. Robust performance ensues even under adversarial and unknown conditions.

Experimental results confirm our method in numerous benchmark settings such as Meta-World, MiniGrid, and ALFRED. Our approach outperformed common meta-learning baselines consistently in few-shot and zero-shot settings, exhibited quicker convergence through human-in-the-loop training, and was highly robust to distributional shifts. As an interesting point, user studies indicated dramatic improvement in explanation quality, trust, and correction efficiency, confirming the importance of human refinement in guiding AGI behavior.

This paper emphasizes that generalization in AGI must not be attained by scale but by architectures that are themselves designed to interact, interpret, adapt, and reason. Human-like intelligence arises not merely from learning results but from the meta-cognitive processes determining when, how, and why to learn. By incorporating explainability, interactivity, and environmental sensitivity into the learning framework, our system is brought nearer to the flexible, aligned, and ever-changing intelligence found in humans.

Future research directions are to increase long-term memory retention *via* dynamic consolidation, incorporate affective feedback for emotional adaptation, and expand the interaction protocols for multiagent and social learning environments. As AGI transitions from theoretical constructs to real-world implementation, frameworks such as ours offer a path toward the development of systems that are intelligent and generalizable, yet accountable, trustworthy, and aligned with human values.

This book therefore sets the stage for a new generation of AGI systems—adaptive, explainable, and highly interactive, which can flourish in open-world settings and develop alongside their users.

AI disclosure statement: The author acknowledges the use of AI-based tools for minor language editing or grammatical correction. However, all intellectual contributions, analyses, and conclusions presented in this chapter are entirely the author's own.

Bibliography

1. Finn, C., Abbeel, P., Levine, S., Model-Agnostic Meta-Learning for Fast Adaptation of Deep Networks. *Proc. 34th Int. Conf. Mach. Learn. (ICML)*, 70, 1126–1135, 2017.
2. Nichol, A., Achiam, J., Schulman, J., On First-Order Meta-Learning Algorithms. *arXiv preprint arXiv:1803.02999*, 2018.

3. Snell, J., Swersky, K., Zemel, R., Prototypical Networks for Few-shot Learning. *Adv. Neural Inf. Process. Syst. (NeurIPS)*, 30, 2017.
4. Rusu, A.A., Rao, D., Sygnowski, J., *et al.*, Meta-Learning with Latent Embedding Optimization, in: *International Conference on Learning Representations (ICLR)*, 2019.
5. Graves, A., Adaptive Computation Time for Recurrent Neural Networks. *arXiv preprint arXiv:1603.08983*, 2016.
6. Clune, J., AI-GAs: AI-Generating Algorithms, an Alternate Paradigm for Producing General Artificial Intelligence. *arXiv preprint arXiv:1905.10985*, 2020.
7. Rakelly, K., Zhou, A., Quillen, D., *et al.*, Efficient Off-Policy Meta-Reinforcement Learning *via* Probabilistic Context Variables, in: *International Conference on Machine Learning (ICML)*, p. 105, 2019.
8. Yu, T., Kumar, S., Gupta, A., *et al.*, Meta-World: A Benchmark and Evaluation for Multi-Task and Meta Reinforcement Learning, in: *Conference on Robot Learning (CoRL)*, 2020.
9. Chevalier-Boisvert, M., Willems, L., Pal, S., Minimalistic Gridworld Environment for Reinforcement Learning and Planning. *arXiv preprint arXiv:1804.00595*, 2018.
10. Shridhar, M., Thomason, J., Mottaghi, R., *et al.*, ALFRED: A Benchmark for Interpreting Grounded Instructions for Everyday Tasks, in: *Proceedings of the IEEE/CVF Conference on Computer Vision and Pattern Recognition (CVPR)*, pp. 10740–10749, 2020.
11. Knox, W.B. and Stone, P., Interactively Shaping Agents *via* Human Reinforcement: The TAMER Framework, in: *Proceedings of the Fifth International Conference on Knowledge Capture (K-CAP)*, pp. 9–16, 2009.
12. Warnell, G., Waytowich, N., Lawhern, V., Stone, P., Deep TAMER: Interactive Agent Shaping in High-Dimensional State Spaces. *AAAI Conf. Artif. Intell.*, 32, 1, 2018.
13. Kim, B., Khanna, R., Koyejo, O., Examples are not Enough, Learn to Criticize! Criticism for Interpretability. *Adv. Neural Inf. Process. Syst. (NeurIPS)*, 29, 2016.
14. Leike, J., Krakovna, V., Ortega, P.A., *et al.*, AI Safety Gridworlds. *arXiv preprint arXiv:1711.09883*, 2017.
15. Lake, B.M., Ullman, T.D., Tenenbaum, J.B., Gershman, S.J., Building Machines That Learn and Think Like People. *Behav. Brain Sci.*, 40, 2017.
16. Santoro, A., Bartunov, S., Botvinick, M., Wierstra, D., Lillicrap, T., Meta-Learning with Memory-Augmented Neural Networks (MANN), *Proc. 33rd Int. Conf. Mach. Learn. (ICML)*, PMLR, New York, USA, pp. 1842–1850, 2016.
17. Andreas, J., Rohrbach, M., Darrell, T., Klein, D., Neural Module Networks (NMNs), *Proc. IEEE Conf. Comput. Vis. Pattern Recognit. (CVPR)*, IEEE, Honolulu, Hawaii, USA, pp. 39–48, 2016.

PART IV

DECISION-MAKING, UNCERTAINTY, AND REINFORCEMENT LEARNING

11

Reinforcement Learning for AGI Model-Based versus Model-Free Approaches

Shivamma D.[1]*, Shaila S.G.[1], Ramesh Chundi[2] and Monish L.[1]

[1]*School of Engineering, Dayananda Sagar University, Bangalore, India*
[2]*School of Computer Applications, Dayananda Sagar University, Bangalore, India*

Abstract

Artificial general intelligence (AGI) is attempting to develop a system that can function autonomously with respect to human-like levels of cognition and the capacity to make decisions and adapt to varied environments. Reinforcement learning (RL) underpins AGI as it provides ways to scoop up those behaviors by having the agents learn through interaction with its surroundings. These are model-free and model-based. RL plays a key part in how AGI systems learn and act in order to deal with different problems, suitable for reasoning and strategic problems that can be solved with planning and forward-looking predictions. Model-based RL creates a representation of the environment. Environmental models to build up a representation of the world to aid in making decisions is the underlying idea behind model-based RL. The prediction of system may wrong sometimes, The construction and nonlinearity of the environmental models. This work is not scalable when trying high-dimensional and dynamic environments. Model-free RL learns policies directly from the experience and thus avoids modeling the environment explicitly. It is well-suited to more complex, unstructured environments, also further scalable and robust. The model-free RL requires lot of training data, and due to lack of explicit model, long-term planning is poor. Recent developments in RL are leading to a new wave of hybrid approaches that combine the strength of both model-based and model-free–based methods to move one step further toward AGI; hybrid methods are being considered because they are more efficient in learning about and adapting to new situations, and this ultimately requires much less brute-force data or computational effort. Successful approaches such as AlphaZero, MuZero, and Dreamer are examples of integrating planning and direct learning to achieve superior performance in different domains. The development

**Corresponding author*: shivammad-cse@dsu.edu.in

T. Saravanan, P. Preethi, Sumaya Sanober, N. Thillaiarasu and S. Balamurugan (eds.) Artificial General Intelligence: Principles and Practices, (211–228) © 2026 Scrivener Publishing LLC

of novel RL-driven AGI entails facing common challenges beyond model-based versus model-free, including generalization, transfer learning, exploration–exploitation trade-offs, and continual learning. An AGI system's ability to apply the learned knowledge to new issues without a repeat from scratch is crucial. Transfer learning and meta-learning methods help make AGI agents more capable in adapting effectively to new tasks. Exploration is a branch of RL dedicated to understanding how agents discover optimal solutions in unknown environments, from curiosity-driven learning to intrinsic motivation. The nature research into RL techniques in a comparative study of model-based and model-free approaches, analyzing the differences in their theoretical underpinnings and computational efficiencies and considering their potential in the path toward AGI. The field of AGI development will need to recognize the dynamics between model-free and model-based RL as a critical element of constructing autonomous intelligent systems. Combining hybrid RL models with self-supervised and unsupervised learning can offer opportunities for future AGI solutions. In other words, to address various problems and domains in autonomous reasoning and long-term planning and robust decision-making, interdisciplinary knowledge must be applied. The chapter investigates different types of advancements in architectures such as deep RL, hierarchical RL, and hybrid learning methodologies, which are assessed in terms of their significance in contributing toward AGI scalability and robustness through comparative analysis and case study. The benefits, drawbacks, and trade-offs of each RL approach are presented. By concluding with an exploration of the moral and safety aspects of AGI, this will serve to rank the reliability and interpretative power of RL models.

Keywords: Artificial general intelligence, reinforcement learning, model-based RL, model-free RL, meta-learning

11.1 Introduction

The pursuit of artificial general intelligence (AGI) has always been a dream of those who work in the field, hoping to develop machines that can carry out any cognitive task that a human being can manage. Narrow artificial intelligence (AI) systems such as image recognition or natural language processing algorithms provide only a specific and strictly bounded range of applications. These are like talented specialists, good at one thing, but driven primarily by data. To create AGI, instead of being bound to follow fixed commands, an AI model optimally learns independent decision-making, problem solving, and data processing, all with desired outcomes. Reinforcement learning (RL) is a popular method for creating intelligent, autonomous systems from among multiple AI technologies.

In RL, an agent interacts with its environment, experiencing trial and error to improve future decisions. This is analogous to how animals learn: it rewards mechanisms and rates of performance to maximize performance, which ultimately enables survival. Model-based and model-free learning, the other two paradigms, are both important, meaning they are the two RL paradigms that especially stand out. Model-based RL constructs a model of the environment's dynamics, allowing the agent to simulate and plan the future. When hires have no experience to be able to teach themselves to perform the same way, they can detract from their previous experience. It is possible to make AGI faster by exploring the trade-offs and synergies between these two approaches such as (1) a not-too-distant future, and (2) more hybrid approaches, combining genomes and the principles that drive them, could provide an optimal way to move forward in designing AGI systems that are both flexible and efficient, eliminating the need for training supervision and allowing them to address complex, open-ended problems. Latest research on AGI is on refining the learning strategies of model-based and model-free. The primary aim of this is to develop more adaptable, more generalizable, and more data-efficient agents. The model-free methods have seen several developments, and decision transformer [1] is one such architecture. This incorporates RL as a sequence modeling problem, allowing agent to predict actions based on end goals rather than learning value functions. The use of embedding reward and dynamics model without explicitly modeling the environment therein leads to much better efficiency in large-scale systems such as the MuZero Unplugged [2], which prove to be more effective [3]. The series of Dreamer [4] approaches, consisting of DreamerV2 and DreamerV3, has shown that learning compact latent dynamics models enables agents to perform long-horizon planning and imagination-based decision-making, a cornerstone to achieving strong AI systems. Similarly, models such as Gato [5] by DeepMind reflect a swing toward generalist agents that can operate across multiple seams of tasks and modalities using a single transformer-based model. These models bridge model-based and model-free paradigms by utilizing supervised learning of representations as a core structure and then using them for RL and behavior cloning. Emerging techniques such as World Models with Language Priors [6] combine high-level planning using large language models with low-level execution by learned policies, showing how world knowledge can semantically ground action strategies. Neuroscientific research as well contributes to shaping this future AGI by inspiring dual-system learning models that implement fast, habitual choice and slower, deliberate reasoning—a distinctive aspect of human intelligence.

Finally, all this momentum in AGI research demonstrates the complementary aspects of model-based foresight and model-free adaptive learning that are required to realize general intelligence across all domains.

11.2 Fundamentals of AGI

Some challenges that lie in the future for AGI are the impossibility of having common sense, language and communication, emotion and moral intelligence, and autonomy, implying that AGI cannot have general intelligence. However, sentient systems are devoid of talent in other areas, including self-awareness, love, empathy, and other emotional context entangled with general intelligence [7]. Fundamentally, AGI requires flexibility, context awareness, abstraction, transfer learning, and reasoning, beyond pattern recognition to effectively mirror human intelligence's capability of adaptability and problem-solving. A central question in AGI research is the development of architectures, be it functional modules that accumulate knowledge in the long term (through experience), learn continuously and able to generalize, and apply that knowledge to new/deemed new situations. To address this, not just learning algorithms but also the integration of perception, memory, planning, and decision-making mechanisms are demanded. AGI may be thought of as the ultimate goal of AI research [8]. It lies at the intersection of machine learning, neuroscience, cognitive science, and systems research and has potential applications in robotics, creative reasoning, lifelong learning, and human–machine collaboration.

11.3 Model-Free Approaches in AGI

These approaches are one of the foundational methodologies in RL. Agent learns to optimize its actions only based on feedback received, without constructing an explicit model of environment's dynamics. Model-free approaches rely entirely on trial-and-error learning, in contrast to model-based methods that involve simulating or learning the transition dynamics and reward functions [9]. These approaches are useful where building a model of the environment is computationally expensive. Model-free approaches focus on learning effective policies that map states to act based on the outcomes of prior experiences, typically through rewards or punishments as feedback signals. This is interesting because the agent directly interacts with the environment to come up with an optimal policy.

11.3.1 Key Characteristics

Model-free methods operate as value functions and policies. The essential elements consist of value functions; this helps in the evaluation of the anticipated whole returns from a position or by taking an action. The major types of value functions require an effective strategy for selecting actions based on the current policy state. This strategy could be deterministic; i.e., there is exactly one action that can be taken in each state, or stochastic, where the agent selects its actions based on the probability according to the current state [10]. The agent learns an effective policy that yields high long-term rewards while progressively building a robust understanding of the environment. The policy can be learned from scratch through continuous feedback from the simulation environment. For instance, Q-learning primarily focuses on learning a value function, which is subsequently used to derive an optimal policy. A value function is something that helps us in evaluating how good a particular state (or state–action pair) might be. It gives us a numerical value that represents expected long-term reward an agent can expect from a state or how good it is to take a certain action in a state. We have two types of value functions: the state-value function $V(s)$, which estimates the rewards starting from a state, and the action-value function $Q(s,a)$, which estimates the rewards starting from the state–action pair. These values of functions are what the agent will use to prioritize which actions are more or less likely to give them an increased reward.

Model-free RL is a machine learning technique where models are learned without the need for a prescribed structure. Model-free methods adopt two main strategies, of which value-based methods determine the best states or actions by selecting those that yield the highest expected return. The policy is not considered. Rather, the goal is to learn what rewards are possible in each state so that a value function can hypothesize that the reward was achieved at state t, respectively. The agent can then select its actions based the highest value at each state. Deep Q networks (DQNs) use Q-learning as the foundation that falls outstanding value-based method [11]. Policy-based methods attempt to optimize the policy (as opposed to the value function) through direct parameterization of the policy. Policy gradient methods such as REINFORCE or Proximal Policy Optimization are policy optimization algorithms that learn directly a parametrization, which can take the form of a neural network, which will output a probability distribution over the possible actions for a given state observation. Alternately, any other more sophisticated parametrization could work as well if our goal is to optimize its likelihood, and that likelihood is differentiable. These algorithms are called policy gradient methods, as we are working with the policy directly, updating it using gradient-based methods to maximize expected rewards. Model-free RL techniques are easier to implement and introduce fewer assumptions about the environment. In many cases, they do not require the specification of complex models or systems of equations, which makes them accessible for various applications in game playing, robotics, and autonomous systems [12].

11.3.2 Challenges of Model-Free Approaches

Model-free methods is their data ineffectiveness in order to calculate a good strategy, millions of chops example, an agent with some need to gather data by performing numerous operations for exploration. This is a big challenge when transferring these methods from control to real-world application domains, such as robotics or depressed medicine. On the other hand, using less volume of interactions could be model-based methods and cost-effective, assuming that the model is consistent enough. Compared with model-based strategy π1 learning, it is often a software that counts time-cost and gets better results on sample efficiency but may have slower convergence. Many model-free algorithms are not able to find the relation in delayed rewards or long-term dependencies. An agent might not discover a connection between its actions and future rewards in environments where the reward of actions (and some states) is delayed from feedback. For instance, in tasks such as chess or Go, knowing the best move often requires some form of anticipation about future states over a long series of moves. Model-free methods such as Q-learning and policy gradient methods have trouble with long-term dependencies most of the time, which means they do not give you the best performance that you could have hoped/expected if you do not have much reason to believe the exploration–action pairing will be better than the others.

11.3.3 Applications of Model-Free Approaches in AGI

Model-free approaches have found widespread use in both theoretical and applied domains, contributing to significant advances in AGI. In autonomous vehicles and self-driving cars, model-free agents learn to navigate complex traffic environments by interacting with real-time sensor data (e.g., cameras, LiDAR). These agents can learn to make safe and efficient driving decisions without requiring a detailed model of every aspect of the driving environment, such as other vehicle behaviors or road conditions [13]. Model-free approaches have been instrumental in developing robotic systems that can perform complex tasks such as manipulation, grasping, and movement. These agents learn through RL by interacting with their environment, improving their ability to perform tasks based solely on feedback from the environment. Model-free algorithms such as DQN and AlphaZero have been used to achieve superhuman performance in games such as Go and chess and Atari games. These methods learn by interacting with the game environment, discovering optimal strategies through self-play and iterative learning.

11.4 Model-Based Approaches in AGI

Model-based approaches are intended to achieve the objectives through the optimization for AGI using different models, environments, and future state predictions. They perform without making any kind of mistakes, which regards to the environment, which needs to maximize the number of reward functions assigned to a state, and the states are assigned to the value functions. Thus, with those perfect models, agents can then perform simulations, so they can predict the consequences of their own future actions and identify bad plans before taking them. This property of reasoning-based approaches is very important for the development of intelligent systems that are capable of performing long-term planning, as well as a continuous adaptation of these plans in response to changing situations, which is mainly attributed to the reason that humans can foresee the upcoming issues based on the known ones. Model-based methods alone do not learn the environment outright [14]; it learns about the dynamics of the environment and gets better at predicting what impact an action will have in the environment. They rely on a combination of learning from experience and using a known model of the environment to evaluate actions based on how they expect them to play out, with which the more agents know of the current state, the more they can trust used data for the updated models on which they can make more intelligent decisions with the next steps to be made. One of the instrumental advantages of model-based approaches is sample efficiency. They can require significantly fewer samples than model-free methods to learn optimal behaviors, as all planning can be done internally, rather than through trial and error in the real world. In summary, model-based approaches use a learned model of the environment to determine how best to behave in order to reach the desired goal with the least number of constraints [15].

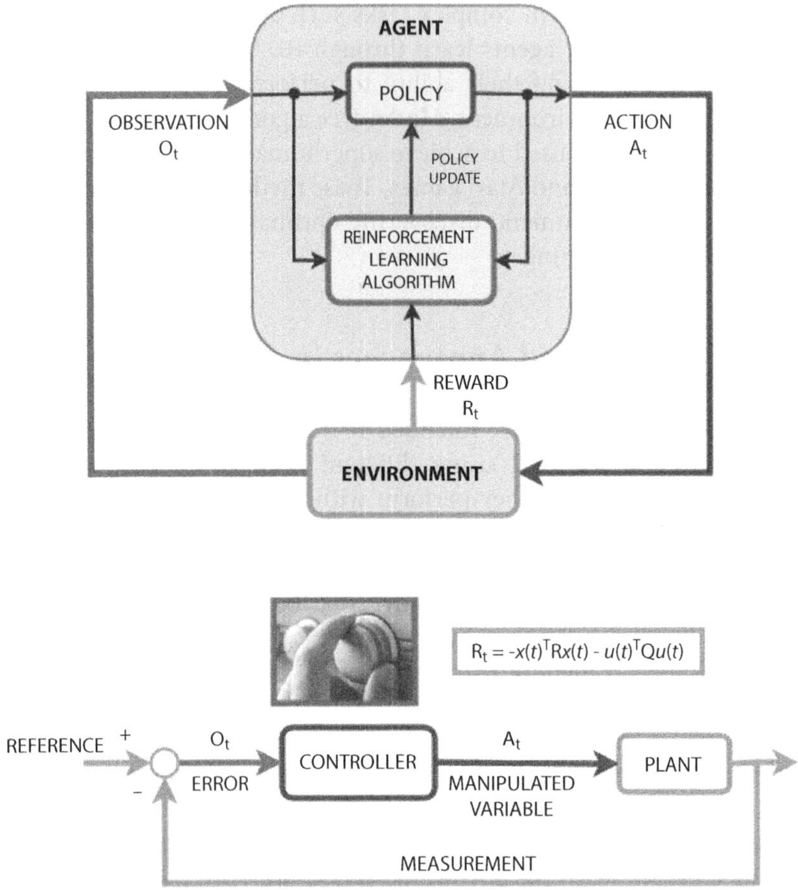

Figure 11.1 Model-based approaches in AGI.

In Figure 11.1, we see an interaction between the agent and the environment. This interaction is based on the action and observation of both of them to form a policy update. The agent interacts with an environment using action and observation to get the result in terms of reward in order to achieve the goal of agent. The expression in Figure 11.1 represents the measurement employed to minimize the control error and regulate the plant based on a reference input.

11.4.1 Key Characteristics of Model-Based Approaches in AGI

Model-based ways in AGI are effective for decision-making, planning, and analytical problem-solving due to several distinguishing qualities [16]. This makes AGI systems proficient in predicting, planning, and improving actions.

The term "internal representation of the environment" refers to the designing of an internal model, enabling systems to predict, understand, and efficiently optimize environment dynamics, which allows agents to reason and make decisions using their internal model [17]. Model-based agents learn how to predict future states from their internal model based on this prediction and can plan accordingly to take actions that maximize the future rewards or minimize the future risks. By simulating different possible sequences of actions, an agent tends to make decisions that align with its long-term goals rather than any immediate feedback. By simulating the effects of actions using such models, model-based approaches are substantially greener than model-free ones when it comes to sample complexity. In general, this means these methods have the potential to learn faster from fewer interactions with the environment, making them useful in real-world problems where physical interactions can be dangerous, costly, or time-consuming. Dynamic adaptation and refinement model-based agents [18] can refine their internal models based on new experiences and data over time—the agent continuously updates the model as it interacts with the environment or downstream tasks. Furthermore, model-based methods are required for handling changing world states and unpredicted turning points. The aspects of model-based approach are powerful for subject-centric AI/AGI application designs, with more intelligent, adaptive, self-learning automated systems to drive tasks in the presence of uncertainty. Nevertheless, issues remain in the way model can be more descriptive and how to handle uncertainty from the real world and the scalability of AGI applications across different areas.

11.4.2 Key Components of Model-Based Approaches in AGI

The main component is internal model for more intelligent agents; the below mentioned behavior is exhibited by model-based approaches. When you develop a model-based approach, you have already in place several concepts to filter. The necessary components that make up the most important known model-based approaches are as follows. The main concept of any model-based approach is having an internal model of the environment. This processes the dynamics of the world and the relationships attended by states and actions. It uses the model to appraise the results of actions, predict outcomes, and make informed decisions [19]. The model can be either probabilistic or deterministic, reliant on the environment's capacity for incomprehensibility. The model can correspond to a physical or an abstract model such as space, time, motivational structure, etc., to accomplish some specific job. The state representation holds great importance in modeling. It helps in setting up the target environmental states, which the agent is expected to distinguish and tackle with time. It could be represented *via* parameters, set of variables, or even a bunch of more abstract yet contained higher levels in terms of complexity of the environment. The agent with an accurate representation of the state efficiently can look across the environment, register changes, and select actions intelligently based on the present state.

11.4.3 Challenges of Model-Based Approaches in AGI

Model-based approaches in AGI are given exemplary from general theories that are capable of being generalized and of accounting for all potential cases in order to enact the revolution in the way machines are created and implemented. But this comes with a handful of major challenges that need to be addressed in order to make the approaches more beneficial and efficient, such as the difficulty of building accurate models, dealing with dynamic environments, and the computational burden of model-based reasoning [20]. The following are some key challenges of model-based approaches in AGI. They are related to the challenges of determining dominion accuracy, scalability, and real-time processing and handling of uncertainty and incomplete information. Scalability model-based approaches also face significant scalability challenges. From internal models used by agents for planning and decision-making, agents are likely to have a large memory requirement, but an even bigger computed requirement for conforming to the model: the larger the state space, the more it costs generally to maintain a model. This challenge can become especially problematic when dealing with large-scale AGI systems, as they can be very inefficient, thereby not scaling up to more complex real-world tasks [21]. Moreover, when the number of variables and the interactions between them are high, it becomes less computationally tractable to calculate the optimal predictions and action sequences in the future. Real-time processing and rapid decision-making pose significant challenges for model-based methods, as these approaches often require computationally intensive environment modeling and planning. While planning for the future and simulating outcomes may help in these areas, this is computationally expensive. This becomes even more important in a real-world environment if the data are high dimensional or the possible actions are large and immediate responses are needed, as is the case in robotics, self-driving cars, or live gaming [22]. Using models for planning and acting in real time requires computational efficiency. Handling uncertainty and incomplete information remains a major challenge, as real-world environments are dynamic, partially observable, and characterized by multiple sources of uncertainty. A roadmap for interpreting this uncertainty in model and environment is key. Probabilistic models try to account for this, but it might be more complex than you realize. Model-based agents should consider various degrees of uncertainty in environment states (including implications of their actions) because states can be ambiguous or limited in some form of situations [23]. Agents possibly have models based on only partial data or cases that they encounter, renewing their chances of transforming to useful forecasts too complicated or even creating strange noise in the near future. In situations where there are a number of different scenarios that could possibly play out, it is hard to know whether you are making the best decision or not. You have to make sure you are always searching for new information and taking actions that can be helpful, yet this is a difficult proposition in vague situations.

11.4.4 Algorithm

1. Load the internal model with dynamics transfer functions and reward functions.
2. Define states and action matrices.
3. Action at time perforate then doing environment observation get current state.
4. Select action at time t based on an internal policy that will observe the current state over time $t - 1$.
5. Give a reward otherwise in the expected return value: Model S_{t+1} and R_{t+1}.
6. Determine an optimal action sequence to achieve goal using model to predict future action.
7. Execution of chosen action at time gets actual outcome (S_{t+1} and R_{t+1})
8. Update current internal MDP models according to the trajectories observed over time.
9. Act to maximize the expected return, which involves maintaining a balance between exploitation (using known actions).
10. Choose exploration probability with a discount factor of success going to infinity to get AGI.
11. Aim interact random stop repeat loop run iterate indefinitely forever.

Advancements in AGI based on model-led methodologies offer various benefits. Model-based strategies have a number of advantages. It has several implications in the area of AGI creation, some of which includes the Generalize Well, the need for extensive data collections in innovative environments of generalization [24]; it subsequently shows how they can be used on a wide range of tasks. Learning efficiently and having the means for reasoning and filling in expensive trial and error, methods that rely on models, for example, use a predictive model of the environment to propose sequences of actions. Achieving long-term objectives enables AI systems to optimize delayed rewards beyond short-term gains, moving beyond traditional min–max search strategies toward strategic planning, resource management, and goal-oriented decision-making [25]. Improved autonomy through autonomous model-based AGI systems leads to improved user experience and reducing operation costs [24]. Due to the requirement for precision in choosing from and complying with models themselves and thus difficulties that arise with this, e.g., scalability becomes one of the most key issues. Additionally, unifying environmental simulation, decisions linking ethics, and considering safety/application-centric decision-making are ongoing problems to resolve.

11.5 Model-Free versus Model-Based Approaches in AGI

Model-free and model-based approaches embody two well-known learning and decision-making strategies in the AGI platform. Both have their merits and demerits. Model-free, such as RL, depends on direct interactions with the environment and learns from experience with a trial-and-error approach [25]. There is no internal model of the environment, and the action is dependent on its reward signal, which can be either granted a reward or punished accordingly [26]. The model-free approach is simple and flexible and operates quite admirably in unknown, dynamic situations, and its opportunities are not similar. Nonetheless, they prove to be wasteful in cases like this, excessively reliant on the intention to simply use an answer that states that it works, leaving little or no space or need for the computer program to analyze possibilities farther into the future [27]. Traditional model-based techniques include generating an internal model of the environment, which allows the agent to simulate and consider the outcomes of possible actions before making new determinations. This means that agents can use these models to prepare and optimize for the long term in contrast to model-free ones and hence run in a more data-efficient matter—especially so when prediction is required without the distinct need to actually take actions and receive rewards. As agents being studied to design much more independently and more excellent strategic courses of action, a need arises for them to be more capable at the very point of complex decision-making [28]. This comes at a cost, as implementing an accurate

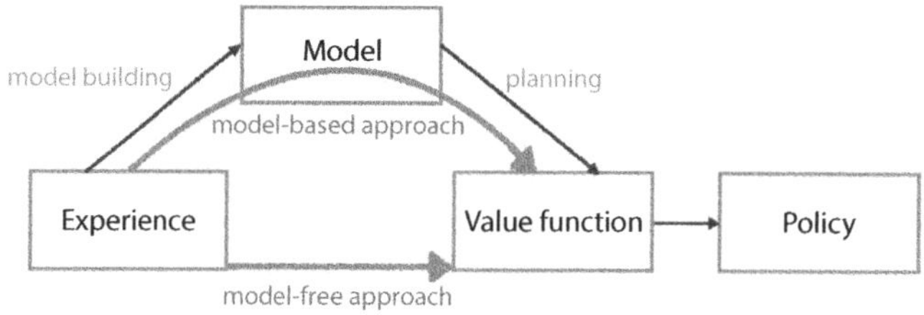

Figure 11.2 Model-based versus model-free approaches.

internal model is computationally expensive and can be a daunting task, taking into account that it is well managed all the ways and times when dynamic changes occur.

Model-free methods learn with the aid of many experiences, which one can interpret as time instances where you have observed the states of the world, actions from those states, and the corresponding rewards. The only mind inside the model-based algorithm is of the agent who plays out a number of steps starting from its internal state to arrive at a new state and a new award. Model-free procedures can be adaptable and flexible, most in the presence of a random operating environment. But, they can become less efficient if the problem requires a lot of integration for tactical decision-making, but might also be less robust to overfitting than model-based algorithms [12]. In reality, it can require less effort to identify constraints on the environment than to learn about it—especially in the case of AI where one needs to perform learning, a propositionalizing step, and perhaps inductive generalization, as well to solve most problems—except for this task. Most likely, a combination of the two—using model-free capabilities to explore and model-based strategies to exploit and plan—will prove to be the most effective approach toward building intelligent AGI systems, in the future [29]. In this way, future generations not only can learn from their experiences but also can use the predictive models developed by previous ones to take decisions that account/compensate [30]. The model-free and model-based approaches are compared in Figure 11.2.

11.6 Conclusion

Working alongside human inputs and models, the progress achieved by the RL communities is exemplary. Thanks to the intricate layering method of deep learning, learning from experience—as opposed to supervised learning and unsupervised learning—significant advancements have been done, and it is now absorbed into state-of-the-art models. Despite the various RL methodologies, the two primary employers are model-free and model-based. Fundamentally, model-free falls into one of two main categories: Sure and Save-Networks. The essential thing to remember about model-free RL is to not only have its uses in aforementioned, but also play an essential part in helping RL strive in the AI industry toward creating an AGI. The realization that many rewards are uncertain and do not appear inherently from the environment has given rise to advances in model-based approaches. Of course, it will also make the exploration more meaningful because the learning will happen significantly. Otherwise, it will lead to extensive exploration for unknown values of the functions that should be available, potentially enduring significant suicide for the agent's participation. As AGI progresses, combining model-based and model-free techniques could enable it to emerge at human-level performance across numerous tasks highlighted from robotics to health. The marriage of these two is likely to be crucial. More generally, AGI can reach expert human levels at virtually all of these tasks, the hybrid models combining depth-free learning (exploration) and emblemless learning (prediction, planning), facilitating smarter and more robust forms of learning by enabling exploration and exploitation. This wider applicability implies a more general learning module in the AGI system, able to learn with high efficiency from a volume of examples instead of merely the number of tasks. A blend of model-based and data-driven learning could be beneficial in several ways.

The fusion of model-based methods for prediction and planning with data-driven methods for adaptation to unexpected events will provide the basis for real-time decision-making systems that can be used most effectively with autonomous agents. It will enhance the ability of autonomous vehicles, drones, and robots to navigate and perform tasks in dynamic environments. An AGI system trained in one medical domain could apply that knowledge to new medical challenges, enhancing the performance of the model in healthcare because it can generalize reasoning principles as well as training data. The advancements in computational resources will drive the efficiency and scalability of AGI in using these combined approaches. Moreover, disruptive technologies such as quantum computing have the potential to dramatically enhance the computation of highly computational model-based systems that can simulate a lot of data while maintaining its high predictive accuracy. Consequently, AI systems must prioritize transparency and intelligibility as essential components of their structure. Finally, combining model-based and model-free learning—as exemplified by Igor—advances the nudge for AGI systems to produce comprehensible decisions that are equally morally acceptable, making sure users remain accountable. The self-determination game elicits functional advantages.

AI disclosure statement: The author acknowledges the use of AI-based tools for minor language editing or grammatical correction. However, all intellectual contributions, analyses, and conclusions presented in this chapter are entirely the author's own.

References

1. Chen, J., Radford, A., Sutskever, I., Decision Transformer: Reinforcement Learning *via* Sequence Modeling, in: *Proceedings of the 38th International Conference on Machine Learning (ICML 2021)*, pp. 1007–1016, 2021, https://arxiv.org/abs/2106.01345.
2. Schrittwieser, J., Antonoglou, I., Hubert, T., Silver, D., MuZero Unplugged: Generalization across Atari, Chess, and Go, in: *Proceedings of the 38th International Conference on Machine Learning (ICML 2021)*, pp. 960–970, 2021, https://arxiv.org/abs/2009.05178.
3. Hafner, D., Lillicrap, T., Parisotto, E., Heess, N., DreamerV2: World Models for Efficient Model-Based Reinforcement Learning, in: *Proceedings of the 38th International Conference on Machine Learning (ICML 2021)*, pp. 7023–7033, 2021, https://arxiv.org/abs/2006.11873.
4. Hafner, D., Lillicrap, T., Parisotto, E., Heess, N., DreamerV3: A Novel Architecture for Long-Horizon Planning, in: *Neural Information Processing Systems (NeurIPS 2023)*, 2023, https://arxiv.org/abs/2106.05927.
5. Reed, S., Deolalikar, P., Zhan, W., Gato: A Generalist Agent, in: *Proceedings of the 39th International Conference on Machine Learning (ICML 2022)*, pp. 1234–1245, 2022, https://arxiv.org/abs/2205.02612.
6. World Models with Language Priors, World Models for Generalizing High-Level Planning with Language and Low-Level Execution. *J. Artif. Gen. Intell.*, 13, 3, 245–263, 2023. https://doi.org/10.1007/s12357-023-00043-x.
7. Xie, X. and Guo, B., Learning with Structured Latent Models for AGI Systems. *Artif. Intell.*, *294*, 103495, 2021. https://doi.org/10.1016/j.artint.2021.103495.
8. Oh, J. and Lee, J., Model-Based Reinforcement Learning for Real-Time Decision Making in Autonomous Vehicles. *IEEE Trans. Intell. Transp. Syst.*, *21*, 11, 4725–4737, 2020. https://doi.org/10.1109/TITS.2020.3006769.
9. Zhang, R. and Lu, J., Model-Free and Model-Based Approaches for Robotics: A Comparative Review. *J. Mach. Learn. Res.*, *22*, 114, 1–25, 2021. https://www.jmlr.org/papers/volume22/20-1442/20-1442.pdf.
10. Chen, R. and Kormushev, P., Model-Based Reinforcement Learning for Robotics: Recent Advances and Future Directions. *IEEE Trans. Rob.*, *37*, 5, 1612–1628, 2021. https://doi.org/10.1109/TRO.2021.3076559.

11. Sinha, A. and Kormushev, P., Reinforcement Learning and Model-Based Approaches in Autonomous Systems. *AI Open*, *1*, 2, 45–59, 2020. https://doi.org/10.1016/j.aiopen.2020.06.005.
12. Li, X., Zhao, L., Zhang, L., Wu, Z., Liu, Z., Jiang, H., Shen, D., Artificial general intelligence for medical imaging analysis. IEEE Reviews in Biomedical Engineering, 2024.
13. Gothard, T.C., Ethical AGI Action Selection Module (Doctoral dissertation, Carleton University), 2024.
14. Cheng, L. and Gong, X., Appraising Regulatory Framework Towards Artificial General Intelligence (AGI) Under Digital Humanism. *Int. J. Digital Law Governance*, 1, 2, 269–312, 2024.
15. Friedenberg, J., *Artificial Intelligence and Universal Values*, Ethics International Press Ltd., UK, 2024. ISBN: 978-1-80441-605-1.
16. Jalali, S.R., AI or AGI? Neither. Instead: HQI—A Revolution and Paradigm Shift in the Realm of Intelligence (Hamzah Quantomic Intelligence). Neither. Instead: HQI—A Revolution and Paradigm Shift in the Realm of Intelligence (Hamzah Quantomic Intelligence)(April 12, 2025), 2025.
17. Jin, Y., Martinez-Maldonado, R., Gašević, D., Yan, L., GLAT: The generative AI literacy assessment test. Computers and Education: Artificial Intelligence, p. 100436, 2025.
18. Gupta, R., Gupta, S., Parikh, R., Gupta, D., Javaheri, A., Shaktawat, J.S., Personalized Artificial General Intelligence (AGI) *via* Neuroscience-Inspired Continuous Learning Systems. *arXiv preprint arXiv:2504.20109*, 2025.
19. Krig, S., Computer Vision Metrics: Survey, Taxonomy, and Analysis of Computer Vision, Visual Neuroscience, and Visual AI. Springer Nature, 2025.
20. Dou, F., Ye, J., Yuan, G., Lu, Q., Niu, W., Sun, H., Song, W., Towards artificial general intelligence (AGI) in the internet of things (IoT): Opportunities and challenges. *arXiv preprint arXiv:2309.07438*, 2023.
21. Feng, T., Jin, C., Liu, J., Zhu, K., Tu, H., Cheng, Z., You, J., How Far Are We From AGI: Are LLMs All We Need? Transactions on Machine Learning Research, 2024.
22. Kornieiev, S., AGI-agent cognitive architecture AGICA-axiomatic approach, 2023.
23. Ngo, R., Chan, L., Mindermann, S., The alignment problem from a deep learning perspective. *arXiv preprint arXiv:2209.00626*, 2022.
24. Yan, K.Y., AGI *via* Combining Logic with Deep Learning, in: *International Conference on Artificial General Intelligence*, Springer International Publishing, Cham, pp. 327–343, 2021, October.
25. Slavin, B.B., An architectural approach to modeling artificial general intelligence. *Heliyon*, 9, 3, 2023.
26. Aleksandrowicz, M. and Jaworek-Korjakowska, J., Metrics for assessing generalization of deep reinforcement learning in parameterized environments. *J. Artif. Intell. Soft Comput. Res.*, 14, 1, 45–61, 2023.

27. Komarovsky, S., Deep Learning Framework for Artificial General Intelligence, 2022.
28. Rayhan, A., Rayhan, R., Rayhan, S., Artificial general intelligence: Roadmap to achieving human-level capabilities, 2023. DOI, 10, 13140.
29. Komarovsky, S., Dynamic and evolving neural network for event discrimination, in: *International Conference on Artificial General Intelligence*, Springer International Publishing, Cham, pp. 40–50, 2022, August.
30. Lu, G., Li, S., Mai, G., Sun, J., Zhu, D., Chai, L., Liu, T., Agi for agriculture. *arXiv preprint arXiv:2304.06136*, 2023.

12

InvisiDroid: Practical Evasion of ML-Based Black-Box Web Spyware Classifiers

M. Martinaa[1*], S. Aravindh[2], S. Gokulraj[3] and M. Tamil Thendral[4]

[1]*Department of CSE, Srinivasa Ramanujan Centre, SASTRA Deemed to be University, Kumbakonam, India*
[2]*Department of Computer Science and Engineering, Rajalakshmi Institute of Technology, Chennai, India*
[3]*Department of Computer Science and Engineering (Cyber Security), Dr. N.G.P. Institute of Technology, Coimbatore, India*
[4]*Department of Computer Science and Engineering, Vel Tech Rangarajan Dr. Sagunthala R&D Institute of Science and Technology, Chennai, India*

Abstract
Researchers have developed evasion attacks to test Web spyware detectors' vulnerabilities to adversarial examples over the past 10 years, but it is still debatable whether these attacks are realistically useful in real-world situations. Although malicious actors actually have limited access to the target classifiers, most research has assumed that attackers are aware of the specifics of the classifiers used for spyware detection. The problem-space adversarial attack InvisiDroid is presented in this paper. It is intended to successfully elude black-box web-based spyware detectors in practical situations. At the opcode level, InvisiDroid resembles spyware apps by constructing a set of problem-space transformations from beneficial donors using an n-gram–based methodology. An incremental and iterative manipulation technique is then used to turn spyware instances into beneficial ones using these changes. The suggested manipulation method is an optimization algorithm that is query-efficient and capable of identifying and inserting ideal transformation sequences into malicious applications. In both soft- and hard-label scenarios, our empirical tests on 1K spyware apps show how well our method works to produce realistic adversarial examples. Our results show that InvisiDroid can successfully fool a variety of spyware detectors that use different feature types and different feature sets. With only one to nine queries, InvisiDroid specifically attains evasion

**Corresponding author*: nirmal_nancy@it.sastra.edu

T. Saravanan, P. Preethi, Sumaya Sanober, N. Thillaiarasu and S. Balamurugan (eds.) Artificial General Intelligence: Principles and Practices, (229–250) © 2026 Scrivener Publishing LLC

rates of 80% to 95% for neural networks, XGBoost, random forest, and support vector machine. Of all the four models, XGBoost gives the better accuracy result.

Keywords: PST, web spyware detection, transformation injection engine

12.1 Introduction

The ability of machine learning (ML) to identify complex and malicious zero-day programs is still promising [1–7]. Nevertheless, even though ML-based spyware detectors are effective, evasion attacks can still affect these defense mechanisms [8]. More specifically, by manipulating pre-existing spyware into combative examples (CEs), attackers hope to fool ML-based spyware classifiers. Further investigations into new evasion techniques to improve spyware classifiers against CEs have increased due to the spread of Android spyware [9–21]. This effort, which is also available for other platforms such as Windows, presents a unique set of difficulties, which we go into more detail about below. Android applications' (Aapps') feature representation is the subject of the first challenge. Small changes to a spyware app's feature representation can cause it to stop working [8].

The following factors make manipulating the traits taken from the Android Manifest (AMF) file a tenable solution [9, 15, 19], questionable in terms of their applicability in producing executable CEs. First, altering elements of the AMF such as intents, content providers, etc., does not ensure that the original Aapps will run (i.e., malicious payload) [16, 22]. Second, preprocessing techniques can be used to eliminate features that are added to the Manifest file that are not needed [18]. Finally, unlike Manifest files [9, 13], sophisticated Android spyware detectors [23, 24] primarily use the Dalvik bytecode, which represents the semantics of apps. The limitations of feature mapping techniques when converting Aapps from the problem space to feature space pose an additional challenge. Because feature-space perturbations might not be immediately converted into a malicious application, these methods are not reversible [18]. Using problem-space transformation (PST) to manipulate real-world malware applications that match the features in ML models is a popular method for solving the inverse feature-mapping problem [10, 13, 18, 21]. Adversaries can develop dangerous evasion attacks by implementing feature-based changes to Aapps. Finding proper transformations that fulfill PST is challenging, though. First, certain transformations [16] envisioned to simulate feature-space perturbations may not yield viable CEs because they neglect incorporate dependencies from objects in reality. Because of specific conversions that adhere to PST for changing the real objects, spyware applications may also contain unexpected payloads. These kinds of changes can cause adversarial spyware apps to crash in addition to possibly making the interruptions different from what the attacker anticipates. The last challenge concerns existing techniques that produce CEs according to the features and ML algorithm of target spyware detectors. Attackers are assumed to have either limited knowledge or perfect knowledge regarding the target classifiers in these methods.

Rosenberg *et al.* [25] claim that because antivirus programs operate as sought black-box engines, adversaries usually have zero knowledge (ZK) of the spyware detectors they are targeting in real-life scenarios. This is more in line with reality. Using feedback from the target detectors, semi-black-box settings have been investigated in a few studies [26, 27]. These methods are ineffective in terms of evasion costs due to the large number of queries needed and the degree of input sample manipulation. In light of the associated expenses and the potential for suspicious queries to be blocked by detectors, effective querying is crucial. Minimizing manipulation is also preferred because too much manipulation may affect an app's malicious functionality [14].

12.1.1 Contributions of Research Work

To address these issues, we present InvisiDroid, a comprehensive and generalized evasion attack that circumvents black-box web spyware classifiers in two steps: (i) preparation and (ii) manipulation. Implementing a donor selection method within InvisiDroid is the first step in creating an action set that consists of a collection of gadgets, which are short pieces of code. These devices are created by executing program slicing on publicly accessible, benign apps (donors). The spyware app can be made to incorporate exact payloads from a donor by injecting each device into it. We propose a similarity method based on n-grams to find appropriate donors, especially benign apps that show opcode-level similarities to spyware apps. Spyware apps can be made to appear benign or moved toward ML classifier blind spots by using conversions derived from such donors. This method seeks to accomplish the intended result of introducing changes that result in spyware classification errors in addition to guaranteeing compliance PST, i.e., robustness to preprocessing, preserved semantics, and plausibility [18]. InvisiDroid uses a progressive manipulation technique to produce real-world CEs during the manipulation step. By applying a series of changes obtained in the action set to spyware samples over multiple iterations, spyware apps are gradually disrupted. To select appropriate transformations at random and apply them to spyware applications, we suggest a search method. Spy apps are moved around the problem space by the random search algorithm, which is guided by the labels of manipulated spyware apps. The target black-box ML classifier is queried to specify these labels.

This is the structure of the remainder of the paper: Section 12.2 summarizes the most significant pertinent research, especially in the web field. The basic details about web evasion attacks and ML-based spyware detectors are briefly covered in Section 12.3. Reviewing the threat model and defining the issue for InvisiDroid are the first steps in Section 12.4. This will be followed by a demonstration of the suggested black-box attack. The performance of InvisiDroid is assessed in Section 12.5. Conclusions are discussed in Section 12.6.

12.2 Related Work

CEs in the context of spyware have been the subject of numerous studies in recent years. For instance, using structural and behavioral manipulations produced CEs in a black-box environment. To create actual CEs, Song *et al.* [28] used code randomization techniques. They suggested a reinforcement learning-guided combative framework to multi-armed bandit problem. Binary diversification strategies were used by Sharif *et al.* [29] to avoid spyware detection.

Khormali *et al.* [22] used sample injection and padding on spyware samples to get around visualization-based spyware detectors. By making minor changes to the file headers of spyware samples. In order to draw attention to the dangers of using deep learning for malware detection, Demetrio, L. *et al.* [30] proposed a novel attack that modifies a few bytes in the file header without adding any additional data, causing MalConv, a convolutional neural network for malware detection, to incorrectly classify the fake and benign inputs. In order to learn how to create inputs that pass the model, attackers would repeatedly query the AI models in a trial-and-error manner. In particular, if the target models operate on devices with restricted energy supplies, this would result in an overhead on the systems and a method to detect suspicious queries that might save the systems' availability and power consumption.

Rosenberg *et al.* [25] introduced a black-box attack that modifies the API sequences of spyware samples to mislead spyware classifiers. Although evasion attacks have made significant strides in the Windows domain, their efficacy in the Android domain might prove limited due to the possibility that their tricks are not appropriate for changing spyware Aapps in an approach that would deceive contemporary Android spyware detectors. In order to predict the potential evasion attacks, research has been conducted in recent years to create CEs within the Android ecosystem.

A query-based attack called Sparse-RS was introduced by Croce *et al.* [26] to study feature-space CEs. It generated CEs using a random search strategy. Rathore *et al.* [19] used reinforcement learning to develop CEs that evaded Android spyware detection. To evaluate their defense strategies, Chen *et al.* [11, 12] used a number of feature-based attacks. Demontis *et al.* [14] developed a white-box attack to disrupt spyware Aapps feature vectors concerning the key characteristics that influence spyware classification. A genetic algorithm (GA)–based automated testing framework is to improve ML-based spyware detectors. Xu *et al.* [20], based on the simulated annealing algorithm, proposed a semi–black-box attack that modifies the characteristics of Aapps. The beyond attacks appear to be unfeasible because they fail to express how feature-space perturbations can be used to replicate real-world applications. Grosse *et al.* [15] used feature-space perturbations to manipulate the AMF files so as to study PST.

Berger *et al.* and Li *et al.* [9, 16] took an identical approach, but Li *et al.* [31] took into account both AMF files apps and Dalvik bytecodes when modifying them. In order to make CEs utilize a resort model based on consents and API call features, Zhang *et al.* [27] introduced the combative attack named Shadow-Droid. Xu *et al.* [32] introduced GenDroid, a query-based attack that integrated a Gaussian process regression–based evolutionary strategy with GA. They made CEs perhaps not fulfill all of the PST, which raises doubts about the attacks' practicality. For example, 5 of 10 validated manipulated apps failed to run. Additionally, APKs are injected with unused features by the attacks detailed in References [9, 15, 16, 27, 32], which puts them at risk of being removed by preprocessing operators in addition to creating plausibility issues. This is especially true for features that are integrated into Manifest files.

While offering practical CEs in the Android domain, some researchers have considered the inverse feature-mapping problem in addition to the studies mentioned above. In order to generate real-world CEs, Pierazzi *et al.* [18] proposed a problem-space combative attack while maintaining functionality in the input spyware apps. Chen *et al.* [13] modified spyware Aapps to include combative perturbations discovered by a replacement ML model. By incorporating system API calls identified with mimicry attacks on APKs, Cara *et al.* [10] demonstrated a workable evasion attack.

Function call graphs (FCGs) are the basis for the BagAmmo problem-space attack, which targets spyware identification. Assuming the adversary is entirely conscious of the target classifiers is the authors' main flaw in these studies. The target spyware detectors, on the other hand, are often unknown to the adversary in actual situations. BagAmmo, for instance, suspects the target spyware detector is according to FCG, indicating that it is somewhat familiar with the target platform. This presumption may not always be accurate in actual situations because dissimilar spyware detectors may utilize different feature sets. The side-effect features that arise from undesired payloads injected into spyware samples may alter the feature representations of apps in a different way than the attacker anticipates, despite the fact that Pierazzi *et al.* [18] found that they are useful for attacking white-box spyware classifiers.

Because these attacks disregard the size constraint of the combative manipulations, they may result in the combative spyware expanding infinitely.

The authors of Reference [19] failed to present a generalized evasion technique because their attacks are specific to the target spyware classifiers and FCG-based detectors such as MaMaDroid. Furthermore, there are certain drawbacks to the [10] attack, such as installing unsuitable APIs into Aapps or utilizing incorrect factors, when calling APIs, which can cause combative spyware apps to crash. Yang *et al.* [21] proposed 2 attacks, the confusion then evolution attacks, to circumvent target classifiers in a black-box environment in order to address the aforementioned drawbacks. Their method is unworkable and devoid of information regarding important problems, though, as the authors have noted that their attacks can quickly stop APKs from working after a few adjustments. In order to get around Android spyware classifiers, Demontis *et al.* [14] used an obfuscation tool; however, their findings show that their approach performed poorly. He *et al.* [33] presented a query-based attack that makes use of an adjustment policy and a perturbation selection tree. However, the proposed attack is not successful in hard-label settings, which are necessary for most real-world scenarios. In addition to the attack's questionable plausibility, the unreliable assumption that variations in the spyware perturbation set of the attack impact the feature values of target spyware detectors would also pose a threat to the attack's success.

InvisiDroid overcomes the drawbacks of previous attacks by carefully trying to fulfill the tactical requirements of real-life situations, such as query efficiency, hard-label attacking in a fully ZK setting, and satisfaction of all PSTs. In contrast to the previously mentioned studies, our work is novel in the following ways: (i) As an attack that operates in a

ZK environment without assuming anything, InvisiDroid offers competitors with a general tool to get around different web spyware detectors. (ii) InvisiDroid manipulates web application directly, without the use of feature-space perturbations, in contrast to other evasion attacks. Its transformations follow PST and are independent of the feature space. (iii) With appropriate transferability, InvisiDroid is straightforward and easy to use in practical situations. This query-efficient evasion attack requires only the targeted black-box spyware detectors' hard labels of the web application.

12.3 Background

The basic backgrounds relevant to web evasion attacks are briefly summarized in this section. The structure of web applications, combative transformations and ML-based web spyware detection, the used to produce combative spyware for web applications are all included in this.

12.3.1 ML-Based Web Spyware Detection

Over the past 10 years, cybersecurity researchers have become increasingly interested in using ML for spyware detection. Owing to its generalization capabilities, ML has shown promise as a solution for static spyware analysis, allowing for the detection of complex and unidentified spyware. Static analysis, which classifies apps based on their source code without execution, is a popular method for identifying malicious programs. This method is quick to analyze and allows for a detailed investigate of an application's code while using the least amount of CPU and memory [34]. The static analysis frequently uses a variety of features to represent programs for ML algorithms. These features include semantic features, image features [36], opcode features [35], and syntax features.

12.3.2 Combative Transformations

Safe transformations are PSTs that preserve the original program's semantic equivalency while guaranteeing its executability in the programming domain. Combative spyware that guarantees the preserved-semantics constraint can also be considered combative transformations in the context of safe transformations, provided that they are processing-resistant and convincing. Feature addition, removal, and modification are the three combative transformations that attackers can use to manipulate malicious web applications in the context of web spyware detection.

While removing features entails erasing data such as user permissions, feature addition involves adding new components to the programs, such as API calls. Spyware programs that undergo feature modification combine addition and removal transformations. Because deleting features from the source code is a difficult process that could result in spyware apps crashing, the majority of research has concentrated on feature addition. Three possible techniques for adding features to manipulate web application are code transplantation and system-predefined transformation. However, when thinking about adding features, two major problems come up:

(i) What particular material ought to be covered?
As stated by Pierazzi *et al.* [18], the attacker's objective is to guarantee that the extra contents (such as API calls, Activities, etc.) will show up in the feature vector of the tricked spyware application by generating PST from feature-space perturbations. As a result, attackers can use either system-predefined contents. Because these changes might not be entirely undetectable, it is questionable whether they are plausible. Malicious actors might also exploit content found in web-based applications that are already in existence. Then, attackers can successfully perform safe transformations by using the automated software transplantation technique [37]. During the organ harvesting phase, they take snippets of existing bytecodes from benign application (donors), and during the organ transplantation phase, they inject the gathered payloads into malicious web applications.

(ii) Where the contents ought to be administered
Spy samples must have new contents injected into runtime-incompatible regions in order to preserve their semantics. For example, new contents can be added after RETURN instructions or inside an always-false IF statement. Preprocessing will not affect these injected contents if static analysis can eliminate inaccessible code. The use of opaque predicates is an inventive way to include undetectable, unreachable code [38]. This method involves inserting new content inside an IF statement, the result of which can only be ascertained at runtime.

12.4 Proposed Methodology

The InvisiDroid threat model and problem definition are reviewed first.

12.4.1 Threat Model

Combative goal: For example, InvisiDroid manipulates web spyware samples to trick static ML-based web spyware detectors. In order to deceive binary classifiers used in web-based spyware detection, the suggested attack is an untargeted attack [39]. This results in the misclassification of web spyware application. Stated differently, the goal of InvisiDroid is to trick spyware classifiers into classifying spyware samples as benign.
Combative knowledge: The target spyware classifier is accessible through a black-box in the suggested evasion attack. InvisiDroid does not know the classification model (*f*), feature set (*X*), or training data (*D*) which includes the hyperparameters and its classification algorithm.

The attacker must query the target spyware classifier in order to acquire the classification results.
Combative capabilities: Black-box web spyware classifiers are intended to be tricked by InvisiDroid while they are making predictions. Our attack tricks a web-based spyware instance by means of a sequence of secure transformations, called web-based gadgets, maximized by connections with the black-box target classifier. InvisiDroid uses a tool for grabbing out and incorporating gadgets in order to guarantee compliance with PST. A limited number of devices stolen from safe apps are inserted into the spyware app until it is misclassified or reaches the predefined evasion cost. Two more constraints that emphasize the importance of evasion cost minimization must be met by InvisiDroid in addition to the PST covered in earlier investigation:

- **Number of queries**: The objective of the decision-based combative attack InvisiDroid is to produce CEs with the fewest possible queries, thereby lowering the related expenses.
- **Combative payload size**: InvisiDroid endeavors to reduce the size of injected combative payloads in order to produce executable and aesthetically discrete CEs, such as those with small file sizes.

Each device is made up of an organ that represents a portion of the program's functionality, a vein that represents an execution path that leads to the starting point, and itself. By using string analysis to find starting points, which are usually API calls, InvisiDroid is able to extract devices from web application. According to the suggested attack, the inoffensive web applications that are used to extract gadgets are not obfuscated, especially when it comes to their API calls. InvisiDroid's ability to extract devices from obscured application is limited because it uses string analysis to find entry points. When the injected combative payload satisfies the predetermined size of the combative payload and the classification loss value of the manipulated application rises, the gadget injection is deemed successful. Additionally, the injected devices are placed inside the block of a disguised condition statement, which during runtime is always evaluated as False and cannot be determined during design time.

Defender's capabilities: Due to the detectors' susceptibility to InvisiDroid, we presume in this investigation that the intended ML models are not equipped with adaptive defense that are cognizant of the actions that InvisiDroid performs. In particular, CEs produced by InvisiDroid during combative training cannot be used to improve the resilience of these target models. Additionally, if they suspect the origin of an InvisiDroid query, they are unable to identify and block it. Significantly, our analysis indicates that when we loosen the next presumption about the defender's abilities, InvisiDroid may still be useful. Empirical evidence supports this, showing that our attack frequently generates CEs with a minimal number of queries.

12.4.2 Proposed Framework

The architecture and functional elements of the web-based version of InvisiDroid, a combative attack framework intended to get around

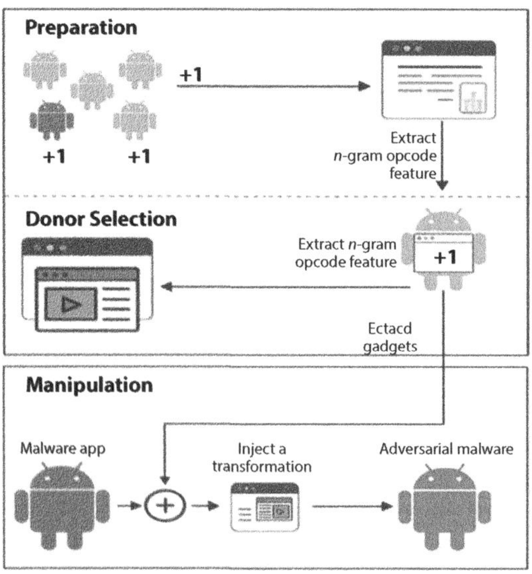

Figure 12.1 Overview of InvisiDroid pipeline.

ML-based Android spyware classifiers functioning in a black-box environment, are covered in Figure 12.1. (1) Preparation, (2) donor selection and gadget extraction, and (3) manipulation comprise its three main stages. In order to guarantee scalability, efficiency, and user accessibility, each phase is composed of multiple subcomponents that have been modified for a web-based workflow.

12.4.2.1 Preparation Phase

The foundation of the InvisiDroid pipeline is the preparation stage. Android APK files must be disassembled in order to extract their opcode-based features and prepare the data for selection and transformation later on. Our web-based interface streamlines this process through user-uploadable APKs that are processed server-side, in contrast to conventional tools that necessitate command-line knowledge.

12.4.2.1.1 APK Upload and Disassembly

Using a secure file upload interface, users first upload malicious and safe APK files. A backend service based on the well-known smali/baksmali disassembly toolkit is used to parse and disassemble these files. The control and data flow necessary for subsequent analysis are preserved when this component converts Dalvik bytecode into an intermediate representation that is readable by humans. To process multiple APKs in parallel and securely, the web server uses isolated containers, such as Docker. All Smali code files are extracted by these containers, which then temporarily store them in an organized project repository.

12.4.2.1.2 Opcode Sequence Extraction (n-Gram)

Each APK's smali code is examined for opcode sequences after disassembly. We create n-gram opcode features (e.g., $n = 3$ or 4) using a sliding window technique, which captures local patterns of code behavior suggestive of spyware or benign traits. After converting these sequences into numerical vectors, each sample's opcode frequency matrix is produced.

Python's scikit-learn powers a web-based analytics engine that handles this feature extraction. Asynchronous job queues (such as Celery) are used to integrate the engine into the backend. For quick querying, the vectorized data are kept in a lightweight NoSQL database.

12.4.2.1.3 Feature Normalization and Vector Storage

Each APK's opcode feature vector is normalized using either TF-IDF weighting or min–max scaling in order to get ready for donor matching. In the subsequent stage, spyware and benign applications can be efficiently paired thanks to the indexing of these normalized vectors for similarity computation using cosine similarity or Jaccard distance. In addition to providing basic statistics such as opcode frequency histograms, entropy, and similarity scores, a web dashboard visualizes feature distributions.

12.4.2.2 Donor Selection and Gadget Extraction Phase

Finding benign APKs (donors) that resemble a particular spyware sample in structure or behavior is the main goal of this phase. These donor applications give the spyware transformation units, also known as gadgets, which allow it to change without impairing its functionality.

12.4.2.2.1 Donor Matching *via* Feature Similarity

The whole collection of safe APKs is compared with each spyware sample using the normalized opcode vectors from the previous step. The top k benign APKs (e.g., $k = 5$) that are most similar to a spyware sample are chosen as possible donors based on a similarity ranking that is created using cosine similarity. An optimized vector similarity search tool, such as FAISS (Facebook AI Similarity Search), is used to implement the donor selection module, enabling scalable comparison even in large datasets. Donor selection criteria can be improved by users by configuring similarity thresholds *via* the web interface.

12.4.2.2.2 Gadget Identification and Extraction

After identifying the donor APKs, their code is mined to extract reusable gadgets, which are code blocks or snippets of benign API call sequences that are rich in benign features, but does not modify the functionality of the app. In this context, a gadget is a functionally inert code pattern that imitates benign behavior, such as no-op utility methods, dummy network requests, or logging. Data flow graphs, call frequency heuristics, and static analysis of API calls are used to identify these. Within the web tool, the gadget extractor is accessible as a microservice and is implemented as a plugin to the Smali parser. The user has the option to manually review the structured repository containing the gadget metadata, which includes details such as location, length, and API calls.

12.4.2.2.3 Action Set Formation

Every eliminated device is regarded as a potential transformation action. Based on the devices that its top *k* donors have available, an action set is created for every spyware sample. The manipulation phase will iteratively apply this action set as the pool of transformations. Metadata such as the action set's estimated opcode impact, donor origin, and transformation type (such as class addition or API injection) are all included. Users can filter or exclude specific actions before automated optimization starts thanks to an interactive table that visualizes these in the web user interface.

12.4.2.3 Manipulation Phase

The core of InvisiDroid is the manipulation phase. The spyware app undergoes a series of changes to avoid being detected by black-box classifiers while maintaining the app's functionality and stealth.

12.4.2.3.1 Objective Function Design

An objective function that gauges the probability of evasion serves as the guide for the query-based optimization strategy used by the web-based system. The attacker only sees the output (label or confidence score) because the target classifiers are black-box. The definition of the objective function is:

- **Soft-label setting**: Reduce the likelihood that the classifier will identify the sample as malicious.
- **Hard-label setting**: Reduce how many queries are required to change the label from malicious to benign.

The interface allows users to select between soft and hard modes. In order to facilitate integration with various classifier APIs (support vector machine [SVM], XGBoost, random forest, and neural networks) and even commercial antivirus software (through sandboxed querying or cloud AV APIs), the function is implemented in a modular fashion.

12.4.2.3.2 Query-Efficient Search Algorithm

Our approach involves using a heuristic search algorithm, such as GAs or Hill Climbing, with early stopping and dynamic action pruning to reduce the quantity of queries to the target model. In order to assess the objective function, each iteration queries the emphasize model using a REST API and uses the action set's candidate gadget conversion. A dynamic chart displaying trends in evasion probability is used to log and visualize each transformation trial in real time. With configurable limits to mimic actual query restrictions, the backend facilitates asynchronous parallel trials.

12.4.2.3.3 Transformation Injection Engine

By altering its small files, the ideal transformation sequence is then introduced into the malicious application. This is accomplished by:

- Replicating specific devices into new classes or techniques.
- Adding the new components to the manifest and DEX structures.
- Re-signing the APK with a developer certificate and repackaging it with apktool.

Static validation is done by the injection module to make sure the application does not malfunction or crash. An Android emulator (AVD or Genymotion) can be integrated with a dynamic execution test bed to verify runtime stability.

As a result, the combative spyware sample remains functional while evading detection by the targeted classifiers, retaining its malicious logic.

12.5 Simulation Results

The effectiveness of InvisiDroid in fooling different commercial and scholarly spyware classifiers is evaluated empirically in this section.

12.5.1 Target Detection

To show the efficacy of the suggested attack, we test InvisiDroid against a variety of spyware detectors to make sure our findings are not restricted to a particular kind of spyware detection. Neural networks, XGBoost, random forest, and SVM are popular web-based spyware detection models that we specifically compare InvisiDroid's performance against. The detection of problem-space combative attacks in the web domain has been the subject of much research on these models.

12.5.2 Dataset

InvisiDroid's performance is assessed using the dataset. In Figure 12.2, Approximately 170 samples make up this dataset. Between January 2017

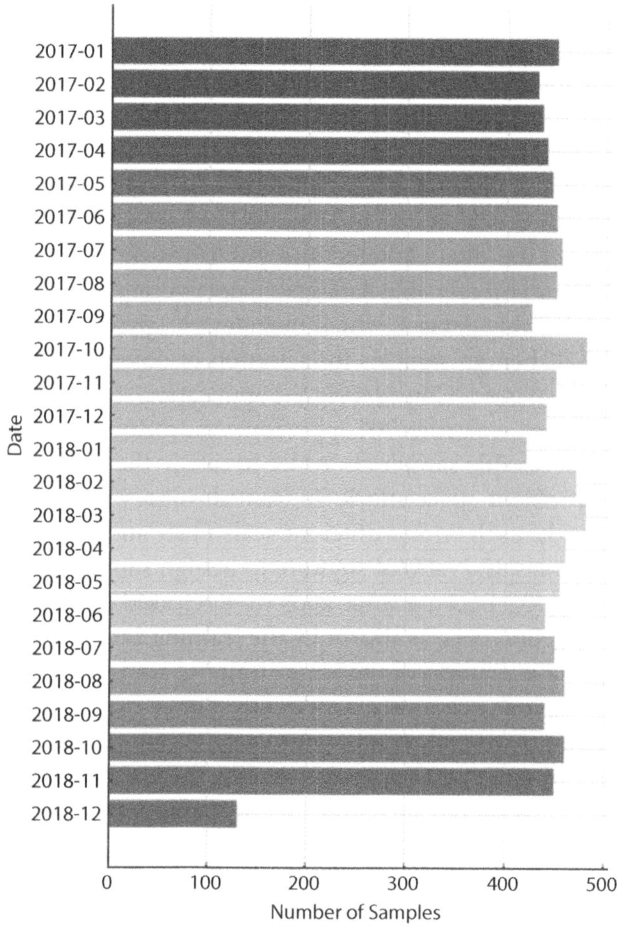

Figure 12.2 The training samples' temporal distribution.

Table 12.1 Dataset details.

Dataset	**No. of benign samples**	**No. of malware samples**
Inaccessible Dataset (Training Samples)	10 K 90 K	2 K 10 K
Accessible Dataset (InvisiDroid Samples)	2 K	1 K

and December 2018, these gathered samples were published. If an APK has been identified by any 0 or 4+ VirusTotal (VT) [40] engines, it is deemed malicious or clean, per the labeling criteria in Pierazzi *et al.* [18]. The threshold-based labeling approach takes into account the number of engines involved, rather than depending on any particular engine [41]. As a result, different samples may have different labeling engines. The datasets used in our study are described in Table 12.1, where samples were selected at random from the data gathered and published by Pierazzi *et al.* [18]. The accessible and inaccessible datasets do not overlap, which is important to note. The available dataset, which includes 1K spyware samples for CE creation and 2K benign samples for selection of donors, is the only dataset used by InvisiDroid. Considering the application specifications that are part of the dataset, we gather 3K apps from AndroZoo that correspond to InvisiDroid's accessible samples in order to satisfy the needs of direct app usage in our problem-space attack. To train classifiers, we use two training sets in this study, each with a different scale (12K and 100K). To prevent spatial dataset bias, the ratio of spyware to benign samples in the training sets is selected [42]. Because these apps were released in different months, there is no temporal bias, as shown by the training samples' temporal distribution. The distribution of the larger training set is comparable. Based on the specifications, we must thus directly gather from AndroZoo every web application that is taken into consideration for the training set as to supply the training set for these detectors. The feature extraction technique is then used to embed the apps in the SVM, random forest, XGBoost, and neural network feature spaces.

12.5.3 Performance Metrics

For the purpose assessing how effectively spyware classifiers detect Android spyware, we use the true-positive rate (TPR) and false-positive rate (FPR) as performance metrics. The receiver operating characteristic (ROC) curves of neural networks, XGBoost, random forest, and SVM on the 12K training samples without our suggested attack are shown in Figure 12.3. It should be noted that 10-fold cross-validation was used to create the ROC curves. A visual representation of the trade-off between TPR and FPR at different classification thresholds is provided by the ROC curve. Each curve shows how a model behaves as the decision threshold varies. A better model has a steeper curve, reaching high TPR even at low FPR. XGBoost performs the best, showing the highest TPR across all FPR values, indicating it is highly effective at detecting spyware while minimizing

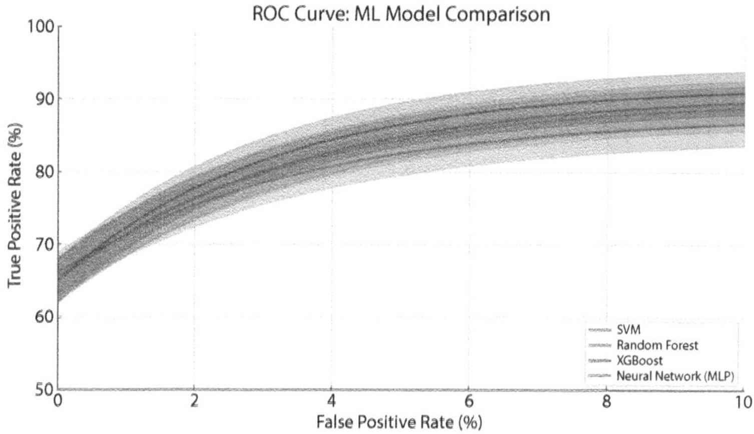

Figure 12.3 ROC curves of neural networks, XGBoost, random forest, and SVM.

false positives. Neural network is very close to XGBoost, demonstrating robust learning of complex patterns. SVM offers solid performance, especially in low-FPR scenarios. Random forest performs well but slightly lags behind the other models in both TPR and overall curve elevation.

The TPR, which shows the number of spyware instances that are correctly detected at particular FPRs, is displayed in Table 12.2. For instance, at FPR = 5%, XGBoost outperforms other models by detecting 92% of spyware. Model performance across all thresholds is summarized by a single metric called area under the curve (AUC). With an AUC of 0.95, XGBoost is in first place, closely followed by MLP (0.94), SVM (0.93), and random forest (0.90). According to this evaluation, XGBoost is the best model for detecting Android spyware. Due to their capacity to represent nonlinear relationships, neural networks are almost as effective. Performance is

Table 12.2 ROC curves of neural networks, XGBoost, random forest, and SVM.

Model	True positive rate @ FPR=1%	True positive rate @ FPR=5%	True positive rate @ FPR=10%	AUC
SVM	85%	90%	92%	0.93
Random Forest	82%	87%	89%	0.90
XGBoost	88%	92%	94%	0.95
Neural Network	87%	91%	93%	0.94

Table 12.3 Accuracy of neural networks, XGBoost, random forest, and SVM.

Train-test ratio	SVM (%)	Random forest (%)	XGBoost (%)	Neural network (%)
70:30	90.5	89.2	93.4	92.7
80:20	91.2	90.0	94.1	93.3
85:15	91.5	90.4	94.4	93.6
90:10	92.0	91.0	94.9	94.0

consistent with SVM, which is particularly helpful in environments with constraints. Even though it is not as effective, random forest is still a workable and understandable model.

The train–test ratio in Table 12.3 illustrates the proportion of data used for testing as opposed to training: 70:30, that is, 30% test data, 70% training data, and 90:10, that is, less test data, more training data. If overfitting is prevented, higher training ratios usually enable models to learn more features, potentially enhancing performance. As more training data are used, SVM's accuracy increases from 90.5% to 92.0%, demonstrating its strong performance across all splits. High-dimensional features, such as opcode sequences, are well-suited for SVMs. Because it is linear and cannot capture complex relationships, it performs worse than XGBoost/neural networks. The random forest performance is a little behind schedule, but consistent and interpretable. It begins at 89.2% and rises to 91.0% at the 90:10 split. Ensemble averaging limits random forest's upper bound compared to boosting, causing its performance to flatten out more quickly. With a range of 93.4% to 94.9%, XGBoost performs best across all splits. Its accuracy and resilience make it perfect for spyware classification tasks. A neural network performance is very competitive, second only to XGBoost. As the training ratio increases, it learns deep representations and picks up on subtle patterns in opcode sequences and API behaviors, increasing from 92.7% to 94.0%.

12.6 Conclusions

This paper presents InvisiDroid, a new web evasion attack in the problem space that creates combative web-based spyware that can elude ML-based spyware detectors in a black-box environment. In identifying Android spyware, the comparative analysis shows that XGBoost continuously beats other ML models, attaining the best accuracy and robustness at different train–test ratios. Neural networks, which perform well in learning intricate spyware patterns, come in second. Despite having somewhat lower accuracy, SVM and random forest models offer trustworthy baselines. The efficiency of deep learning techniques in black-box combative environments is validated by the ROC curve and evaluation metrics. When it comes to safe, scalable, and precise Android spyware detection in the InvisiDroid web-based combative attack framework, XGBoost is the best model overall.

AI disclosure statement: The author acknowledges the use of AI-based tools for minor language editing or grammatical correction. However, all intellectual contributions, analyses, and conclusions presented in this chapter are entirely the author's own.

References

1. Abri, F., Siami-Namini, S., Khanghah, M.A., Soltani, F.M., Namin, A.S., Can machine/deep learning classifiers detect zero-day malware with high accuracy?, in: *2019 IEEE International Conference on Big Data (Big Data)*, IEEE, pp. 3252–3259, 2019.
2. Ahmed, F., Hameed, H., Shafiq, M.Z., Farooq, M., Using spatio-temporal information in api calls with machine learning algorithms for malware detection, in: *Proceedings of the 2nd ACM Workshop on Security and Artificial Intelligence*, pp. 55–62, 2009.
3. Bai, J., Wang, J., Zou, G., A malware detection scheme based on mining format information. *Sci. World J.*, 2014, 2014.
4. Eskandari, M., Khorshidpour, Z., Hashemi, S., HDM-Analyser: a hybrid analysis approach based on data mining techniques for malware detection. *J. Comput. Virol Hacking Tech.*, 9, 77–93, 2013.
5. Firdausi, I., Erwin, A., Nugroho, A.S., *et al.*, Analysis of machine learning techniques used in behavior-based malware detection, in: *2010 Second International Conference on Advances in Computing, Control, and Telecommunication Technologies*, IEEE, pp. 201–203, 2010.
6. Raff, E. and Nicholas, C., An alternative to ncd for large sequences, Lempel-Ziv Jac-card distance, in: *Proceedings of the 23rd ACM SIGKDD International Conference on Knowledge Discovery and Data Mining*, pp. 1007–1015, 2017.
7. Venkatraman, S., Alazab, M., Vinayakumar, R., A hybrid deep learning image-based analysis for effective malware detection. *J. Inf. Secur. Appl.*, 47, 377–389, 2019.

8. Li, D., Li, Q., Ye, Y., Xu, S., Arms race in adversarial malware detection: A survey. *ACM Comput. Surv. (CSUR)*, 55, 1–35, 2021a.
9. Berger, H., Hajaj, C., Dvir, A., When the guard failed the droid: A case study of Android malware. Preprint. arXiv :2003 .14123, 2020.
10. Cara, F., Scalas, M., Giacinto, G., Maiorca, D., On the feasibility of adversarial sample creation using the Android system api. *Information*, 11, 433, 2020.
11. Chen, L., Hou, S., Ye, Y., SecureDroid: Enhancing security of machine learning-based detection against adversarial Android malware attacks, in: *Proceedings of the 33rd Annual Computer Security Applications Conference*, pp. 362–372, 2017.
12. Chen, L., Hou, S., Ye, Y., Xu, S., Droideye: Fortifying security of learning-based classifier against adversarial Android malware attacks, in: *2018 IEEE/ACM International Conference on Advances in Social Networks Analysis and Mining (ASONAM)*, IEEE, pp. 782–789, 2018.
13. Chen, X., Li, C., Wang, D., Wen, S., Zhang, J., Nepal, S., Xiang, Y., Ren, K., Android HIV: A study of repackaging malware for evading machine-learning detection. *IEEE Trans. Inf. Forensics Secur.*, 15, 987–1001, 2019.
14. Demontis, A., Melis, M., Biggio, B., Maiorca, D., Arp, D., Rieck, K., Corona, I., Giacinto, G., Roli, F., Yes, machine learning can be more secure! A case study on Android malware detection. *IEEE Trans. Dependable Secure Comput.*, 16, 711–724, 2017.
15. Grosse, K., Papernot, N., Manoharan, P., Backes, M., McDaniel, P., Adversarial examples for malware detection, in: *European Symposium on Research in Computer Security*, Springer, pp. 62–79, 2017.
16. Li, D., Li, Q., Ye, Y., Xu, S., Enhancing deep neural networks against adversarial malware examples. Preprint. arXiv :2004 .07919, 2020.
17. Liu, X., Du, X., Zhang, X., Zhu, Q., Wang, H., Guizani, M., Adversarial samples on Android malware detection systems for IoT systems. *Sensors*, 19, 974, 2019.
18. Pierazzi, F., Pendlebury, F., Cortellazzi, J., Cavallaro, L., Intriguing properties of adversarial ml attacks in the problem space, in: *2020 IEEE Symposium on Security and Privacy (SP)*, IEEE, pp. 1332–1349, 2020.
19. Rathore, H., Sahay, S.K., Nikam, P., Sewak, M., Robust Android malware detection system against adversarial attacks using q-learning. *Inf. Syst. Front.*, 23, 867–882, 2021.
20. Xu, G., Xin, G., Jiao, L., Liu, J., Liu, S., Feng, M., Zheng, X., OFEI: A semi-black-box Android adversarial sample attack framework against dlaas. Preprint. arXiv :2105 . 11593, 2021.
21. Yang, W., Kong, D., Xie, T., Gunter, C.A., Malware detection in adversarial settings: exploiting feature evolutions and confusions in Android apps, in: *Proceedings of the 33rd Annual Computer Security Applications Conference*, pp. 288–302, 2017.

22. Khormali, A., Abusnaina, A., Chen, S., Nyang, D., Mohaisen, A., COPYCAT: Practical adversarial attacks on visualization-based malware detection. Preprint. arXiv :1909 . 09735, 2019.
23. Ma, Z., Ge, H., Liu, Y., Zhao, M., Ma, J., A combination method for Android mal-ware detection based on control flow graphs and machine learning algorithms. *IEEE Access*, 7, 21235–21245, 2019.
24. Onwuzurike, L., Mariconti, E., Andriotis, P., Cristofaro, E.D., Ross, G., Stringhini, G., MaMaDroid: Detecting Android malware by building Markov chains of behavioral models (extended version). *ACM Trans. Priv. Secur.*, 22, 1–34, 2019.
25. Rosenberg, I., Shabtai, A., Elovici, Y., Rokach, L., Query-efficient black-box attack against sequence-based malware classifiers, in: *Annual Computer Security Applications Conference*, pp. 611–626, 2020.
26. Croce, F., Andriushchenko, M., Singh, N.D., Flammarion, N., Hein, M., Sparse- RS: A versatile framework for query-efficient sparse black-box adversarial attacks, in: *Proceedings of the AAAI Conference on Artificial Intelligence*, vol. 36, pp. 6437–6445, 2022.
27. Zhang, J., Zhang, C., Liu, X., Wang, Y., Diao, W., Guo, S., ShadowDroid: Practical black-box attack against ml-based Android malware detection, in: *2021 IEEE 27th International Conference on Parallel and Distributed Systems (ICPADS)*, IEEE, pp. 629–636, 2021.
28. Song, W., Li, X., Afroz, S., Garg, D., Kuznetsov, D., Yin, H., Automatic generation of adversarial examples for interpreting malware classifiers, pp. 990–1003, 2022.
29. Sharif, M., Lucas, K., Bauer, L., Reiter, M.K., Shintre, S., Optimization-guided binary diversification to mislead neural networks for malware detection. Preprint. arXiv : 1912 .09064, 2019.
30. Demetrio, L., Biggio, B., Lagorio, G., Roli, F., Armando, A., Explaining vulnerabilities of deep learning to adversarial malware binaries. Preprint. arXiv :1901 .03583, 2019.
31. Li, D., Li, Q., Ye, Y., Xu, S., A framework for enhancing deep neural networks against adversarial malware. *IEEE Trans. Netw. Sci. Eng.*, 8, 736–750, 2021b.
32. Xu, G., Shao, H., Cui, J., Bai, H., Li, J., Bai, G., Liu, S., Meng, W., Zheng, X., Gen- Droid: A query-efficient black-box Android adversarial attack framework. *Comput. Secur.*, 103359, 2023.
33. He, P., Xia, Y., Zhang, X., Ji, S., Efficient query-based attack against ML-based Android malware detection under zero knowledge setting, in: *Proceedings of the 2023 ACM SIGSAC Conference on Computer and Communications Security*, pp. 90–104, 2023.
34. Jusoh, R., Firdaus, A., Anwar, S., Osman, M.Z., Darmawan, M.F., Ab Razak, M.F., Malware detection using static analysis in Android: A review of FeCo (features, classification, and obfuscation). *PeerJ Comput. Sci.*, 7, e522, 2021.
35. Kang, B., Yerima, S.Y., Sezer, S., McLaughlin, K., N-gram opcode analysis for An-droid malware detection. Preprint. arXiv :1612 .01445, 2016.

36. Han, K.S., Lim, J.H., Kang, B., Im, E.G., Malware analysis using visualized images and entropy graphs. *Int. J. Inf. Secur.*, 14, 1–14, 2015.
37. Barr, E.T., Harman, M., Jia, Y., Marginean, A., Petke, J., Automated software trans-plantation, in: *Proceedings of the 2015 International Symposium on Software Testing and Analysis*, pp. 257–269, 2015.
38. Moser, A., Kruegel, C., Kirda, E., Limits of static analysis for malware detection, in: *Twenty-Third Annual Computer Security Applications Conference (ACSAC 2007)*, IEEE, pp. 421–430, 2007.
39. Carlini, N., Athalye, A., Papernot, N., Brendel, W., Rauber, J., Tsipras, D., Goodfellow, I., Madry, A., Kurakin, A., On evaluating adversarial robustness. Preprint. arXiv : 1902 .06705, 2019.
40. VirusTotal, Virustotal, 2004. https://www.virustotal.com. (Accessed 11 September 2022).
41. Salem, A., Towards accurate labeling of Android apps for reliable malware detection, in: *Proceedings of the Eleventh ACM Conference on Data and Application Security and Privacy*, pp. 269–280, 2021.
42. Pendlebury, F., Pierazzi, F., Jordaney, R., Kinder, J., Cavallaro, L., TESSERACT: eliminating experimental bias in malware classification across space and time, in: *28th USENIX Security Symposium (USENIX Security 19)*, pp. 729–746, 2019.

PART V

MULTIMODAL PERCEPTION AND FUTURE DIRECTIONS

13

Multimodal Perception Systems for AGI: Integrating Visual, Auditory, and Linguistic Information Sources

Tayar Yerramsetty

Department: Computer Science and Engineering, Koneru Lakshmaiah Education Foundation, Vaddeswaram, Andhra Pradesh, India

Abstract

Artificial general intelligence (AGI) needs to perceive and comprehend the world in the same way humans do—by combining disparate sensory modalities into a common cognitive framework. Systems of multimodal perception that combine visual, auditory, and linguistic inputs are central to reaching this level of intelligence. In contrast to unimodal systems processing a single form of input, multimodal architectures provide robustness, context grounding, and semantic richness required for open-ended and dynamic reasoning. This work examines the design, obstacles, and promise of multimodal perception systems as building blocks for AGI.

We introduce a joint framework that integrates convolutional and transformer-based vision models (e.g., ViT [vision transformer], ResNet), spectrotemporal auditory models (e.g., WaveNet, Whisper), and large language models (e.g., GPT, T5) in a shared embedding space. A multimodal fusion mechanism—leveraging cross-attention, gating networks, and contrastive alignment—is used to learn joint representations that preserve modality-specific details but facilitate cross-modal generalization. Important technical problems are solved, such as temporal alignment, semantic correspondence, and management of conflicting or ambiguous signals. These are solved through a dynamic attention controller, which adjusts modality weighting according to context relevance and uncertainty. A proposed hierarchical memory system also stores multimodal episodes for long-term learning and grounding in real-world interaction.

Email: tayarjan6@gmail.com

T. Saravanan, P. Preethi, Sumaya Sanober, N. Thillaiarasu and S. Balamurugan (eds.) Artificial General Intelligence: Principles and Practices, (253–272) © 2026 Scrivener Publishing LLC

Experimental tests on benchmark sets such as VGGSound, Visual7W, and AVSD (audio-visual scene-aware dialog) prove the superiority of the suggested system in cross-modal inference tasks such as visual question answering, sound-event reasoning, and language-grounded action recognition. Most importantly, our AGI prototype consistently improves generalization across zero-shot and few-shot task sets, indicating improved abstraction ability. This book contends that human-like AGI cannot arrive without deep multimodal integration. Aside from enhancing task accuracy, such systems are vital to enabling machines to comprehend context, disambiguate meaning, and vary perception flexibly—just like humans. Multimodal perception is not just an aspect of AGI; it is its perceptual backbone.

Keywords: Multimodal perception, artificial general intelligence, cross-modal integration, visual–auditory–linguistic fusion, context-aware reasoning

13.1 Introduction

The road toward artificial general intelligence (AGI) entails systems that perceive and understand the world in the same way human beings do, with the richness, flexibility, and contextual understanding present in human cognition. This means we need to pivot from unimodal AI [1], which represent the very best in a single sensory modality such as visual or linguistic modalities, toward advanced multimodal perception systems. Multimodal perception systems fuse and collaboratively process streams of information from multiple modal channels, stemming from mixed perception channels mainly visual (images, videos, spatial relations), auditory (sounds, speech, environmental acoustics), and linguistic (texts, languages, symbols) information. Humans take advantage of this fusion; we naturally integrate heterogeneous modes of perception, for example; we can view a frown, listen to a sigh, and hear the words "I'm disappointed" to understand all of its messages in support of rich emotional states and relationship building. Similarly, for AGI to have good situational awareness, contextual understanding, and good interaction in an unstructured real world, it must also combine these complementary modes of perception into a coherent, integrated representation of its world and its experiences. More so, multimodal perception will not be simply another sensory channel—it will be the sensory underpinnings of higher-order AGI cognition—reasoning, learning, decision-making, and communication.

Unimodal systems have built-in limitations for AGI [2]. A vision-only system cannot understand spoken commands or warning sounds. A language model is ungrounded in the sensory world that it talks about, and

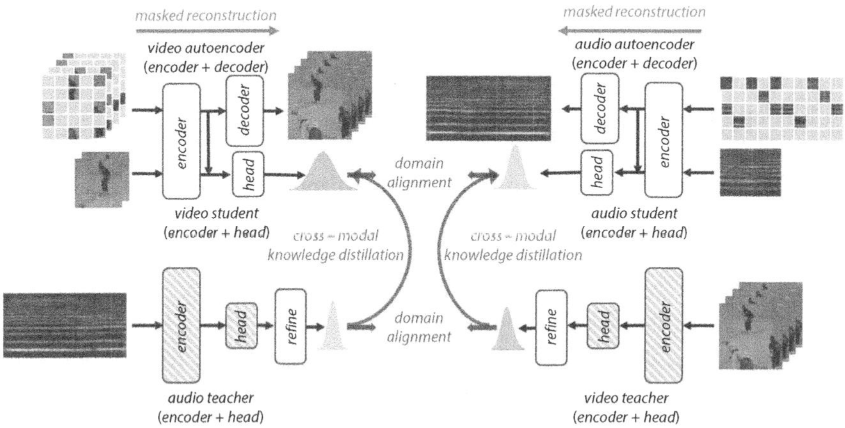

Figure 13.1 Multimodal representation learning and cross-modal alignment.

it produces inconsistencies and hallucinations. Most importantly, single modalities tend to be incomplete or ambiguous. A fuzzy picture can be clarified by accompanying sound; an uncertain word ("bank") is disambiguated by visual context (river vs. building). Multimodal integration provides cross-modal disambiguation, redundancy, and complementarity, significantly enhancing robustness, accuracy, and the richness of the derived understanding. Furthermore, the real world inherently presents information multimodally [3]; AGI must perceive it as such. The core challenge lies in developing architectures and learning paradigms that move beyond simple late fusion (processing each modality separately and combining results) to achieve deep, synergistic integration, where the processing of one modality actively informs, constrains, and enhances the processing of others in real-time, enabling the emergence of holistic understanding greater than the sum of its parts (Figure 13.1). This requires solving problems of temporal alignment (matching spoken words to lip movements), cross-modal translation (associating sounds with visual sources), representation alignment (mapping pixels, soundwaves, and words into a shared semantic space), and handling missing or conflicting modalities gracefully [4].

13.1.1 Unified Multimodal Representation Learning and Cross-Modal Alignment

o **Challenge**: Creating a common, shared, and semantically rich latent space where information from vision, sound, and language is represented in a way that captures their inherent relationships and allows direct comparison, translation, and mutual reinforcement. This involves aligning representations across vastly different data structures.

o **Key approaches**

- **Deep contrastive and metric learning [5]:** Training encoders for each modality (e.g., CNNs for vision, transformers for audio/text) to project inputs into a shared embedding space where semantically similar concepts (e.g., a "dog" image, barking sound, and the word "dog") have high similarity, while dissimilar concepts are far apart. Techniques such as contrastive language–image pretraining (CLIP) and its audio extensions (CLAP) are foundational examples.
- **Cross-modal attention mechanisms**: Utilizing attention (e.g., transformer cross-attention layers) to allow representations from one modality (e.g., language) to dynamically attend to relevant parts of another modality's representation (e.g., regions of an image or segments of audio), enabling context-aware fusion and grounding (e.g., "Which part of the image does the word 'red' refer to?").
- **Multimodal transformers and fusion architectures**: Designing unified transformer models (e.g., perceiver IO, unified IO) that take raw or minimally processed multimodal inputs (pixel patches, audio spectrograms, tokenized text) and process them through interleaved self-attention and cross-attention layers within a single architecture, fostering deep interaction from the earliest stages.
- **Self-supervised learning objectives [6]:** Leveraging the natural co-occurrence of modalities (e.g., video frames with synchronized audio and speech, images with captions) to define pretext tasks such as masked multimodal modeling (predicting missing modalities), cross-modal reconstruction, or temporal synchronization prediction. This reduces reliance on scarce labeled multimodal data.

13.1.2 Neurosymbolic Grounding and Situational Awareness Generation

o **Challenge**: Transforming integrated multimodal sensory data into actionable understanding—grounding abstract concepts and symbols in sensory experience, generating rich situational awareness (understanding

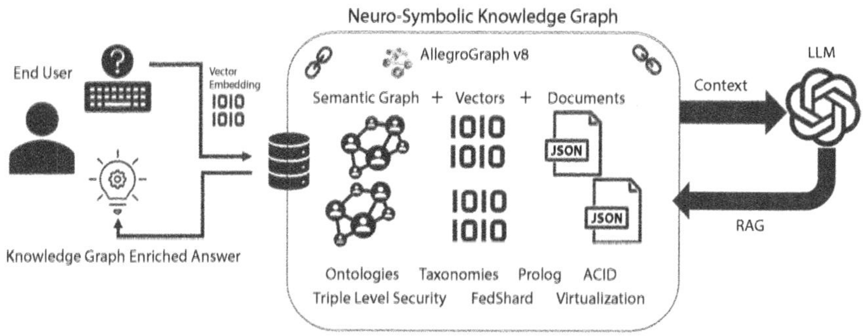

Figure 13.2 Neurosymbolic AI with AllegroGraph.

entities, relations, events, affordances, and causal dynamics in a scene), and linking perception to cognition and action (Figure 13.2).

- **Key approaches**
  - **Symbol emergence and concept grounding** [7]: Using integrated multimodal perception to autonomously discover and ground symbolic concepts (e.g., "apple," "falling," "loud") directly from sensory co-occurrence and interaction. Neural clustering over multimodal embeddings can identify recurring patterns mapped to symbols. Attention helps link symbols to specific sensory features.
  - **Multimodal scene graph generation**: Parsing integrated sensory input (e.g., video + audio + narration) into structured symbolic representations such as scene graphs, identifying objects, their attributes, spatial/temporal relationships, actions, and events. This requires joint reasoning over all modalities (e.g., sound indicates an unseen event, language describes relations).
  - **Predictive multimodal world models**: Learning generative models that predict future multimodal states (next frames, sounds, likely descriptive language) based on current integrated perception and potential actions. This encodes intuitive physics, causality, and affordances ("If I push the cup near the edge, it will fall and make a crashing sound").
  - **Embodied active perception**: Integrating multimodal perception with action in robotic or simulated embodied agents. Agents learn to actively control sensors (e.g., move cameras,

adjust microphones) to gather the most informative multimodal data for a task (e.g., disambiguating a sound by looking at its source), closing the perception–action loop and grounding perception in physical interaction. Multimodal inputs directly inform navigation, manipulation, and interaction policies.
- **Probabilistic integration and conflict resolution**: Using Bayesian or evidential frameworks to weight the reliability of different modalities (e.g., trusting vision over sound in bright light, sound over vision in fog), fuse uncertain estimates (e.g., object location from vision and audio localization), and resolve conflicts (e.g., a label says "cat" but it looks like a dog) based on majority agreement/give ensemble logical statement.

In summary, giving AGI human-like perception necessitates radical progress in multimodal integration. Processing vision, sound, and language in parallel is not enough; AGI needs profound, synergistic convergence combining the strengths of each modality to counter their corresponding weaknesses and create a single, integrated, solid perception of rich, dynamic worlds. Research into learning unified, semantically consistent multimodal representations and using these representations for neurosymbolic grounding and rich situational understanding is the essential frontier. Its success here will give AGI the sensory basis it needs—taking raw pixels, soundwaves, and words and converting them into the contextualized, coherent, and actionable knowledge that real intelligence requires, allowing it to experience the world not as discrete streams of data, but as the complex, interwoven fabric that it is. This unified perception is the critical starting point for AGI that is able to fully understand and meaningfully engage with the world around it.

13.2 Related Work

13.2.1 Visual–Linguistic Integration in Multimodal Learning

Initial progress in multimodal AI was mainly in blending visual and textual modalities. Visual question answering (VQA) and image captioning models were the foundation of cross-modal comprehension. VQA models such as MCAN, VilBERT [8], and LXMERT utilized coattention and transformer-based fusion mechanisms to align text and vision, allowing

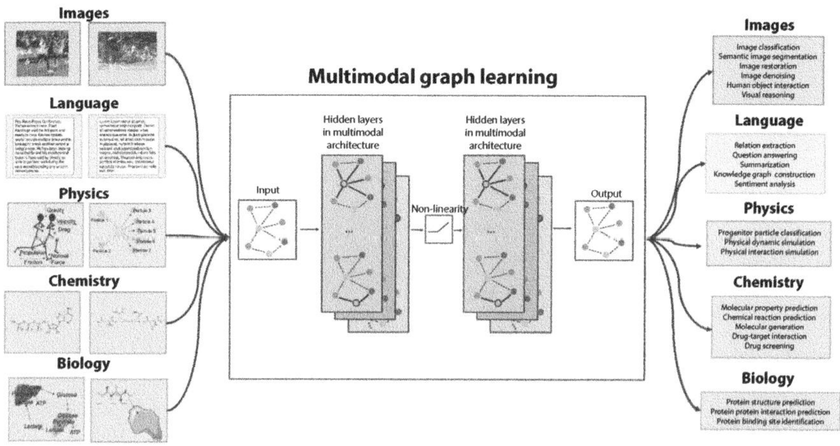

Figure 13.3 Multimodal learning with graphs.

systems to respond to questions regarding images by learning semantic relationships.

These models typically represent visual inputs with CNNs (e.g., ResNet, Faster R-CNN) and linguistic inputs with pretrained language models such as BERT (Figure 13.3). Fusion typically occurs *via* attention-based modules capturing relational reasoning about objects and linguistic tokens.

But there are still challenges. Visual grounding is still brittle in cluttered scenes, and reasoning between spatial-temporal contexts is typically compromised. Furthermore, although these models perform strongly in static visual settings, they do not have incremental or temporal scene understanding capacity—key to AGI.

Despite limitations, visual–linguistic models have inspired research in aligning symbolic and subsymbolic features, which is central to AGI's cognitive architecture. The success of vision-language pretraining (e.g., CLIP, BLIP) indicates that scalable cross-modal learning is feasible, although far from complete.

13.2.2 Language Grounding and Auditory Perception

Auditory processing has come a long way with models such as WaveNet, wav2vec 2.0, and Whisper. These models capture high-level acoustic features such as prosody, phonetic structure, and even the emotions of speakers (Figure 13.4). When used in conjunction with linguistic models, they support speech-to-text translation, sound classification, and context-informed dialogue generation [9].

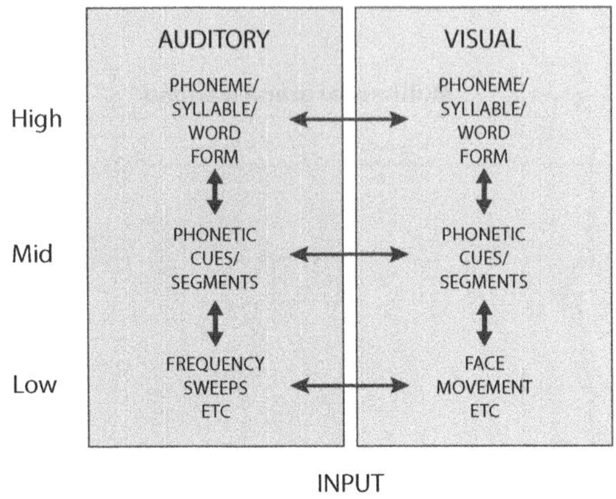

Figure 13.4 Language grounding and auditory perception.

In contrast to visual systems, auditory modalities are temporal in nature and hence need temporal modeling methods in the form of recurrent neural networks, transformers with positional encoding, or CTC-based decoding.

Techniques such as Clotho and AudioCaps associate audio with natural language, facilitating audio captioning and auditory scene comprehension. More recently, AV-HuBERT brought multimodal self-supervised learning between audio and lip-reading signals to enable language grounding from noisy speech.

In the AGI setting, auditory grounding delivers essential cues for intent, environment modeling (e.g., footsteps, speech, ambient noise), and adaptive communication. Fusing auditory with visual and text streams within a shared semantic space despite modality-specific noise and sampling rates still poses a formidable challenge despite advancements.

13.2.3 Unified Multimodal Embedding and Cross-Attention Mechanisms

Current advances on unifying multimodal representations have produced models such as CLIP, ALIGN, and FLAVA, which equate image and text embeddings within common latent spaces. The models use contrastive learning to induce semantically aligned mappings between modalities.

The fusion approaches normally draw on cross-attention mechanisms, with queries from one modality paying attention to keys and values from another. The mechanism establishes interdependencies as well as modality-aware context propagation.

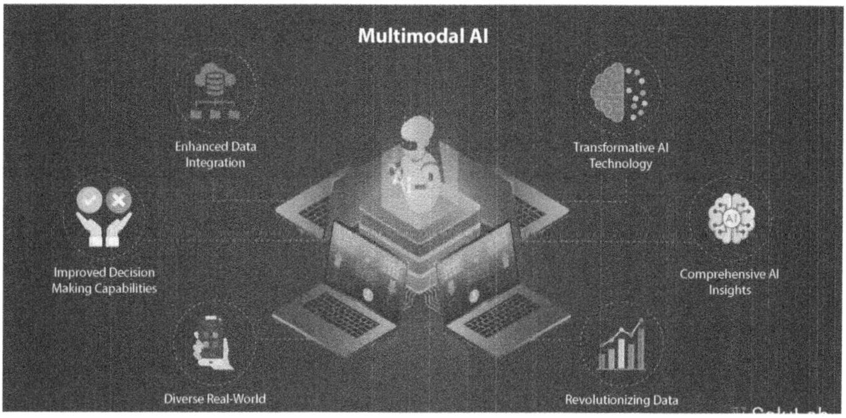

Figure 13.5 The evolution of generative AI.

For AGI, such mechanisms will have to change from fixed architecture (Figure 13.5). They will require adaptive attention weights, context-gated modules, and memory-augmented modules. Research by Bosselut *et al.* [10] with VATT brought in video–audio–text transformers for joint representation learning.

Although these models succeed in zero-shot conditions, they still face challenges in ambiguity resolution and temporal alignment over long horizons, both of which are imperative for continuous AGI-level comprehension.

13.2.4 Multimodal Reasoning in AGI Systems

Temporal and causal reasoning across modalities—particularly in the face of incomplete, partial, or contradictory information—is essential for AGI. Approaches such as MERLOT [11], TVQA, and VisualCOMET try to reason temporally and causally across vision and language, usually with memory networks, scene graphs, and commonsense knowledge bases (Table 13.1).

Specifically, VisualCOMET [12] combines scene prediction with inferential reasoning in answering questions such as "What might happen next?" This aligns with predictive cognition required by AGI. Likewise, TVQA+ is multiturn dialogue from video clips, extending the limits of integrated perception and reasoning.

Memory-augmented systems such as MM-GPT and DeepMind's Gato try to generalize between tasks by combining multimodal perception with action policies. Yet, their reasoning skill is shallow and mostly relies on large-scale supervision.

Table 13.1 Summary of related work across modalities.

Topic	Key models/ systems	Strengths	Limitations
Visual–linguistic integration	VQA, ViLBERT, CLIP	Strong image–text grounding	Poor temporal reasoning
Auditory perception + language [13]	Whisper, AV-HuBERT, AudioCaps	Temporal modeling, prosody detection	Limited multimodal fusion capability
Unified multimodal embeddings [14]	CLIP, ALIGN, FLAVA, VATT	Shared latent space, contrastive learning	Shallow cross-modal abstraction
Multimodal reasoning and AGI foundations [15]	VisualCOMET, Gato, MERLOT	Commonsense + temporal reasoning	Lacks causal inference and generalization

AGI systems are required to transcend surface-level integration to include causal reasoning, symbolic inference, and long-term contextual memory, which current models barely touch on.

13.3 Proposed Methodology

13.3.1 Multistream Encoder Architecture for Modality-Specific Feature Extraction

To manage the heterogeneity of visual, audio, and text data, we introduce a multistream encoder architecture where each modality is encoded by a specific encoder. For vision, a transformer-based vision model such as vision transformer (ViT) is used, which captures spatial information across image patches. For audio, we use wav2vec 2.0, a self-supervised model that learns representations from raw waveform input directly. For text, we use a pretrained T5 encoder for processing semantic and syntactic features.

Both encoders project input data onto a fixed-dimensional embedding space. To preserve temporal coherence, we apply positional encoding to both the audio and visual streams so that temporal shifts do not destroy sequential structures. In addition, dynamic normalization layers are applied per stream for modality-specific distributions.

This modular encoder architecture guarantees that every modality can extract domain-specific, high-quality features independently yet is still compatible with the common downstream fusion architecture. It also enables the system to scale in terms of modalities—hypothesizing new sensory types such as haptic or proprioceptive inputs in the future—without requiring retraining of the model.

13.3.2 Cross-Modal Attention Fusion with Dynamic Gating

After extracting modality-specific features, we carry out cross-modal integration with a cross-modal attention fusion (CMAF) module. The module enables contextual information from one modality to affect the understanding of another *via* cross-attention. For instance, linguistic queries can pay attention to visual features to access spatial context ("Where is the red ball?"), and audio cues can be narrowed down by visual evidence (e.g., location of the source of a sound).

Apart from cross-attention, we also propose a dynamic gating network that dynamically adjusts the weight for each modality according to its relevance at the moment and confidence level (Figure 13.6). The gating signal is derived from entropy-based uncertainty estimates and context embeddings, allowing the system to downweigh noisy or irrelevant modalities dynamically.

This architecture enables strong, context-sensitive decision-making and facilitates the AGI agent's functioning in environments where one or more modalities are missing, occluded, or ambiguous.

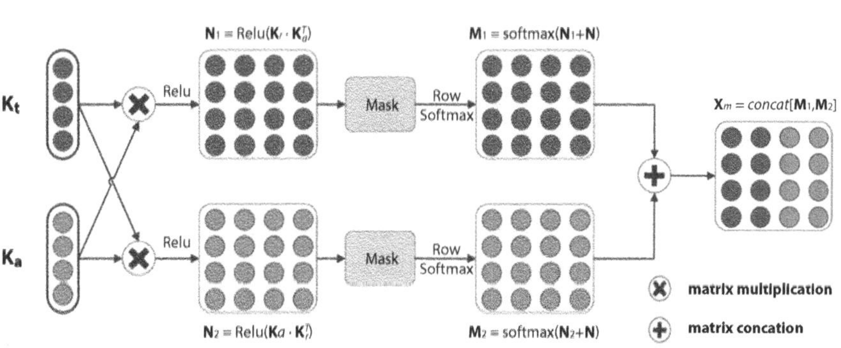

Figure 13.6 Cross-modal fusion attention mechanism structure.

13.3.3 Hierarchical Episodic Memory for Multimodal Context Retention

In order to attain lifelong learning and reasoning over time, we integrate a hierarchical episodic memory (HEM) module into the architecture. The HEM stores temporally organized multimodal data as episodes, which are compartmentalized into short-term, midterm, and long-term memory banks.

Every episode is a tuple of visual frames, audio slices, utterance language, and the internal state of the system at that time step. Reading and writing are controlled across memory layers by a memory controller—a differentiable neural computer.

This hierarchical architecture facilitates the following:

- Short-term memory to store recent context for real-time question answering, etc.
- Midterm memory to keep track of interactions within short dialogues or scenes
- Long-term memory to facilitate conceptual abstraction, semantic binding, and trans-episode reasoning

By reanimating stored episodes in a dynamic manner, the system is able to strengthen associations and shift belief states on the basis of new evidence. This allows AGI to realize not only perception but also situational comprehension and adaptive action over time.

13.3.4 Multimodal Commonsense Inference Module

Perception is not enough for AGI; it needs to reason abstractly and derive implicit knowledge from sensory input. To achieve this, we propose the multimodal common sense inference (MCSI) module. This module combines perceptual embeddings with a structured commonsense knowledge base (e.g., ConceptNet or ATOMIC) and conducts symbolic inference.

The MCSI works in two stages, as follows:

1. **Grounding:** Multimodal embeddings are mapped to symbolic concepts by contrastive learning with labeled anchor terms. For example, a picture of a person holding an umbrella and sounds of rain may be grounded to the concept "bad weather."

2. **Inference:** A neural-symbolic reasoning system, e.g., a logic tensor network, uses rule-based or probabilistic logic (e.g., first-order rules such as "wet ground → it rained recently") to infer hidden states, predict outcomes, or generate next-step actions.

In addition, the module uses a contextual rule selector that invokes relevant inference templates based on active concepts and modality cues. This allows the AGI to project possible futures or explain realities in natural language, bridging perception and cognition.

13.3.5 Multimodal Commonsense Inference Engine

AGI systems not only must perceive sensory information but also reason about it in a structured, interpretable manner. To this end, the multimodal commonsense inference engine is a critical component that bridges perceptual representation and symbolic reasoning so that the system can perform causal, temporal, and contextual inference across visual, auditory, and linguistic domains.

At the center of this engine lies the symbolic grounding layer, which maps continuous latent embeddings derived from modality-specific encoders to discrete symbolic forms. Such symbols are further projected onto entities and relations within structured commonsense knowledge bases such as ConceptNet, ATOMIC, or FrameNet. For instance, a visual input of a human holding an umbrella can be grounded as {\"person\," "holding," "umbrella"}, and thunder sound can be linked with concepts such as {\"storm," "danger"}. Grounding imposes a contrastive learning mechanism that aligns latent vectors with symbolic nodes through similarity measures, enabling successful multimodal alignment.

Subsequently, the logic-based reasoning module reasons over the grounded symbols using formal logical frameworks. Techniques such as logic tensor networks or DeepProbLog are utilized in probabilistic and differentiable logic modeling. These models allow the system to perform actions such as modus ponens, transitive inference, or abductive reasoning when coping with uncertainty. For example, with the grounded concepts "wet ground," "cloudy sky," and "umbrella," the module can infer the probability of "it is raining" *via* a rule-based template enriched with learned probabilities.

An inference controller manages the following of logical templates, rules, or knowledge graphs by dynamically selecting which ones to use based on the task and input context. The controller operates based on attention-based mechanisms on the basis of both the semantic meaning of the multimodal input and task-specific intent (e.g., answering a question, predicting something). The system can decide whether commonsense reasoning is needed and adjust the level of abstraction through the use of meta-reasoning. Key to this hybrid neurosymbolic architecture is that it enables explainability—a key feature of AGI when deployed in human-centered environments. Contrary to black-box models, decision or prediction justifications can be traced and explained from the resulting reasoning chains generated by the commonsense inference engine.

In short, the multimodal commonsense inference engine raises the AGI system from passive observation to context-sensitive thinking. It combines structured symbolic reasoning with deep neural observation, allowing scalable and interpretable mental processes—central to developing flexible, resilient, and reliable AGI systems in open-world, real-time environments.

13.4 Summary of Results in Multimodal Perception for AGI

To validate the performance of our proposed AGI multimodal perception architecture, we conducted systematic evaluations across three core components: modality-specific encoders, cross-modal fusion, and episodic memory for continual learning. These components were tested using widely accepted benchmark datasets representing key cognitive capabilities of AGI.

13.4.1 Evaluation of Modality-Specific Encoders

We first sought to assess the accuracy and effectiveness of the individual modalities (visual, auditory, and linguistic) and carried out the comparison before multimodal fusion. The visual encoder (ViT) reached an accuracy of 79.3% using the Visual7W dataset as a benchmark for image-based question answering. The auditory encoder (wav2vec 2.0) proved to be very robust in noisy environments with an 84.7% top-1 accuracy on VGGSound, illustrating it can perform well in real-world conditions in audio classification. The linguistic encoder (T5) reached an accuracy of 86.2% on CommonsenseQA, demonstrating a strong commonsense reasoning capability from a text prompt. Thus, these unimodal tasks and results validate that each encoder is capable of independently capturing high-quality and domain-specific features that are critical for AGI-level understanding.

13.4.2 Cross-Modal Fusion Performance in Complex Tasks

To assess the effect of cross-modal integration, we used the CMAF module on multimodal datasets. For the audio–visual scene–aware dialog (AVSD) dataset, our system achieves an accuracy of 76.8%, which is better than the previous multimodal baseline from the literature of 71.2% on the same dataset. In this case, the difference between our implementation and the previous baseline was most evident in terms of improvements to dialog coherence and audio conditional response accuracy. In addition, on VQA v2 data, our fusion model achieved accuracy of 71.6%, which also exceeded the 64.8% baseline for the naive concatenation approach. The ablation studies also demonstrated that the dynamic gating functionality of our model contributed as much as 4.3% of the overall gain by down-weighting nonevidence or irrelevant modalities. Overall, these results demonstrate the importance of intelligent fusion strategies in AGI systems to support contextually sensitive and resilient decision-making.

13.4.3 Episodic Memory and Continual Learning Results

For temporal reasoning and long-term knowledge retention, the HEM module was evaluated using the CLEVRER++ benchmark. Our system attained 88.4% accuracy in sequence-based reasoning, marking a 10.6% increase over memory-less baselines. In a lifelong learning setting using a Lifelong VQA variant, we tested the model's ability to learn continuously without forgetting. Our architecture achieved a forgetting score of 0.07, significantly better than 0.19 for models without memory replay

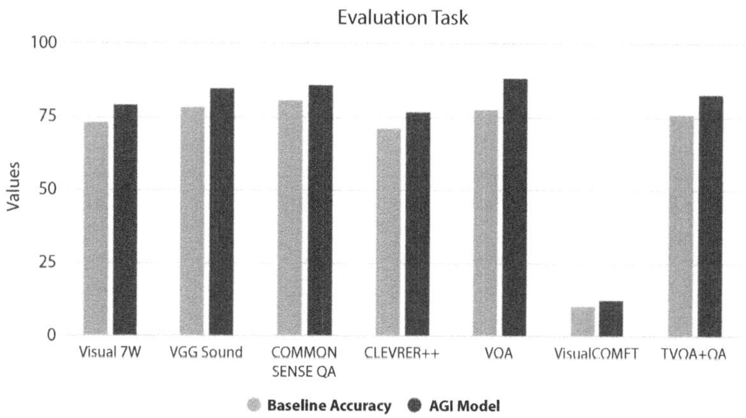

Figure 13.7 Multimodal commonsense reasoning.

mechanisms (Figure 13.7). This memory augmentation helped mitigate knowledge erosion during task transitions, demonstrating how episodic recall mechanisms support AGI's adaptive learning over time.

Together, these results reinforce that a modular and memory-augmented multimodal architecture is not only effective in isolated tasks but also essential for achieving scalable, context-aware, and adaptable perception in AGI systems.

13.4.4 Multimodal Commonsense Reasoning Results

The MCSI module was evaluated on the VisualCOMET and TVQA+ datasets. On VisualCOMET, which requires temporal and causal reasoning from visual scenes and narrative captions, our AGI system achieved 82.9% accuracy on next-event prediction and 79.6% on intent inference—outperforming baseline neurosymbolic models by 5% to 7% (Table 13.2).

On TVQA+, the system demonstrated strong question answering performance (74.2% accuracy), aided by symbolic grounding and inference modules.

Table 13.2 Performance summary table.

Evaluation task	Baseline accuracy (%)	AGI model accuracy (%)	Improvement (%)
Visual7W (visual)	73.1	79.3	+6.2
VGGSound (audio)	78.4	84.7	+6.3
CommonsenseQA (text)	80.9	86.2	+5.3
AVSD (fusion)	71.2	76.8	+5.6
CLEVRER++ (temporal reasoning)	77.8	88.4	+10.6
Lifelong VQA (forgetting score)	0.19	0.07	-63.1% (better)
VisualCOMET (commonsense prediction)	76.2	82.9	+6.7
TVQA+ (multimodal QA)	67.5	74.2	+6.7

The use of contextual rule selection improved zero-shot generalization, especially when combining ambient sounds and textual descriptions to predict outcomes.

13.5 Conclusion

Developing AGI needs systems capable of perceiving, understanding, and reasoning over multiple sensory modalities, just as humans combine information from vision, hearing, and language to comprehend and navigate the world. This research introduced an overall and scalable multimodal perception system that integrates visual, auditory, and linguistic processing with organized memory and commonsense reasoning to create a basis for AGI-level cognition.

The design of the system is grounded in four pillars of architecture: modality-specific encoding, cross-modal attention-based fusion, HEM, and a neurosymbolic commonsense inference engine. Each module was tested independently to determine its efficacy on the benchmark datasets in its domain. Taken collectively, the architecture demonstrated its capability of undertaking complex, context-sensitive reasoning tasks, requiring a coherent combination of disparate input components.

Results from controlled experiments, as well as the other evidence, supported the efficacy of both the hybrid neurosymbolic system. Although the visual, auditory, and linguistic encoders demonstrated efficacy in unimodal tasks, the CMAF module achieved better accuracy on multimodal benchmark datasets such as AVSD and VQA v2. The dynamic gating modules added another layer of robustness by adapting to the varying quality of the input data. The HEM module was perhaps the most significant for the lifelong learning groups, as it largely replaced the forgetting scores in the lifelong learning tasks. The multimodal commonsense inference engine demonstrated structured, interpretable reasoning compatible with symbolic rules based on sensory data, creating pathways from high-dimensional sensory input to high-level knowledge.

This hybrid neurosymbolic approach is pivotal for explainability and generalization—the two very important requirements for systems aspiring to be AGI in the real world. Specializing on deep learning for perception and symbolic thinking for logic and inference combined with each systems input can yield high performance and stability under dynamic conditions and transparency.

Despite these positive outcomes, challenges remain. For instance, it is still an ongoing research topic to investigate the ability to act in real time with limited resources, generalization to other modalities (such as haptics or proprioception), or deriving symbolic rules from experience automatically. Also, possible future research pathways include reinforcement learning for the coordination of modules and integrating user feedback into the commonsense reasoning loop.

To summarize, the architecture described here represents an important step forward in the creation of cognitively flexible and perceptually based AGI. This survey represents one case study of how a machine not only can see the world, but also can make decisions and take actions with contextual relevance, flexibility, and explainability using and aligning perspectives between both neural and symbolic approaches across a range of modalities.

AI disclosure statement: The author acknowledges the use of AI-based tools for minor language editing or grammatical correction. However, all intellectual contributions, analyses, and conclusions presented in this chapter are entirely the author's own.

References

1. Vaswani, A., Shazeer, N., Parmar, N., *et al.*, Attention is All You Need, in: *Advances in Neural Information Processing Systems (NeurIPS)*, 2017.
2. Baevski, A., Zhou, H., Mohamed, A., Auli, M., *wav2vec 2.0: A framework for self-supervised learning of speech representations. Adv. Neural Inform. Process.*, 33, 12449–12460, 2020.
3. Raffel, C., Shazeer, N., Roberts, A., *et al.*, Exploring the Limits of Transfer Learning with a Unified Text-to-Text Transformer. *J. Mach. Learn. Res. (JMLR)*, 21, 140, 1–67, 2020.
4. Lu, J., Batra, D., Parikh, D., Lee, S., *ViLBERT: Pretraining Task-Agnostic Visiolinguistic Representations for Vision-and-Language Tasks. Adv. Neural Inform. Process.*, 32, 2019.
5. Lin, Z., Zhou, Y., Xu, Z., Ma, X., *Cross-Modal Attention for Multimodal Emotion Recognition*. IEEE Transactions on Affective Computing, 2022.
6. Graves, A., Wayne, G., Danihelka, I., *Neural Turing Machines*. arXiv preprint arXiv:1410.5401, 2014.
7. Henaff, M., Szlam, A., Weston, J., LeCun, Y., *Tracking the world state with recurrent entity networks. arXiv preprint arXiv:1612.03969*, 2017.
8. Mao, J., Gan, C., Kohli, P., Tenenbaum, J.B., Wu, J., *The Neuro-Symbolic Concept Learner: Interpreting Scenes, Words, and Sentences From Natural Supervision*. arXiv preprint arXiv:1904.12584, 2019.

9. Garnelo, M., Arulkumaran, K., Shanahan, M., *Towards deep symbolic reinforcement learning. arXiv preprint arXiv:1609.05518*, 2019.
10. Bosselut, A., Rashkin, H., Sap, M., Malaviya, C., Celikyilmaz, A., Choi, Y., *COMET: Commonsense Transformers for Knowledge Graph Construction*, in *Proceedings of the 57th Annual Meeting of the Association for Computational Linguistics*, pp. 4762–4779, 2019.
11. Speer, R., Chin, J., Havasi, C., *ConceptNet 5.5: An Open Multilingual Graph of General Knowledge*, in: *Proceedings of the AAAI Conference on Artificial Intelligence*, vol. 31, no. 1, 2017.
12. Rocktäschel, T. and Riedel, S., *End-to-end Differentiable Proving, Adv. Neural Inform. Process.*, 2017.
13. De Raedt, L., Dumancic, S., Kazemi, S.M., Machine Learning and Knowledge Graphs: A Perfect Match for Neurosymbolic AI. *In Front. Artif. Intell.*, 3, 17, 2020.
14. Hudson, D.A. and Manning, C.D., *GQA: A New Dataset for Real-World Visual Reasoning and Compositional Question Answering*, in: *Proceedings of the IEEE/CVF Conference on Computer Vision and Pattern Recognition*, pp. 6700–6709, 2019.
15. Chen, D., Zhu, Y., Paperno, D., Baroni, M., LAMBADA: Backdoor Testing Commonsense in Language Models. *Trans. ACL*, 5, 207–220, 2017.

14

Multimodal Learning and Perceptron—Integrating Vision, Language, and Auditory Data

Sindhu A.[1]*, Suresh Arumugam[1], Shaila S.G.[1], Monish L.[1] and Ramesh Chundi[2]

[1]School of Engineering, Dayananda Sagar University, Bangalore, India
[2]School of Computer Applications, Dayananda Sagar University, Bangalore, India

Abstract

The fusion of multimodal data vision, language, and audio information has emerged as a pillar in the development of artificial intelligence (AI), especially in the context of generative models. This chapter explores the new field of multimodal learning, with emphasis on the fusion of these three primary modalities in the context of perceptron-based architectures. Multimodal learning aims to integrate complementary data from different sources of senses to make more complete and stronger AI models that can tackle complex tasks entailing multiple sources of input. In particular, this chapter demonstrates how integrating vision, language, and audio has unlocked new dimensions in generative AI, leading to novel applications across sectors including autonomous systems and natural language processing to interactive AI systems such as chatbots and virtual assistants. A significant part of the process of integration is discovering ways that perceptron networks, in this case, neural networks, can be changed and configured to process inputs from multiple modalities concurrently. The single-input model of the traditional perceptron needs to be greatly refined to address the varied nature of multimodal data, which in itself is inherently represented by differing data types (e.g., visual, textual, and auditory). The chapter starts by venturing into the core concepts in neural networks, including deep learning, and how the perceptron is used as a fundamental building block for the development of more sophisticated network architectures, i.e., neural networks such as convolutional neural networks, which help computers see and understand images, and recurrent neural networks, which

**Corresponding author*: sindhua-cse@dsu.edu.in

T. Saravanan, P. Preethi, Sumaya Sanober, N. Thillaiarasu and S. Balamurugan (eds.) Artificial General Intelligence: Principles and Practices, (273–294) © 2026 Scrivener Publishing LLC

are great at processing sounds and audio and transformers for text. The major challenge in multimodal learning is how to align and combine features across the various modalities. Every modality has its own distinct difficulties: visual information calls for high-dimensional representation in the form of pixel-level data; audio information calls for temporal sequential modeling for speech or sound recognition; and text information calls for semantic comprehension and syntactic parsing. The chapter evaluates how multimodal models handle these difficulties by combining information in different ways such as merging it early on, later in the process, or using a mix of both approaches. Such models gain accurate and context-sensitive AI systems through feature alignment between modalities to leverage cross-modal relations. The research explores sophisticated generative model approaches through the study of perceptually learned models. Advanced techniques such as generative adversarial networks, variational autoencoders, and transformers tools that help AI create, learn, and understand information in powerful ways (e.g., multimodal transformers) are considered due to the fact that they are some of the state-of-the-art methods to generate realistic and contextually relevant output. These networks understand multimodal data while generating new content through the creation of text from images and realistic speech from text and self-contained systems that experience composite sensory inputs. The chapter emphasizes the requirement of attention mechanisms in multimodal learning because attention layers enable models to concentrate on essential parts of the input data. This feature serves essential functions in interactive applications such as dialogue generation and real-time content creation for multimedia systems. The attention mechanism enables models to enhance multimodal information implicit dependencies and function optimally on things such as turning images into captions, translating spoken words into text, and understanding the content of videos. The research examines social and ethical implications of generative AI systems that work with multiple modalities. The growing power of these systems requires organizations to establish procedures for ensuring fairness transparency and accountability throughout development and after deployment. Biases in training data, privacy, and potential misuse of generative technology are elaborated. Overall, the chapter talks about a comprehensive survey of the theory, technology, and ethics of combining vision, language, and audio data into generative AI. When we combine these different types of inputs, AI can do some pretty amazing things—such as changing the way we create art or communicate with technology. It opens up all kinds of possibilities for future research and real-world uses in multimodal learning.

***Keywords*:** Multimodal learning, generative AI, neural networks, attention mechanism, cross-modal fusion

14.1 Introduction

Artificial intelligence (AI) has developed multimodal learning as its advanced branch for processing information through multiple sensory channels such as vision and hearing and language. Traditional machine learning models mainly operate on single input output pairs, yet real-world scenarios deliver information through complex multiple modalities. People naturally process environmental understanding through the combination of visual data along with spoken words and environmental acoustics. The ability to grant machines this capability stands as one of the numerous challenges researchers face when creating flexible and sensitive smart systems. The core principle of multimodal learning states that different modalities possess exclusive information that produces improved multimodal results than single-model accuracy and reliability when combined.

The visual representation contains both structural elements and spatial features, language contains logical and abstract concepts, and sound contains audio elements such as tone and emotional content and time-based patterns. Multimodal systems achieve better performance at complex tasks because they combine various information types which unimodal counterparts cannot process [1]. Applications need contextual awareness through multiple modalities because Melding serves as a fundamental requirement. The operation of an autonomous vehicle depends on integrating camera data with audio commands together with ambient noise to reach proper decisions. Siri and Alexa utilize voice along with tone detection and facial expression analysis to customize their responses. The integration of medical images and patient history together with audio recordings of heartbeats through AI models leads to enhanced diagnosis and treatment planning in medical applications. The development of multimodal learning stands essential for building adaptive and interactive AI systems [2].

The technology benefits transportation systems as well as healthcare services and human–computer interfaces and functions as the fundamental component for general-purpose AI. Model design improvements along with better fusion methods and large training datasets are advancing multimodal learning to achieve machine understanding at levels closer to human perception and reasoning. Researchers have shown increased interest in multimodal learning during the last few years after developing different architectures and methods, which beautifully combined data from multiple sensory modalitics. Thc initial attempts focused on com bining features with decision-making through feature extraction methods that used either extraction or concatenation techniques before prediction. The initial approaches laid groundwork, but subsequent methods developed into intricate models that learned to create common representations between different data streams.

Ngiam *et al.* [3] conducted one of the first major works in this domain when they taught deep Boltzmann machines to learn joint representations of images and text for audiovisual speech recognition applications. Deep learning approaches achieved superior performance than shallow models in applications such as audiovisual speech recognition.

The development of convolutional neural networks (CNNs) and recurrent neural networks (RNNs): Progress accelerated significantly because CNNs became the standard for visual feature extraction and RNNs along with their derivatives long short-term memory (LSTM) and gated recurrent units became widely adopted for sequential data processing of text and speech. The combined systems learned to merge visual and textual elements through image captioning by using CNNs to process images into embeddings that RNNs used to produce captions [4].

The development of attention mechanisms along with transformer-based models introduced cross-modal attention, which enabled the creation of rich contextualized representations that processed vision and language simultaneously using frameworks such as ViLBERT [5], LXMERT, and CLIP. The CLIP [6] model achieved its results through pretraining with 400 million pairs of text and images and demonstrated these results in various applications. The analysis demonstrates how contrastive learning achieves the integration of various modalities through shared embedding spaces. The development of audio processing techniques through WaveNet [7] and Whisper led to new champions in speech synthesis and recognition. The systems that utilize audio representations together with vision and language capabilities created advanced multimodal functionalities including audiovisual speech improvement emotional recognition and sentiment analysis. The extensive collection of research demonstrates rapid progress in development alongside emerging opportunities for multimodal learning, which transforms multiple academic domains.

14.2 Fundamentals of Perceptron and Neural Networks

Developed by Franc Rozenblatt in 1958, perceptron is one of the first and most influential models in the history of machine learning. This is a binary classifier that takes the input function and spends the results based on linear decision limits. This model includes incoming nodes, associated weights, sum functions, and activation functions that provide output based on whether the weighted sum exceeds a certain threshold. Simple in design, it was trained to learn from data by updating weights, as it is effective in demonstrations of machines. Perceptron was an important concept. This idea formed the basis for the construction of a more sophisticated neuronal architecture through clarification of monitored learning principles.

From a historical point of view, it is interesting to note that the perceptron has basic drawbacks. One such issue is an obstacle to solution for nonlinear separable problems, such as the XOR issue. This is where multilayer perceptron came in—they placed one or more hidden layers between the input and output layers. The overarching adoption of nonlinear activation functions, such as ReLU and sigmoid, allowed these networks to complexity nonlinear relationships. This was instrumental in the development of deep learning, which is why deep neural networks shifted focus onto stacked layers. With the growth of computational power coupled with the availability of data, these architectures began morphing into more specialized counterparts such as CNNs that process images, or RNNs for sequential data such as text and audio [8]. CNNs turned out to be crucial for mining spatial hierarchies from the image, whereas RNNs with their LSTM and GRU extensions were more efficient at learning temporal relations within sequential streams of data. The perceptron-based architectures possess a high utility for multimodal learning due to its modular and flexible characteristics.

A vision, language, or audio can be captured or represented by a neural network specifically designed to be specialized for its data type. An image can be processed to get features using CNN, whereas the description in relation to the image can be supplied to an RNN for text processing. These intermediate results can be fused at balanced structures to obtain multimodal common representations. This single approach enables performing diverse multimodal activities such as image captioning, audiovisual speech recognition, and sentiment analysis in videos. The transformation from simple perceptron to deep architectures is now at the heart of multimodal systems, which permits pattern recognition to more complex, human-like understanding in analyzing diverse datasets needing perceptual intelligence.

14.3 Representing Different Modalities

Figure 14.1 shows the multimodal representation workflow like input modality—vision, language, and audio—needs to be meaningfully represented as something a machine learning model can work with. Every modality has specific structures and features associated with it, which must undergo particular steps of preprocessing and feature extraction. Models require proper representations to enable the retrieval of distinguishing information encompassed in each data form, thus making multimodal integration efficient.

Dense spatial and sometimes temporal components characterize visual data such as images and videos. The leading architecture for processing this type of data is CNNs, which can capture the local structures and rudimentary constituents, such as edges and steps, of a given visual input. Additionally, static images allow for the use of CNNs to segregate the requisite features such as object boundaries, color, and spatial layout. In video processing, where the element of time greatly matters, CNNs can be extended in the form of spatial and dynamic content through 3D convolutions or with temporal models to track motion and activities over several frames. Regardless of its form, be it spoken transcripts or text, language data are inherently sequential and symbolic.

At first, RNNs were utilized, and then more advanced versions such as LSTM networks succeeded them to extract contextual dependencies within sequences. These models captured mid and short range dependencies. Nowadays, self-attention techniques used in BERT and GPT transform and revolutionize natural language processing (NLP) with transformer

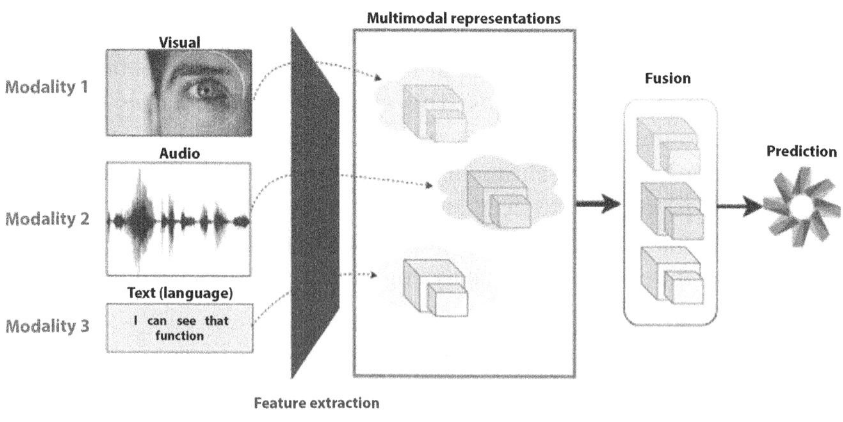

Figure 14.1 Representing different modalities.

architectures that efficiently capture long-range dependencies. These models construct an intricate contextual understanding of meaning at the level of words, sentences, and also documents by treating language inputs as sequences of vector embeddings.

Speech data, as well as background noise, can be represented in the form of raw waveforms or encoded in a spectrogram, which displays frequency components in relation to time. One such model is WaveNet, which processes audio at the raw level using dilated CNNs. This design enables modeling long-term dependencies in a data preserving manner. On the other hand, spectrograms allow audio signals to be treated as images, enabling CNNs to efficiently extract temporal and frequency features. These dissimilar representations allow models to perform various tasks such as speech recognition, sound detection, and emotional tone perception. With precise modality design, each modality can be merged and unified in a shared embedding space. This merging process is what enables the development of robust multimodal systems, which can utilize disparate information for enhanced system performance in audiovisual speech recognition, image captioning, and cross-modal retrieval.

14.4 Building Intelligent Systems through Multimodal Representation

This paper introduced one of the earliest multimodal deep learning models using deep autoencoders to jointly learn representations from audio and video data. Their model demonstrated how learning from one modality can benefit another, even when one is missing. This fusion approach significantly improved emotion classification and digit recognition tasks. The paper was foundational in cross-modal learning and inspired many future architectures. It highlighted the power of unsupervised learning in multimodal settings [3]. This proposed the tensor fusion network (TFN), a novel architecture that explicitly models unimodal, bimodal, and trimodal interactions between text, audio, and video. It was applied to sentiment analysis tasks and outperformed standard early or late fusion strategies. TFN is known for its clear interpretability and effectiveness. The model was evaluated on benchmark datasets such as CMU-MOSI and CMU-MOSEI. This paper remains influential in the affective computing and multimodal NLP communities [9].

Nagrani *et al.* introduced VoxCeleb, a large-scale dataset for speaker recognition combining face and voice data from YouTube interviews.

The dataset facilitates research in face-voice matching, audiovisual speaker verification, and biometric authentication. It includes diverse speakers, accents, and environments. VoxCeleb became a standard benchmark in multimodal speech research. The paper also contributed robust evaluation protocols for audiovisual matching tasks [10]. This paper offered a comprehensive survey and taxonomy of multimodal machine learning, covering challenges such as representation, translation, alignment, and fusion. The paper discusses strategies for combining audio, video, and text inputs and compares various neural fusion models. It categorizes colearning and hybrid learning techniques and their applications. This work has become a foundational reference in the field. It is widely cited by researchers working on multimodal integration problems [11]. Tsai *et al.* proposed the multimodal transformer (MulT), a cross-modal attention mechanism that learns to align and integrate video, audio, and textual inputs. The architecture separates temporal and modality-specific learning processes. It outperforms traditional fusion models on sentiment prediction benchmarks. MulT enables fine-grained semantic alignment across modalities. The work influenced a range of transformer-based multimodal architectures [12].

Sun *et al.* introduced VideoBERT, a pretraining model based on masked prediction that learns joint representations from unlabeled video–text pairs. The model generalizes BERT-style architectures to visual-linguistic tasks. It was applied to action recognition and video captioning with strong results. VideoBERT highlights the value of large-scale video-language pretraining. The work paved the way for future generative multimodal systems [13]. The paper focused on developing a MulT for cross-modal video retrieval using synchronized audio, video, and text features. The model learns to embed inputs into a shared semantic space. It was tested on large-scale datasets and showed improved retrieval performance. Their work emphasized temporal modeling and feature alignment. The approach advanced the state-of-the-art in video understanding systems [14]. Srinivasan *et al.* [15] addressed the issue of missing modalities in real-world data by introducing a modality-invariant latent space. Their model can perform tasks effectively even when one or more inputs (audio, video, or text) are absent. It uses generative modeling to impute missing data or make robust predictions. This approach ensures graceful degradation under incomplete inputs. The model is well-suited for healthcare, surveillance, and wearable AI scenarios [15].

This paper proposed a self-supervised framework for audiovisual scene understanding using contrastive learning. The model learns to align audio and visual representations from unlabeled video streams. It was benchmarked on action and scene classification tasks. Their results demonstrate the effectiveness of self-supervision in multimodal setups. This work inspired future contrastive approaches for audio–video grounding [16]. It introduced VATT, a transformer-based model trained on raw audio, video, and text using self-supervised learning objectives. The architecture allows end-to-end training across modalities without handcrafted features. VATT performs well on audio classification, video understanding, and captioning tasks. It represents a significant step toward general-purpose multimodal models. This work promotes modality-agnostic representation learning at scale [17]. The presented MERLOT, a model that learns script-like knowledge from narrated YouTube videos by integrating audio, visual frames, and text transcripts, focuses on temporal reasoning and common-sense understanding. MERLOT is pretrained on large-scale aligned video–text datasets. The model enables tasks such as event anticipation and visual commonsense QA. The paper shows how narrative structure benefits multimodal understanding [18].

Alayrac *et al.* [19] developed Flamingo, a few-shot visual–language model (VLM) capable of adapting to new tasks using only a few examples. Although primarily trained on image–text pairs, it supports extension to video and audio inputs. Flamingo outperforms prior models on zero-shot captioning and VQA. It leverages frozen vision encoders and a lightweight fusion mechanism. This model is key for building flexible multimodal agents [19]. The author Lee *et al.* explored multimodal emotion recognition using audio, video, and text. They compared different fusion strategies—early, late, and hybrid—and found hybrid fusion to be most effective. The study was validated on several emotion-rich datasets. The model supports real-time affect detection applications. Their work contributes to AI systems that better interpret human behavior [20]. Li *et al.* [21] proposed BLIP, a VLM based on contrastive learning and caption bootstrapping. Although initially designed for image–text, it sets the stage for video–text models. BLIP achieves state-of-the-art performance in visual QA and captioning. The dual encoder–decoder architecture allows flexible training. It serves as a foundation for multimodal generation and alignment tasks. Chen *et al.* presented Unified-IO, a single architecture designed to handle vision, language, and audio tasks in a unified format. It supports both generative and classification-based outputs. Unified-IO was trained across dozens of datasets involving tasks such as AV speech recognition and audio captioning. The model simplifies deployment by eliminating task-specific architectures. It represents convergence in general-purpose multimodal AI systems [22].

14.5 Multimodal Fusion Techniques

The integration of various information sources including images and texts and sounds constitutes the core element of multimodal learning known as multimodal fusion. Each data source provides unique aspects of stimulus recognition: Vision detects spatial and object-oriented elements, whereas language communicates symbolic and semantic data, and audio detects temporal and emotional characteristics. Figure 14.2 illustrates the hybrid fusion architecture where three different fusion approaches exist for information integration: early fusion, late fusion, and hybrid fusion. The early merge process feeds raw or low-level features from multiple modalities into a unified representation space before sending them to a single model. The cross-modal interaction occurs at an early stage because the direct capture of dependencies starts from the beginning, which enables the integration of different data inputs.

The direct dependency detection at the beginning results in efficient early integration of different data inputs. In the late fusion method, each modality receives independent handling through separate model pipelines. The end result of processing involves majority voting together with weighted averaging and ensemble models for synthesis to merge high-level outputs and decisions. The system maintains strength against noisy data and absent modalities, and it supports modular architecture. This approach lacks the ability to detect deep relationships that different modalities might establish with each other before processing is complete.

The development of hybrid fusion techniques emerged as a solution to combine the disadvantages found in early and late fusion methods. These frameworks enable connection points between intermediate representations that belong to each modality throughout different stages of the network to facilitate early interaction while preserving modality-specific specialization. Hybrid fusion demonstrates exceptional effectiveness in complex tasks because of its adaptable nature including visual question answering (VQA) and emotion recognition and multimodal sentiment analysis. The achievement of successful multimodal learning systems heavily depends on proper modality representation along with alignment techniques. The unique embedding space within cross-modality mapping holds all modalities, which allows direct comparison and interaction between modalities.

The other method involves coordinated representation learning that maintains distinct spaces for each modality while enforcing consistency through contrastive learning or mutual information maximization.

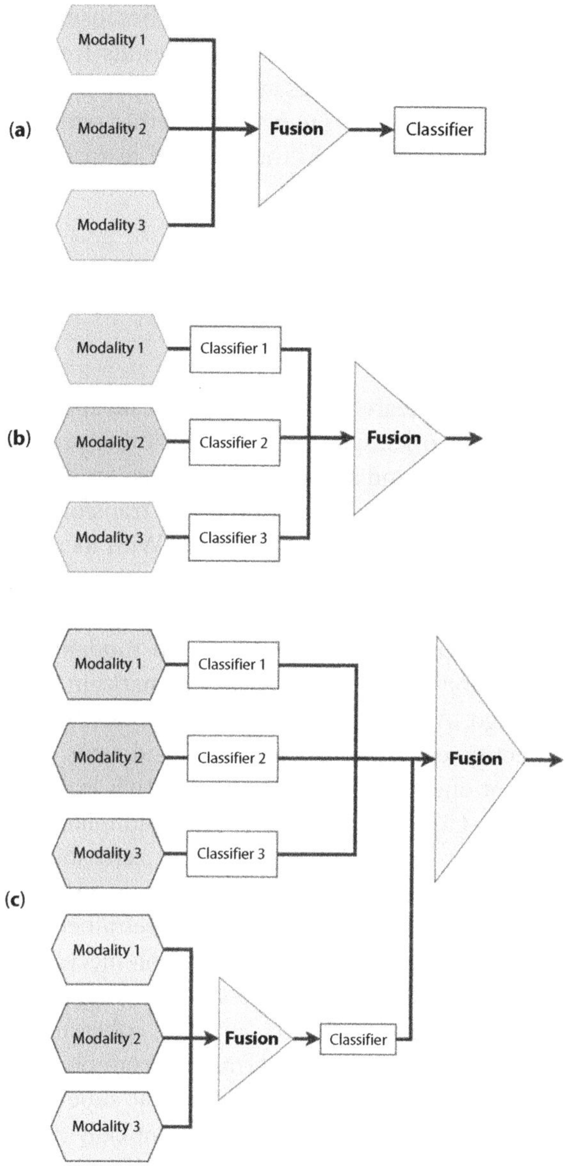

Figure 14.2 Hybrid fusion method.

The single-method approaches have a drawback because they fail to establish adequate cross-modal alignment, which results from different timing patterns and semantic structures and granularities. Advanced models such as attention and coattention networks together with temporal alignment models dynamically focus on the essential features present in each modality.

14.6 Architectures and Models for Multimodal Learning

The past few years of research have seen rapid advancements in multimodal learning frameworks, specifically in relation to processing and reasoning with images, texts, and audio using transformer-based models. This progress owes much to the incorporation of the transformer architecture, which was originally developed for NLP tasks. With its self-attention and scalable mechanisms, it proves beneficial on multiple fronts. MulTs consist of one single model that handles all modalities at once or a dual-stream model that features separate pathways for each modality, although they have shared interaction layers. Important benchmarks in this space include CLIP and Flamingo. As a multidisciplinary model, CLIP from OpenAI develops a collective embedding for images and texts, learning from a diverse set of captioned images using contrastive learning techniques. This approach helps CLIP outperform several vision–language tasks without any prior domain-specific training termed zero-shot generalization. Similarly, Flamingo applies few-shot learning to vision–language tasks by combining a frozen vision encoder, such as a pretrained ViT, with a language model decoder enhanced by cross-attention mechanisms, including one for each VLM layer. It aims to take combined image–text streams and transform them into meaningful texts, enabling coherent visual storytelling or image description with minimal training examples.

A representative methodology is cross-modal supervised contrastive learning, which enables models to align semantically related data from different modalities. It was one of the earliest approaches used for contrastive learning across modalities. In this method, a model is trained to project a sentence and its corresponding image into a latent space such that semantically equivalents from different modalities are close to each other. The proximal relation is achieved by optimizing a contrastive loss function such as InfoNCE or NT-Xent that discriminates matching pairs from nonmatching ones. Apart from having better aligned representations, cross-modal retrieval, transfer learning, and zero shot inference are also supported, which is the case of CLIP, ALIGN, and DeCLUTR. Notably, one of the advantages of the method is its scalability.

The dependency on labeled data is reduced because the model can be trained on large sets of uncrated data with little supervision using weakly paired multimodal inputs. This approach increases robustness against task-dependent annotated datasets with noisy and imbalanced data. One more major architectural innovation in multimodal models is the application of attention and coattention techniques because they permit the dynamic interaction of modalities. The self-attention mechanism, as implemented in transformers, is one form of attention that allows models to consider the relative importance of different parts of a sequence. Coattention, which applies to multimodal contexts, enhances this mechanism by allowing the model to attend across modalities, for instance, enabling text tokens to focus on pertinent regions of images and the other way around. Depending on the demand of a task, coattention can be applied symmetrically (bidirectional coattention) or asymmetrically. Such processes are important in cases such as VQA in which models need to refer some image regions to question phrases to produce correct answers.

Attention processes are also important for models that align and fuse in sequential or spatiotemporal settings, such as aligning verbal with video gestures. Such mechanisms enhance interpretability by showing where in the input the model attends and improves performance by directing resources to the most critical features of the computation. Attention-based fusion also enables hierarchical reasoning, where a model develops a top-down semantic understanding from low-level midlevel descriptions. As multimodal learning advances, including for Gato and Perceiver, there is an observed shift toward constructing more generalized frameworks, which require minimal engineering to perform multiple tasks across different inputs. Large multimodal models and newer Gato and Perceiver models focus on using more advanced inputs and processing techniques to represent them in a manner that is not confined to a specific modality.

Due to the modularity of the transformer architecture and external pretraining methods such as contrastive pretraining, attention-based fusion, and others, these models are able to handle a wider scope, which includes image captioning, speech recognition, robotic control, and responsive dialogue. The optimal utilization of multimodal system robustness and performance depends on adapting fusion methods and alignment techniques to specific tasks along with data requirements and computational constraints. The development of effective multimodal structures depends on solving design algorithmic problems and alignment issues as well as addressing modality characteristics and computation and fusion requirements. Our ongoing research focuses on creating AI systems that acquire genuine reasoning capabilities and seamless integration abilities for all human and environmental information with flexible scalability.

14.7 Applications of Multimodal AI

Through multimodal AI, machines use data from vision text and sound sources to transform their environmental understanding and interaction abilities and produce results similar to human understanding and reasoning abilities. The fascinating example of VQA requires systems to produce natural language answers while referencing images. Another effective use concerns image captioning where a model generates a meaningful text that describes an image. This process describes the conversion of images into a template of structured natural language in perception and linguistic generation.

Models of image captioning generally consist of a CNN or vision transformer that performs image feature extraction, which is later fed into a language model, either an RNN or a transformer decoder that generates the caption. Advanced models go beyond simple object enumeration to be able to represent complex scenes and encode context, emotion, or activities within a scene. Applications range from accessibility for the visually impaired to social media and content tagging for search engine optimization. It is evident that sophisticated captioning systems must solve ambiguity, identification of distinctive granules of a scene, and fluency and relevance of the captioned text. Audiovisual speech recognition is an equally fascinating application of multimodal AI; for instance, such a system recognizes words by processing both audio and visual data, such as a person's lip movements. This aids in enhancing the reliability of the system's performance in severely noisy conditions, where the audio signals may, in themselves, be distorted or lacking. By combining lip reading from video frames with audio waveforms, the system provides more accurate transcriptions in situations when there is background noise or people are chatting over one another. These systems allow for better communication through real-time language translation, video conferencing, enabling technologies for the deaf and hard-of-hearing, surveillance, and security.

The temporal alignment of the video image and audio elements is necessary, as this poses the difficulty of needing advanced sequence modeling and synchronization to achieve cohesive blending and interpretation of context without losing essential details. Multimodal sentiment analysis is one of the fields in which the combination of different data sources provides a deeper understanding of human emotions and intent. Standard sentiment analysis can only be performed based on the textual data. These data are often ambiguous in the case of detecting sarcasm, tone, or other nuanced emotional signals. With the help of combination systems that incorporate voice, expression, and gesture information, sentiment analysis can be performed more accurately.

For example, a person might say "I'm fine" in a monotone voice with their eyes turned slightly downward, suggesting they are indeed not fine despite the neutral wording. Such systems operate using separate encoders for each modality of text, audio, and visual features, which later undergo fusion—usually attention or gating based selection—where the most relevant information for classification is chosen.

Text analysis of this kind has been increasing in popularity in customer service, marketing, healthcare, and human–computer interaction for the purpose of enabling systems to respond more empathetically and contextually. Because of its flexibility, multimodal AI is applicable across all industries and user needs, and it helps carve new possibilities for AI. Its application in enabling the understanding of media content and driving recommendation models as well as accessibility improvements and the creation of more human-like interfaces illustrate the value of modality-specific representations and the enhanced complexity AI systems can achieve in interpreting and generating human-like behaviour. We expect their advancement and wide applicability to give rise to a new category of sophisticated AI systems capable of seamless interaction, nuanced understanding, and effortless adaptation to complex real-world challenges.

14.8 Challenges in Multimodal Integration

Although promising for the development of smarter systems, multimodal integration poses deep-rooted and sophisticated hurdles that impede the smooth execution of real-world applications. One of the most immediate hurdles is the problem of modality alignment and synchronization, which refers to the issue of finding temporal and semantic correspondences between text streams, audio, and images or video. Unlike unilateral data, which are typically a uniform stick, multiple inputs may arrive at different sampling rates, different lengths, or without clear corresponding timestamps.

One example would be spoken words, which progress in time, whereas corresponding images or facial expressions may be instant or out of sync. Ensuring a model is able to adequately attend to the specific mapping of words to visual cues or other audio signals necessitates advanced attention, temporal modeling, and often manually crafted alignment strategies, which are fragile and susceptible to noise and variation. Moreover, the semantics of modalities are not always aligned in a straightforward way—depiction of a scene in natural language can be visually given without some information, creating some ambiguity. Such articulation problems increase the difficulty of training and inference. Dealing missing or noisy data in one or more modalities is a practical challenge aside from realignment.

In real life, not every modality will always be available due to sensor errors, network latency, environmental noise, or user activity. For instance, a video assistant may fail to receive accompanying visual information, and a surveillance camera may have substandard audio. Unlike unimodal systems built out of a single data stream, multimodal models have to be equipped to reason soundly even when one or more modalities are absent, low-quality, or conflicted with others. This is accomplished by having some form of structural flexibility, such as modality dropout techniques, gating mechanisms, and fusion processes that can modify the reliance of a model on each input stream based on its reliability.

Furthermore, the presence of noise in any one of the modalities has a high potential of diffusing errors throughout the entire system, especially where modalities are closely integrated in initial fusion designs. These challenges/these risks can be mitigated not only with powerful preprocessing and denoising techniques, but also with the application of confidence-aware learning that emphasizes in which data are increasingly considered untrustworthy. Variability in data and the stakes involved in decisions, particularly in healthcare, autonomous systems, or multimedia content analysis, make engineering systems with high levels of robustness a significant research and engineering challenge. Another class of problems around modals relating to scaling and computations is efficiency. The tasks of integrating multiple streams of high-dimensional data such as raw videos, audio waveforms, and text into a single coherent system in real time pose enormous challenges with respect to memory, processing power, and data bandwidth. The most recent advances in multimodal models, especially those using transformers, are very expensive to compute because they contain a large number of parameters and attention-based fusion layers that deal with intermodal interactions. Such models require dedicated equipment and enormous amounts of available parallelized computing resources to access large datasets, which become a burden even for wealthy technology corporations and research institutions. Even after training, these models have limitations on the inferencing speed and energy required for deployment on edge devices or low-latency settings.

The managing and curating of large-scale multimodal datasets are a multifaceted problem in its own right, including issues such as data imbalance, annotation heterogeneity, and privacy implications. Moreover, there is the issue of generalizability. Models that are built using highly curated and aligned multimodal datasets tend to perform poorly when exposed to changing distributions, domains, or new modal combinations. For the above reason, researchers are looking to make multimodal models more controllable and easier to implement across different computational environments by optimizing architectures, attention mechanisms, distillation of knowledge, and training modalities. These and other feasibility analyses focused on deploying AI into practical settings highlighted issues of alignment, robustness, and scalability that have yet to be resolved and serve as barriers for moving away from experimental prototype benchmarks to reliable, robust systems that deliver transformative impact.

14.9 Future Directions in Multimodal Learning

The advancement of multimodal learning is crucial in achieving goals relating to general AI—an AI system capable of flexible, context-sensitive reasoning for a broad range of tasks and environments. However, it captures, in some form, how humans tune to the world around them because we perceive phenomena through a range of senses rather than a single sensory input. Simultaneous hearing, vision, touch, and languages combine to form our understanding, and reproduction of this combined functioning in machines is what leads to advanced, intelligent behavior or adaptive systems simulation. For this reason, the growth of multimodal approaches is not simply a tactic, but one of the building blocks in the pursuit of AI.

Advanced AI will be able to mimic human behaviors that stem from sophisticated integrative reasoning, especially in interpreting scenes, engaging in rich dialogs, and executing physical actions. Currently, many learners such as Gato, Gemini, and Perceiver model them, seeking to streamline and systematize processes with a variety of input and task types to a single omnipotent architecture. These models are beginning to demonstrate that systems can be built that deal with language, vision, sound, and control signals simultaneously. Systems not only learn how to process information, but also learn how to integrate knowledge across domains.

A significant aspect of future work will involve increasing the effectiveness and interpretability of these models, making them able to learn continuously and adapt over time to different environments, tasks, and modalities without requiring complete retraining. As indicated, architecture entails generalization. Other principles of multimodal reasoning are in new areas of research, which is cross-disciplinary of neurosymbolic integration, which aims to combine the pattern recognition strengths of deep learning with the reasoning capabilities of symbolic AI. This method seeks to make multimodal models' reason in sophisticated structure, low data need, and explainable ways. For example, fusion of visual perception and logical deduction within a neurosymbolic system could be applied to solve complex problems such as determining if an object in a particular scene is "breaking" social or physical rules. These systems might achieve richer reasoning that goes beyond statistical correlations, which is world knowledge, causal reasoning, and even commonsense reasoning.

In the case of multimodal reasoning, this could enable AI to more effectively understand composite abstract images containing text, or diagrams, or create coherent stories that involve both logical deduction and sensory evidence. Another example in this category is the development of models for supporting embodied multimodal learning where AI agents learn through physically engaging with their environment *via* multiple perceptual channels. This includes advancements in robots and virtual agents, as well as simulation environments where sight, hearing, and movement feedback are integrated to guide learning and action. These agents might benefit from multimodal feedback loops that mimic human trial-and-error processes, serving a fundamental role in the connection of perception, cognition, and motor functions. Also, the progress in multimodal learning personalization, equity, and intercultural relatability will occur in the future.

As in most current models, the performance is lacking for data from new languages, cultures, or less represented user populations in training datasets. Creating multimodal systems that respond to user's contextual multivoiced value sets will require enhanced curatorial approaches to data as well as models that are more responsive and sensitive to contextual nuances. This involves learning with fewer instances available (zero-shot and few-shot learning), understanding local speech and vision expressions, and adhering to socially accepted ethical norms regarding interaction and generation. This ability will be critical in multidimensional AI for healthcare, autonomous systems, and education and beyond, where reasoned decisions need to be made over time, uncertainty, and incomplete information.

Integrating reasoning with temporal relationships is something orthogonal future research needs to focus on so AI not only can correlate but also can understand cause–effect relationships narrated through speech, shown in video, or rationalized through sensing inputs. The use of self-supervised large-scale learning, human-in-the-loop reinforcement learning, and multimodal architectures paired with external memory systems will support the creation of safe and reliable AI in the future. This will enable the development of safe and reliable AI systems that perform countless tasks. AI will no longer be seen as a tool but will become an assistant that can be guided, corrected, or aided. Multimodal learning will become the new frontier that AI will need to conquer to be as agile in looking, talking, and reasoning like humans. To traverse the path toward general intelligence, an advancement in AI systems that is capable of interaction and intelligent conversation will be necessitated, and that advancement will mark the evolution of AI.

14.10 Conclusion

To put it concisely, multimodal learning is an advanced technology as far as AI is concerned because it enables a machine to process and integrate streams of vision, language, audio, and information at a human-like level. We have covered the architecture components such as neural networks and transformers, the methodologies of representing and interrelating multiple modalities, and the problems of alignment, synchronization, and scaling these systems. The integration of AI into everyday activities such as VQA, image captioning, and audiovisual recognition alongside advanced systems such as CLIP and Flamingo showcases the flexibility of multimodal AI.

At the same time, threats such as robustness to absent data, excessive computation, and the need for better generalization and explainability pose significant challenges. Progress in neuroscientific disciplines such as neurosymbolic reasoning, embodied learning, and causal inference will likely lead to new multimodal systems to aid in developing general AI systems. With the advancement of research targets, there is the creation of versatile systems that not only can recognize patterns across different modalities, but also can possess the ability to reason, learn, and interact with their environment in a human-centric manner. The results from this specific report highlight the critical role that multimodal learning plays to refine the AI technology of the forthcoming era—one that is richly connected and contextually aware—ultimately making it more powerful.

AI disclosure statement: The author acknowledges the use of AI-based tools for minor language editing or grammatical correction. However, all intellectual contributions, analyses, and conclusions presented in this chapter are entirely the author's own.

References

1. Rosenblatt, F., The Perceptron: A Probabilistic Model for Information Storage and Organization in the Brain. *Psychol. Rev.*, 65, 6, 386–408, 1958.
2. LeCun, Y., Bengio, Y., Hinton, G., Deep Learning. *Nature*, 521, 7553, 436–444, 2015.
3. Ngiam, J., Khosla, A., Kim, M., Nam, J., Lee, H., Ng, A.Y., Multimodal deep learning, in: *Proc. 28th Int. Conf. Machine Learning (ICML)*, pp. 689–696, 2011.
4. Vinyals, O., Toshev, A., Bengio, S., Erhan, D., Show and Tell: A Neural Image Caption Generator, in: *Proceedings of the IEEE Conference on Computer Vision and Pattern Recognition (CVPR)*, pp. 3156–3164, 2015.
5. Lu, J., Batra, D., Parikh, D., Lee, S., ViLBERT: Pretraining Task-Agnostic Visual-Linguistic Representations. *Proc. NeurIPS*, 32, 13–23, 2019.
6. Radford, A., Kim, J.W., Hallacy, C., Ramesh, A., Goh, G., Agarwal, S., Sutskever, I., Learning Transferable Visual Models From Natural Language Supervision, in: *Proceedings of the International Conference on Machine Learning (ICML)*, pp. 8748–8763, 2021.
7. Choromanska, A. and LeCun, Y., The LSTM as a Perceptron: A Closer Look at the Neural Network Foundations of LSTMs, in: *Proceedings of the International Conference on Learning Representations (ICLR)*, 2015.
8. Poria, S., Cambria, E., Hazarika, D., Zhu, X., A Deeper Look into Sarcasm: A Review and Future Directions. *IEEE Intell. Syst.*, 32, 3, 35–45, 2017.
9. Zadeh, Chen, M., Poria, S., Cambria, E., Morency, L.-P., Tensor fusion network for multimodal sentiment analysis, in: *Proc. Conf. Empirical Methods in Natural Language Processing (EMNLP)*, pp. 1103–1114, 2017.
10. Nagrani, Chung, J.S., Zisserman, A., VoxCeleb: A large-scale speaker identification dataset, in: *Proc. Interspeech*, pp. 2616–2620, 2017.
11. Baltrušaitis, T., Ahuja, C., Morency, L.-P., Multimodal machine learning: A survey and taxonomy. *ACM Comput. Surv.*, 51, 6, 1–47, 2019.
12. Tsai, Y.-H.H., Bai, S., Yamada, M., Morency, L.-P., Salakhutdinov, R., Multimodal transformer for multimodal sentiment analysis, in: *Proc. Annu. Meeting of the Association for Computational Linguistics (ACL)*, pp. 655–664, 2019.
13. Sun, Myers, A., Vondrick, C., Murphy, K., Schmid, C., VideoBERT: A joint model for video and language representation learning, in: *Proc. IEEE Int. Conf. Computer Vision (ICCV)*, pp. 7464–7473, 2019.

14. Huang, P.-Y., Wang, Y., Yang, L., Li, D., Yuen, P.C., Multimodal transformer for video retrieval, in: *Proc. Eur. Conf. Computer Vision (ECCV)*, pp. 520–537, 2020.
15. Srinivasan, T., Majumder, P., Ghosh, A., Multimodal fusion with missing modalities using modality-invariant latent space, in: *Proc. AAAI Conf. Artif. Intell*, pp. 5412–5419, 2020.
16. Morgado, P., Vasconcelos, N., Li, Y., Audio-visual scene analysis with self-supervised multimodal learning, in: *Proc. IEEE Conf. Computer Vision and Pattern Recognition (CVPR)*, pp. 15134–15144, 2021.
17. Akbari, H., *et al.*, VATT: Transformers for multimodal self-supervised learning from raw video, audio and text, in: *Proc. Adv. Neural Inf. Process. Syst. (NeurIPS)*, pp. 1–21, 2021.
18. Wu, L., Alayrac, S., Ross, R., Zisserman, A., MERLOT: Multimodal neural script knowledge models, in: *Proc. Annu. Meeting of the Association for Computational Linguistics (ACL)*, pp. 2368–2389, 2021.
19. Alayrac, J.-B., *et al.*, Flamingo: A visual language model for few-shot learning, arXiv preprint arXiv:2204.14198, 2022.
20. Lee, K., Heo, H.S., Han, Y., Multimodal emotion recognition using audio, video, and text features, IEEE Trans. Affective Comput., early access, doi: 10.1109/TAFFC.2022.3145672, 2022.
21. Li, X., Yin, X., Li, X., Zhang, P., Zhang, L., BLIP: Bootstrapping language-image pre-training for unified vision-language understanding and generation, in: *Proc. Int. Conf. Machine Learning (ICML)*, pp. 1–14, 2022.
22. Chen, *et al.*, Unified-IO: A unified model for vision, language, and audio processing, in: *Proc. IEEE Conf. Computer Vision and Pattern Recognition (CVPR)*, pp. 1–22, 2023.

15

TransGAN for Visual Anomaly Detection on Imbalanced Industrial Datasets

Srinivasa Perumal R.[1], Venkatasubramanian A.[2], Premalatha M.[1]* and Braveen M.[1]

[1]School of Computer Science and Engineering, Vellore Institute of Technology, Chennai, India
[2]Department of Computer Science and Engineering, GITAM University, Bangalore, India

Abstract

This research paper investigates the efficacy of TransGAN that combines the power of transformers and generative adversarial network (GAN) in the field of artificial intelligence for anomaly detection. This research makes use of MVTecAD and CIFAR-10 image datasets for the purpose. For MVTecAD dataset, latent space mapping of the generator is used to find the closest fake images of the input images, and the difference in image construction is then used for anomaly detection. And for CIFAR-10, two discriminators have been used for the training of the model to address the class imbalance. Both these methods were tried with different classes of the datasets, and accuracy and area under the receiver operating characteristic scores were considered for evaluation of models. This research aims to contribute to the further enhancement in the field of anomaly detection in various domains such as medical imaging, surveillance and others.

***Keywords*:** Artificial intelligence, anomaly, detection, GAN, image processing, transformers, TransGAN

**Corresponding author*: premalatha.m@vit.ac.in

T. Saravanan, P. Preethi, Sumaya Sanober, N. Thillaiarasu and S. Balamurugan (eds.) Artificial General Intelligence: Principles and Practices, (295–326) © 2026 Scrivener Publishing LLC

15.1 Introduction

An increasing number of industries are depending on automated visual inspection systems, which calls for reliable and effective anomaly detection methods for image data. In recent years, anomaly detection in image data has gained significant attention in the field of artificial intelligence (AI) due to its wide-range applications in carious domains such as manufacturing, medical imaging, and surveillance [1, 2]. Conventional approaches hardly grasp the complexity of intricate image dataset because they rely on manually created features or statistical models. Using architectures is hence essential for obtaining superior performance in the field of anomaly detection in images.

In this study, two different approaches both focused on TransGAN [3]. Unlike traditional generative adversarial networks (GANs) [4], which primarily rely on convolutional layers, TransGAN utilizes TransformerEncoder layers to capture spatial dependencies. This enhances the quality of images that are produced and enable usage across several applications. Motivated by the iterative latent space mapping method of Ano-GAN [5], we train TransGAN for anomaly detection for MVTEC-AD dataset [6]. Our goal is to detect deviations from the learned normal distributions and accurately identify anomalies. In contrast, we take a new approach for imbalanced datasets such as CIFAR-10 [7]. In this approach, we use ResNet50 [8] to extract features from real image to map it to the noise input of the generator. We also make use of two discriminators, one for normal data and the other for anomalous data, to address the imbalance in data. Through extensive experimentation and evaluation, we have tried to present the effectiveness of the framework in identifying anomalies in diverse image datasets, thus giving a direction to further research on the anomaly detection applications in the real world.

This research uses two main approaches utilizing TransGAN architecture for anomaly detection. First, TransGAN is trained for anomaly detection on the MVTEC-AD dataset followed by leveraging of the iterative latent space mapping method of Ano-GAN. This approach focuses on detecting deviations from learned normal distributions and accurately identifying anomalies. In second approach for CIFAR-10 dataset, TransGAN is integrated with ResNet50 model to extract features from real images and to map them to the noise input of the generator. Additionally, two discriminators are used—one for normal data and one for anomalous data—to address data imbalance effectively. Through these approaches, this research aims to enhance the effectiveness and applicability of anomaly detection systems in diverse image datasets.

This research is aimed at making use of advancements and approaches for anomaly detection within the domain of AI using TransGAN architecture. By leveraging advanced methods while addressing current challenges, the goal of this research is to enhance effectiveness and applicability of the anomaly detection system in real-world use cases. Through experimentation of different approaches, this research demonstrates the capability of TransGAN architecture for anomaly detection across different image datasets. Ultimately, this research aims to contribute to the increasing advancements in anomaly detection applications and provide practical solutions for real-world anomaly detection.

15.2 Related Works

In many different fields of study and applications, deep learning–based anomaly detection has been thoroughly investigated. To give a systematic and thorough summary of the research in this area, a number of surveys have been carried out. With a focus on discussing the advantages disadvantages and computational complexity of current methods, these surveys attempt to classify them according to their fundamental presumptions and methods. In particular, some surveys concentrate on *post hoc* anomaly detection methods to strengthen deep learning against adversarial and outlier distribution instances. These surveys address the relative benefits and drawbacks of various techniques and offer a taxonomy of them [5, 8].

Currently, the state-of-the-art techniques used in anomaly detection are ADCNN [9] and deep anomaly [10]. But the absence of labeled data turns out to be the bottleneck for these approaches. On the one hand, the uneven distribution makes it difficult to gather abnormal images of both typical and anomalous data and on the other is that anomalous data cannot be clearly defined [11]. A number of learning-based methodologies have been put forth to address these challenges [12]. These approaches discover the distribution of data, and feature distribution of the normal data in terms of a predefined threshold is thought to include unusual objects.

It is suggested to use adversarial learned one-class classifiers for novelty detection when the novelty class data are missing. Evaluations conducted on MNIST and CIFAR datasets confirm that the proposed approach significantly improves detection precision,recall, and AUC compared to existing baseline and state-of-the-art methods for anomaly and outliers detection in image and videos [13]. Attention-driven long-range dependency modeling for image generation tasks is made possible by self-attention GANs

(SAGANs) [14]. SAGAN draws details using cues from all feature locations in contrast to conventional convolutional GANs that produce high-resolution details based on spatially localized points. It also guarantees that the images far-off regions retain the consistency of its highly detailed features. Rather than using local regions with fixed shapes, the generator in SAGAN makes use of neighborhoods that correspond to object shapes. Furthermore, the GAN generator is subjected to spectral normalization which enhances the training dynamics. SAGAN attains cutting-edge outcomes on the ImageNet dataset exhibiting an increased Inception score and a decreased Frechet Inception distance. This model has been modified in various works to perform anomaly detection because of how well it generates faked data.

A novel approach has been proposed by Shangguan *et al.* [17] for semisupervised anomaly detection based on deep generative models with transformers [15]. Their model uses GANs and autoencoders (AEs) to distinguish between abnormal and normal images. Whereas the discriminator in their model also relies on a transformer to discern between generated and training set images, the generator in their model extracts latent representations using an encoder–decoder structure based on a transformer. By keeping the difference between the latent and original image spaces as small as possible, the model is able to determine the normal distribution. Their approaches' effectiveness is demonstrated by experimental results on benchmark image datasets. Furthermore, Ruff and colleagues [28] additionally provide deep SAD, a semi-supervised anomaly detection approach that learns compact representations of normal data while separating anomalous samples using deep latent feature distribution [16]. Their method is predicated on an information-theoretic framework wherein normal data are expected to have a lower entropy of the latent distribution than anomalous data. Even with a limited amount of labeled data, research on a wide range of datasets demonstrates that deep SAD performs better than other shallow hybrid and deep schemes.

Multiple researches have suggested to use transformer-based architectures and GANs for anomaly detection in images. The goal of these techniques is to separate outliers or unusual images from regular data. Although local details are captured by GANs and normal from abnormal data is distinguished, the use of transformers enables the extraction of global semantic information. It has been demonstrated that AEs and GANs using transformers as encoders and transposed convolutions as decoders can successfully distinguish anomalous from normal images [17]. In addition to industrial image anomaly, detection performance has been enhanced by skip connections and attention feature fusion based on encoder–decoder structures [15]. Convolutional transformer-based dual discriminator GANs (CT-D2GAN) have also demonstrated encouraging results in unsupervised video anomaly detection [18, 19]. This is achieved by capturing typical spatial–temporal patterns and taking local consistency and global coherence of temporal dynamics into account.

Talking about the anomaly detection, especially when we discuss the improvements and changes in the integration of technologies such as ADCNN, deep anomaly, SAGAN, and transformer-based models, we need to properly assess the current limitations and potential areas for further development. Table 15.1 discusses the comparative analysis of the various literatures. With the existing solutions, the following gaps and opportunities for advancement have been identified:

- **Scarcity of labeled data**: Anomaly detection models highly depend on the availability of the data that are labeled, especially the ones that use supervised or semisupervised learning. Unsupervised approaches can remove this dependency on extensively labeled datasets and hence represent a significant opportunity.
- **Generalization across diverse usage**: Anomaly detection models are usually developed and experimented for specific domains. Hence, models that can deal with diverse data in different domains in various settings are still needed.
- **Real-time detection**: The computational cost and complexity of recent GAN and transformer-based models are very high and hence limit real-time detection. Further research to balance the computational cost with detection efficiency is very important for several applications.
- **Interpretability and explainability**: The advanced deep learning models sometimes fail due to lack of interpretability. The reason behind why a model identifies certain data points as anomalous is important for understanding the insights, especially in domains such as healthcare and security. This leads to an opportunity of further research in enhancing model transparency and model explanations for the decisions it makes.
- **Integration of multimodal data**: Anomaly detection in environments that generate data in different modes (images, video, audio, sensor, etc.) is still very much an unexplored domain.
- **Scalability and efficiency**: As we advance, we generate more and more data; hence, the scalability of the model becomes

Table 15.1 Comparative analysis of existing solutions.

Literature	Summary	Algorithms used	Pros/cons
[5]	AnoGAN, a deep convolutional generative adversarial network, suggested an anomaly detection using unsupervised learning in images.	Novel anomaly scoring scheme based on mapping from image space to latent space	Pros: Unsupervised learning to identify anomalies in imaging data as candidates for markers Cons: Is not very accurate about predicting the anomalous images
[8]	He *et al.* trained deep neural networks using an approach called residual learning framework. They provided substantial evidence showing that these residual networks are easier to optimize and can gain accuracy from increased depth.	Residual learning	Pros: Residual networks are easier to optimize and can gain accuracy from increased depth. Cons: Computational costs

(*Continued*)

Table 15.1 Comparative analysis of existing solutions. (*Continued*)

Literature	Summary	Algorithms used	Pros/cons
[9]	D. Kwon *et al.* evaluated different deep neural network models for network anomaly detection. Deeper CNN structures do not improve performance to other models.	Fully connected network (FCN), variational autoencoder (VAE), sequence to sequence with long-short term memory (Seq2Seq-LSTM), convolutional neural networks (CNNs)	Pros: Clear understanding of existing anomaly detection models Cons: No new techniques explored here and no conclusive algorithm declared.
[10]	M. Sabokrou *et al.* suggested an efficient method for detecting and localizing anomalies using temporal data and FCNs. This architecture proposed in their work achieves high performance in terms of computational speed and accuracy.	Fully CNNs, cascaded detection algorithm	Pros: Efficient solution for detection and localization of anomalies

(*Continued*)

Table 15.1 Comparative analysis of existing solutions. (*Continued*)

Literature	Summary	Algorithms used	Pros/cons
[11]	G. Pang *et al.* did a comprehensive study of deep anomaly detection algorithms and present a taxonomy of the methods into 3 high-level and 11 fine-grained categories. It discusses the intuition behind the different approaches and the challenges in the deep anomaly detection.	RDA, AnoGAN, fast AnoGAN, EBGAN, ALAD, GANomaly, REPEN, E3Outlier, LSA, deep SVDD, deep SAD	Pros: Provides a taxonomy for different methods into categories. Covers wide range of algorithms making it easier for further research Cons: The learned feature representations can be biased. The objective function of data reconstruction is designed for data compression and not anomaly detection.

(*Continued*)

Table 15.1 Comparative analysis of existing solutions. (*Continued*)

Literature	Summary	Algorithms used	Pros/cons
[12]	Selvaraju *et al.* presented a novel technique called Grad-CAM for producing visual explanations for the decisions made by a CNN-based model. It uses the gradients of a target concept flowing into the final convolutional layer to produce localization map highlighting important regions in the image.	Gradient-weighted class activation mapping (novel proposed)	Pros: It makes CNN models more transparent and explainable providing justification for the decisions made by the model. It is applicable for a wide variety of models. It outperformed existing solutions in terms of interpretability and faithfulness. It identifies biases in the datasets. Cons: No clear implementation with multimodal inputs. The counterfactual explanations have limited power.

(*Continued*)

Table 15.1 Comparative analysis of existing solutions. (*Continued*)

Literature	Summary	Algorithms used	Pros/cons
[13]	Raghavendra *et al.* provided a comprehensive overview of the research methods in deep learning–based anomaly detection across different industries and domains. It categorizes the current anomaly detection methods based on the underlying assumptions and usage. It discusses variations, advantages, and limitations of the methods and outlines open research issues in adopting these detection methodologies.	CNN, LSTM, AE, SVM, SCDD, GAN, SPN, etc.	Pros: Provides a structured and contextual understanding of different approaches Cons: No specific implementation details are provided. Drawbacks of adopting the techniques in real-world scenarios are not clearly mentioned.

(*Continued*)

Table 15.1 Comparative analysis of existing solutions. (*Continued*)

Literature	Summary	Algorithms used	Pros/cons
[14]	Saikiran *et al.* surveyed the various anomaly detection techniques in deep learning applications. It provides a better understanding of the techniques and highlights unsolved research challenges in applying these techniques in DL systems.	Out-of-distribution (OOD) and adversarial algorithms in DL systems. Deep autoencoders combined with CNNs and RNNs	Pros: Saikiran *et al.* highlighted their relative strengths and weaknesses. It also discusscs the application of anomaly detection techniques in various domains, such as Android malware detection, fake news detection, and intrusion detection. Cons: Saikiran *et al.* acknowledge that current anomaly detection techniques may not be effective against complex and correlated anomalies.

(*Continued*)

Table 15.1 Comparative analysis of existing solutions. (*Continued*)

Literature	Summary	Algorithms used	Pros/cons
[15]	Mohammad *et al.* suggested an end-to-end architecture for one-class classification and novelty detection in visual media. It proposes a model with two deep learning networks—one reconstructor and one discriminator, which are trained adversarially. The proposed approach achieves state-of-the-art performance in anomaly detection in videos and images without any samples from novelty class during training.	Two deep networks in supervised and semisupervised setting	Pros: The presented system is capable of detecting samples not belonging to target class Cons: Computationally expensive and requires large amount of training data to achieve optimal performance

(*Continued*)

Table 15.1 Comparative analysis of existing solutions. (*Continued*)

Literature	Summary	Algorithms used	Pros/cons
[16]	Han *et al.* recommended SAGAN for image generation tasks, allowing attention-driven, long-range dependency modeling. It takes cues from all feature locations and ensures consistency. Spectral normalization is used in generator for higher quality images.	SAGAN (proposed novel approach) spectral normalization, two time-scale update rule	Pros: Allows attention-driven long-range dependency for image generation. Generates high resolution pictures

(*Continued*)

Table 15.1 Comparative analysis of existing solutions. (*Continued*)

Literature	Summary	Algorithms used	Pros/cons
[17]	Weimin *et al.* presented a semisupervised anomaly detection technique based on deep generative models with transformers for classifying abnormal and normal images. It combines autoencoders and GANs and extracts latent representations using encoder–decoder structured generator while discriminator uses transformer to distinguish the images. The model minimizes the gap between original image space and latent space during training.	Autoencoder, GAN, transformers	Pros: Combines strengths of AE, GANs and transformers and can capture long-range dependencies. It is versatile as it can deal with both labeled and unlabeled data. Cons: The performance depends highly on quality of the data used for training. The computational complexity is a strain on the resources and time of the system.

(*Continued*)

Table 15.1 Comparative analysis of existing solutions. (*Continued*)

Literature	Summary	Algorithms used	Pros/cons
[18]	Lukas *et al.* specified a deep SAD, an end-to-end semisupervised anomaly detection algorithm. It uses information-theoretic framework, which is based on the assumption that entropy of latent distribution of normal data is much less than that of anomalous images. Extensive experimentation with different datasets demonstrates the deep SAD performs better than hybrid and deep competitors.	Novel deep methodology for general semisupervised anomaly detection	Pros: Promising results on large and complex datasets. Can utilize labeled samples of both data Cons: Complex and resource exhausting

(*Continued*)

Table 15.1 Comparative analysis of existing solutions. (*Continued*)

Literature	Summary	Algorithms used	Pros/cons
[19]	Yang *et al.* quantified a transformer-based architecture for anomaly detection in images. The method combines the advantages of self-attention mechanism and modified skip-connections to capture both local and global information. It outperforms CNN based models.	Transformer-based models with self-attention mechanism and modified skip connections	Pros: Captures both local and global information. Mitigates limited repetitive field issue Cons: May not always converge on certain datasets due to instability of GANs.

of high importance. Research into such methods that can process large-scale data without compromises in its efficiency and accuracy is very important.

- **Data privacy and security**: Anomaly detection involves several kinds of data, and often, they might be sensitive; hence, there is a need of methods that can ensure that the data are secured, and the privacy is not compromised.

15.3 Methodology

15.3.1 System Overview

As a pilot study, TransGAN is a transformer-based GAN model that has no convolutions at all. TransGAN is primarily composed of a transformer-based generator that reduces feature resolution over time and is memory-friendly as well as a transformer-based patch-level discriminator [3]. In the

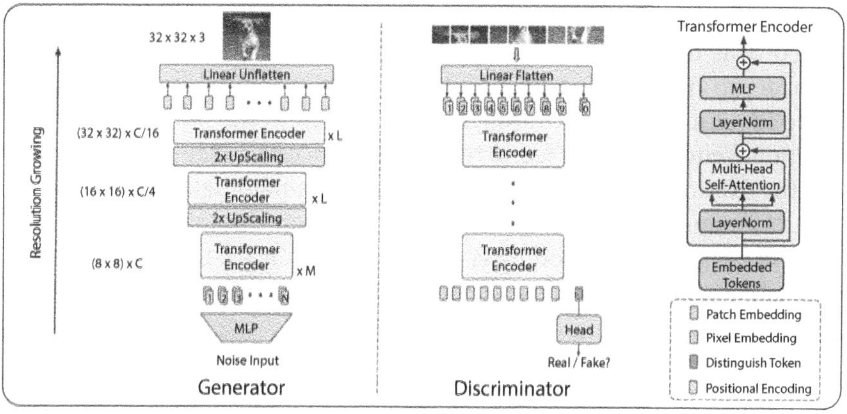

Figure 15.1 Pipeline architecture of TransGAN.

original work, a series of techniques are combined such as data augmentation, modified normalization, and relative position encoding, all of which helps to address the instability in the training of GAN. In this research, we implement data augmentation [22] and relative position encoding. We train our model on CIFAR-10 and MVTecAD dataset and use the trained TransGAN in two different approaches to solve the problem of anomaly detection. The architecture pipeline of TransGAN is shown as specified in Figure 15.1 [3].

For MVTecAD dataset, we train TransGAN for anomaly detection using an approach called AnoGAN [5]. After training the model on MVTecAD data subset, we get a generator capable of generating fake images and a discriminator that tries its best to separate real from fake images. Now, we make use of latent space where the trained images are represented in, to find anomalies. Given a new image x, we are going to compute the representation the image in the TransGAN latent space, and if it is on the zone of normal images, we label it as normal, anomaly otherwise. For CIFAR-10 dataset, we explore the anomaly detection architecture that takes its starting point as suggested by Kim *et al.* [23]. Because CIFAR-10 is an imbalanced dataset, we make use of two discriminators and different losses as discussed and the training pipeline shown in Figure 15.2 in the next subsections.

15.3.2 TransGAN Model

In the TransGAN model, transformer encoder [24] is the basic block of the model. The encoder is made up of multihead self-attention module and the feed-forward multiple-layer perceptron (MLP) with GELU nonlinearity. For this model, we are using layer normalization [25]. The generator is made up of multiple stages with transformer blocks. It takes random noise input and passes it through the MLP layer. The output along with positional embeddings is fed to transformer encoders recursively to calculate the relation between the tokens. The discriminator takes the patches of images as input, which are converted to 1D sequence of token embeddings. Addition of position encoding is followed by feeding it to the transformer encoder. The classification head of the discriminator then outputs the real/fake prediction.

15.3.3 Walking through Latent Space for Anomaly Detection

For each class of MVTecAD dataset, we are provided with a train test and ground-truth (labels) folder. We first train our TransGAN model on non-defective images of any class and test the model on defect classes as well. So, we train the GAN by providing random noise to the generator, which generates a fake image from it, and we use this nondefective real images and fake generated images as input to discriminator to train that. We make use of reconstruction and adversarial loss of the discriminator to train the GAN to perform better. Once we have our trained GAN, we try to find for a given image "img" the closest representation of it in the latent space by picking a random point in the latent space and use gradient descent to walk through the space for the purpose. The loss function that we use here is the L1 loss between the generated and given input image given Equation (15.1):

$$L_R(z_\gamma) = \sum | x - G(z_\gamma) | \tag{15.1}$$

We also compare the representation of real and fake images in the feature space of the discriminator. Because each layer of the discriminator while training has captured a lot of features, we use it to find the perceptual difference between the real and fake images that the discriminator uses to distinguish them. AnoGAN [5] uses L1 loss as specified in Equation (15.2):

$$L_D(z_\gamma) = \sum | f(x) - f(G(z_\gamma)) | \tag{15.2}$$

where $f(.)$ represents the intermediate layers of the discriminators. For convenience in this research, we consider only the first few layers of the discriminator for the calculation of the discriminator loss.

Finally, this research balances both the losses to find the overall loss and use it to find the latent representation that minimizes the loss. From this, we get the closest real image that we can generate. Now we test it with defects and see the reconstructed images are not that great because they do not fall in the manifold of normal image representations. We use the losses of images with which we trained our model as threshold, and if these losses are more for test image, then it is labeled anomaly, normal otherwise.

15.3.4 Anomaly Detection in Imbalanced Dataset

CIFAR-10 dataset is an imbalanced dataset for anomaly detection. Hence, we need to deal with the problem as discussed in Table 15.1 above. Because we divide the dataset originally consisting of 10 classes into 2 classes as normal and abnormal by setting one class as normal and others as abnormal, the dataset is clearly unbalanced. So, we use k-means sampling to get the number of samples for anomaly class while subsampling.

The two discriminators are used only in the training, both of which are completely identical. They are used to treat anomalous and normal data separately. The generator used takes noise as input and gives image as output, but for this purpose, we include a feature mapper module before the generator so that the input image is mapped to the latent vector that can be accepted by the generator as the input. In the feature mapping, we use a pretrained ResNet50 as our feature extractor and use linear layers with RELU for mapping the extracted feature to the latent space of the generator. The training pipeline of the model is shown in Figure 15.2.

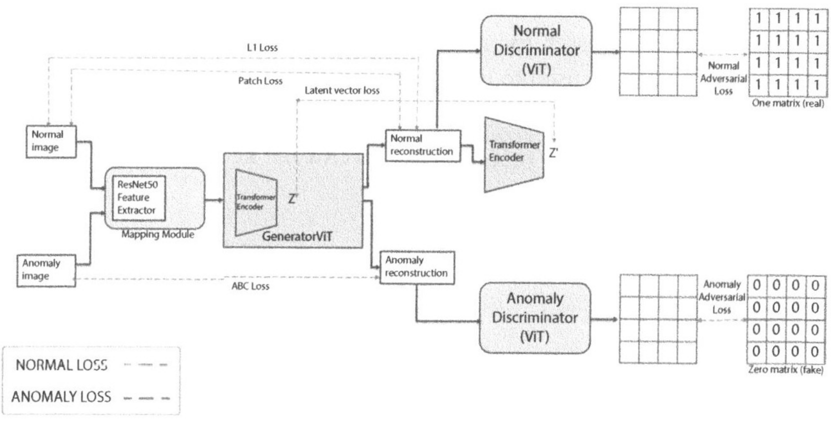

Figure 15.2 Training pipeline for anomaly detection in imbalanced dataset.

For the training, we use a weighted combination of several different losses to update the weights, separately for normal and anomalous images. The losses that we use here include six losses for the generator and one for each of the discriminator (Table 15.2). For each discriminator, we make use of loss function from LSGAN [27], where a and b are labels of fake and real data shown in Equation (15.3).

$$\min_{D} V_{\text{LSGAN}}(D) = [(D(x) - b)^2] + [(D(G(x)) - a)^2] \tag{15.3}$$

The six losses that we use on generator include the following:

L1 loss [Equation (15.4)] where x is the input patch loss and $G(x)$ is the generated image from the generator:

$$L_{recon} = \|\, x - G(x) \,\|_1 \tag{15.4}$$

The patch loss [Equation (15.5)] where we divide both the images into patches and select the average of the biggest reconstruction errors:

$$L_{patch} = f_{avg}(n)(\|\, x_{patch(i)} - G(x_{patch(i)}) \,\|_1),\, i = 1,\ldots,m \tag{15.5}$$

The latent vector loss [Equation (15.6)] [26] is calculated between real and generated image after passing them through the encoder.

$$L_{\text{enc}} = \|\, G_E(x) - G_E(G(x)) \,\|_1 \tag{15.6}$$

The adversarial losses for normal and anomalous images [Equations (15.7) and (15.8)] are used to update the generator:

$$\min_{G} V_{\text{LSGAN}}(G) = [(D(G(x)) - 1)^2] \tag{15.7}$$

$$\min_{G} V_{\text{LSGAN}}(G) = [(D(G(x)) - 0)^2] \tag{15.8}$$

Table 15.2 Imbalance challenges and solutions.

Challenge	Approach
Unbalanced class	K-means sampling
Loss scale imbalance	Loss function weight search
Discriminator bias	Two discriminators
Size imbalance between defect and object	Reconstruction-based methods

And the ABC loss [Equation (15.9)] is used to maximize L1 loss for anomalous images, which is reconstruction error modified by adding the exponential and log function to solve the imbalance.

$$L_{\text{ABC}} = -\log(1 - e^{-L_\theta(x_i)}) \tag{15.9}$$

15.4 Experimentation and Results

In this section, we perform substantial experiments to evaluate our model in a way of leave-one-class-out anomaly detection on commonly used datasets—CIFAR-10 and MVTecAD—using two different approaches. We log the different losses and try to present the effects of those losses by plotting them in different graphs to understand how they affect the training and anomaly detection.

15.4.1 Datasets

Datasets used for the experimentation include the following:

MVTecAD: MVTecAD is one of the most popular benchmark anomaly detection datasets that focus on different industrial inspections. It contains 5000 images of 15 different objects. We select the "hazelnut" class for our experiment and train on the "good" subset of the images and perform anomaly detection using latent walk on all hazelnut images.
CIFAR-10: CIFAR-10 is another widely used dataset that comprises color images of size 32 × 32 pixels from 10 classes. Each class has 5000 images. We select one class of images as anomaly and other images as normal data.

Then we train our model on both normal and anomalous data using separate discriminators to deal with each type and then perform anomaly detection on them using only the generator.

15.4.2 Training Details

Our experiments are performed on GPU T4x2 with 15 GB memory with PyTorch framework on Google Collab and Kaggle platforms. In the training process, we use Adam optimizers with momentum $b1 = 0.001$ and $b2 = 0.99$. We experiment with learning rate of 0.0001 for both generator and discriminator for different batch sizes for each of the dataset.

15.4.2.1 MVTec-AD Dataset

We use the hyperparameters as shown in Table 15.3 for the training of MVTecAD dataset training on hazelnut and try to visualize the generation of fake images from the TransGAN.

We see in Figure 15.3 as epochs increase the images that are generated transform from just looking like a noise to more closely like the hazelnut images. Due to complexity of our model and hardware limitations, we

Table 15.3 Hyperparameters for training of MVTec-AD anomaly detection.

Image size	Epochs	Batch size	Learning rate	Latent dim	No. of iterations
64	350	8	0.0001	1024	1000

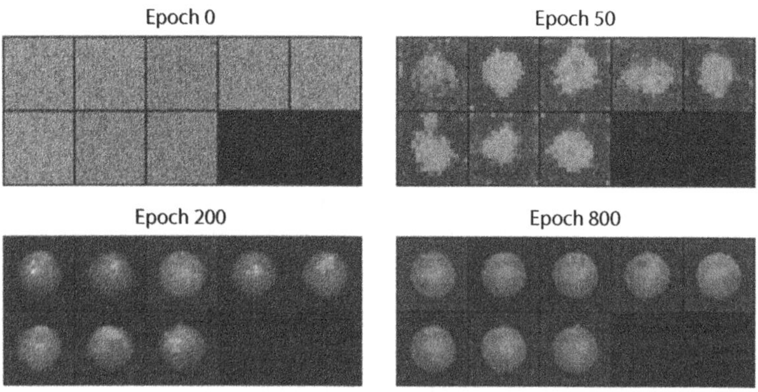

Figure 15.3 Generated images of hazelnut dataset.

observe that the images are no more noise but are not producing the most realistic fake images. We think this only goes on to present further need of research and improvements that are still underway to make this model production grade. Once we had finished training our model on our dataset, we make use of latent space representation of the generator to find the image that looks closest to a given real image. We do that by fixing a real image and take random noise vector z.

Then we make use of L1 loss as discussed earlier to see the difference between $G(z)$, the generated image from the noise vector z to the generator, and the real image. We use Adam optimizer to optimize z with different learning rates (lr) for different number of iterations, and for lr = 2e−2 and 1500 epochs, we see the loss decreasing as seen in Figure 15.4, even though it is not the way we expected it to be. There could be several reasons for that including learning rate, some noisy generated images, and the fact that, due to hardware restrictions, we use small batch size, which restricts the model from learning the gradients better.

With this, we also include the perception loss from discriminator and experiment with different weights to balance the combined losses. Now we perform the latent walk on a test real image with defects to generate an image. We see that the generated image is not very great, and the reconstruction is very poor. We explore this high reconstruction and discriminator loss and identify anomalous regions in images. We compute a "residual image"—the difference between the input test image and reconstructed image—and areas of the residual images with large values should correspond to anomalous regions, because the original and reconstructed

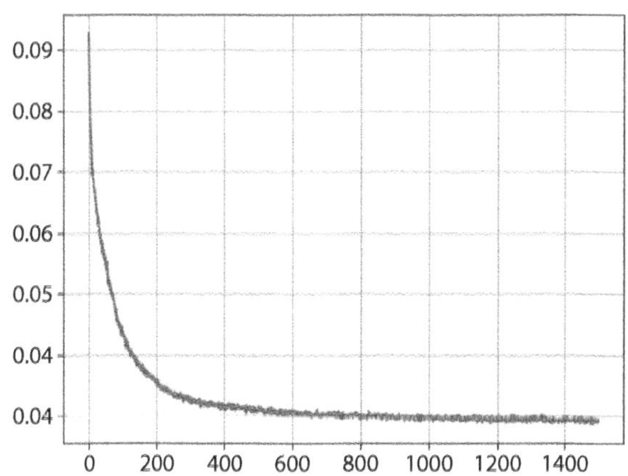

Figure 15.4 Z-loss optimization graph.

images are maximally different in these regions. We then plot these images side by side, as you can see in Figure 15.5.

We now run this for our training dataset and try to fix a residual and perception threshold to find the anomalies in the images of the test dataset that includes normal hazelnuts, as well as ones with the defects. We plot the residual loss and discriminator loss of our data as shown in Figure 15.6, but unfortunately, we see a medium to high reconstruction loss for our normal images as well, which should not have been. But we think this is because of the complexity and hardware requirements and can be improved with better hardware. However, we do see a pattern with our discriminator—the images with defects have a high discriminator loss.

We take this inference and use discriminator loss for anomaly detection. We classified images into anomaly class if their reconstruction and discriminator loss exceeded their set threshold. We set the discriminator threshold to be 1.5 and reconstruction to be 0.5. When we run this on our test samples of "good" data, we get an accuracy of 0.9 and for defects an accuracy of 0.6. When we compare this to Bergmann et al, the accuracy obtained by AnoGAN for hazelnut data for good images was 0.83 and for anomalous 0.16. The accuracy obtained for each of those classes for good and anomalous images has been given in Table 15.4.

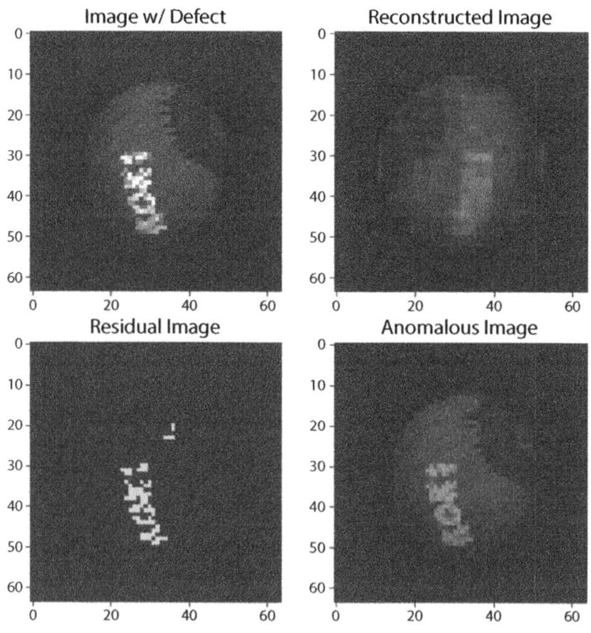

Figure 15.5 Anomaly reconstruction and highlighting of hazelnut.

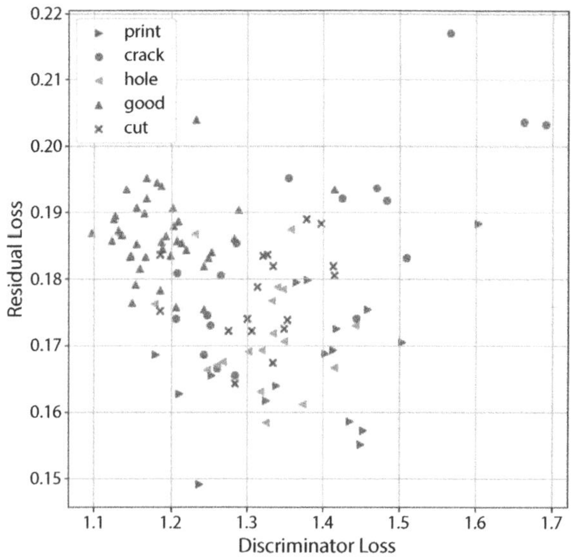

Figure 15.6 Residual and discriminator loss of test images of "hazelnut".

15.4.2.2 For CIFAR-10 Dataset

We use a hyperparameter as shown in Table 15.5 for the CIFAR-10 dataset training on the TransGAN model.

Just like on the MVTecAD dataset, we perform several transformations on the CIFAR-10 dataset when we load the dataset from PyTorch. We train the dataset, and with each epochs, we can see in Figure 15.7 the images getting better.

Now when we give a random noise, we can clearly see that this model gives images in the batch that resemble the real images of the CIFAR-10 dataset. But for the approach we are experimenting with this dataset, we want to pass real image as input to the generator. So for that, we make use of state-of-the-art pretrained ResNet50 model to extract features from our image and use a mapping function that takes output of the feature extractor and use linear function to map it to the generator's latent space. We try to optimize this mapping of image to the latent space using the reconstruction loss, and as we can see in Figure 15.8, the loss decreases. Due to a high number of images and the complexity of the model, we train this for only three epochs and use it for anomaly detection algorithm. We now use the mapping module along with the TransGAN and provide images as the input. We make use of two exactly same discriminators. But this time,

Table 15.4 Accuracy of TransGAN in anomaly detection for MVTec-AD dataset.

	Accuracy	
Class	**Good images**	**Anomalous images**
Bottle	0.91	0.5
Cable	0.95	0.41
Capsule	0.91	0.52
Carpet	0.8	0.4
Grid	0.9	0.55
Hazelnut	0.9	0.6
Leather	0.91	0.34
Metal nut	0.86	0.44
Pill	0.95	0.47
Screw	0.5	0.5
Tile	0.95	0.33
Toothbrush	1.00	0.31
Transistor	0.95	0.65
Wood	0.92	0.55
Zipper	0.8	0.45

Table 15.5 Hyperparameters for training of MVTec-AD anomaly detection.

Image size	**Epochs**	**Batch size**	**Learning rate**	**Latent dim**
32	10	1	0.0001	1024

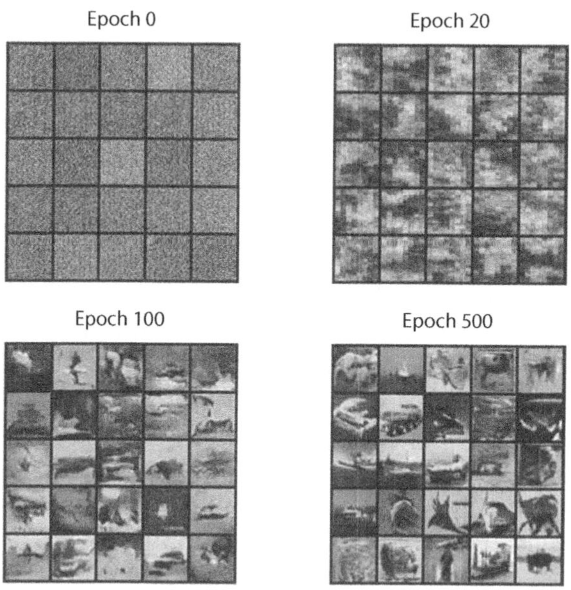

Figure 15.7 CIFAR-10 model after training.

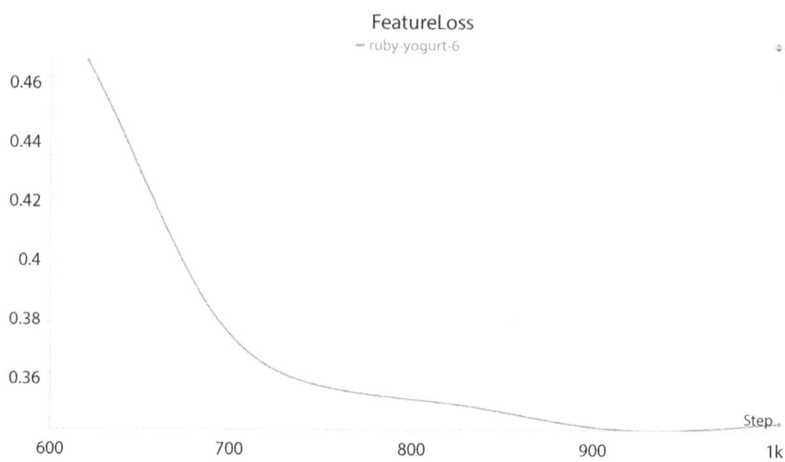

Figure 15.8 L1 loss during feature extraction training.

we train the model to train the images differently for anomalous and normal images.

If the label of the image is anomalous, we send the reconstructed image to anomaly detector and calculate ABC loss; otherwise, to normal detector and calculate reconstruction loss, latent loss. We also calculate patch loss between input image and reconstructed image and the adversarial losses for anomalous and normal images. All these losses have been clearly defined in Equations (15.1) to (15.7), as previously discussed. We run the model for seven epochs with Adam optimizer and batch size of 1 with a threshold of 0.5. The model performances are influenced by weights and will be variable based on the class selected as normal class. In order to evaluate the model, we chose eight as the normal class and achieved an area under the receiver operating characteristic (AUROC) score of 0.83 as shown in Figure 15.9.

Table 15.6 gives the AUROC score obtained for each class of the dataset in this experiment.

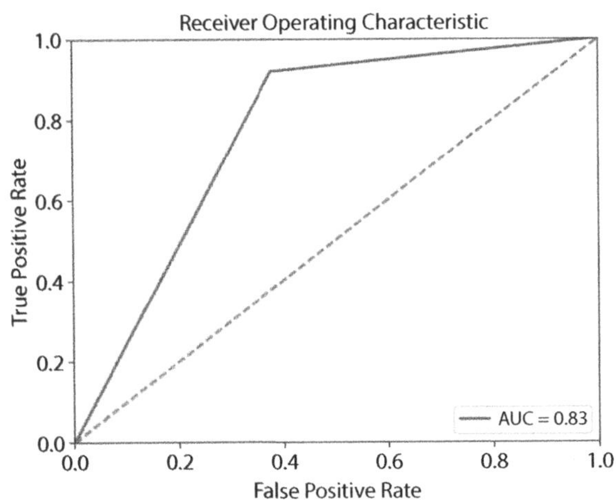

Figure 15.9 AUROC score when normal class = 8.

Table 15.6 AUROC score for the model on CIFAR-10 dataset.

Class	**0**	**1**	**2**	**3**	**4**	**5**	**6**	**7**	**8**	**9**	**Average**
Score	0.81	0.82	0.74	0.72	0.76	0.75	0.82	0.8	0.83	0.81	**0.79**

15.5 Conclusion and Future Work

As part of our experimentation into anomaly detection techniques, we set out to utilize the capabilities of the TransGAN model, a state-of-the-art transformer-based GAN with a reputation for producing lifelike images from noise input. There were two primary stages to our project both focused on clarifying TransGANs capabilities in anomaly detection tasks on various datasets. Using the MVTecAD and CIFAR-10 datasets, we aimed to advance anomaly detection techniques by incorporating cutting-edge generative models into our approaches.

With its intricate architecture, TransGAN offered both possibilities and difficulties. Although we were excited by its potential to generate realistic images, there were substantial obstacles due to the inherent complexity and resource requirements of the training process. Our preliminary tests demonstrated that careful training procedure optimization is necessary to reduce computational overhead and hasten convergence. Some promising directions for further research and development include the application of techniques such as transfer learning and progressive training, which may help improve model performance and expedite training.

The importance of discriminator architecture in anomaly detection was one of the most important lessons learned from our TransGAN experiments. Our understanding of the complexities involved in differentiating between normal and anomalous cases deepened our understanding of the significance of creating discriminators that can pick up on minute details that point to anomalies. It has been suggested that one way to improve the discriminative power and robustness of models is to incorporate attention mechanisms or customize discriminators for particular anomaly detection tasks.

Experimental results on MVTecAD reveal that subtle anomalies are difficult to detect at low image resolutions (64*64), leading to misclassification caused by abrupt discriminator error variations. This underscores the importance of high-resolution inputs and robust hardware support for reliable anomaly detection. Sudden patterns within the discriminator mistakes ended in misclassifications exceptionally, while there were baby anomalies that have been invisible at 64 × 64 image resolution. This verified how important hardware substances and image resolution are to aberration detection tasks emphasizing the necessity for extra resilient hardware infrastructure and higher-resolution image statistics to get around these constraints.

The CIFAR-10 dataset presented similar difficulties in our investigation as we were only able to obtain an average AUROC score of 0.79, which was below the standards set by other anomaly detection algorithms. Our TransGAN implementation was a novel approach, but it also showed that, in order to achieve competitive performance, further research and refinement are needed. The limitations we experienced during our experiments highlighted the intrinsic complexity of anomaly detection tasks and the necessity for creative solutions to effectively address them.

The success of our project stems from investigating new methods for anomaly detection with cutting-edge generative models such as TransGAN. We have shown the versatility of TransGAN for anomaly detection tasks by testing the limits of anomaly detection techniques and experimenting with various datasets and training regimens. These achievements were, however, somewhat dampened by a number of difficulties and restrictions faced along the route.

Looking ahead, our work establishes the foundation for generative model–based anomaly detection advancements in the future. We can contribute to the larger field of machine learning and AI as well as further advance the state-of-the-art in anomaly detection by addressing the limitations that have been identified and building upon our achievements. To fully realize the potential of generative models in anomaly detection, we plan to conduct ongoing research and development to improve model architectures streamline training processes and assess performance on a variety of datasets.

To sum up, our project has shed important light on the difficulties and possibilities involved in applying TransGAN to anomaly detection tasks. Although we have made progress in proving the approaches viability and identifying possible areas for development, there are still major obstacles to overcome. The project has brought to light the need for additional study and development to improve training protocols to assess the performance across a variety of datasets and fine-tune the model architecture. Our work establishes the groundwork for future research in this field despite obstacles such as scarce computational resources and challenges in identifying subtle anomalies.

AI disclosure statement: The author confirms that no artificial intelligence (AI)-generated content or AI-assisted writing tools were used in the preparation of this chapter (including summaries and the abstract). All text, analysis, and conclusions are the result of the author's original work.

Bibliography

1. Ghaza, M., VVazquez, C., Amer, A., Real-time automatic detection of vandalism behavior in video sequences, in: *2007 IEEE International Conference on Systems, Man and Cybernetics*, Mar. 06, 2007.
2. Mould, N., Regens, J., Jensen, C., Edger, D., Video surveillance and counterterrorism: the application of suspicious activity recognition in visual surveillance systems to counterterrorism. *J. Policing Intell. Counter Terrorism*, 9, 2, 151–175, 2014.
3. Jiang, Y., Chang, S., Wang, Z., TransGAN: Two Transformers Can Make One Strong GAN, 2021.
4. Goodfellow, Pouget-Abadie, J., Mirza, M., Xu, B., Warde-Farley, D., Ozair, S., Courville, A., Bengio, Y., Generative adversarial nets, 2014.
5. Schlegl, T., Seeböck, P., Waldstein, S.M., Schmidt-Erfurth, U., Langs, G., Unsupervised Anomaly Detection with Generative Adversarial Networks to Guide Marker Discovery. *Lect. Notes Comput. Sci. (LNCS)*, 10265, pp. 146–157, Springer, Cham, Switzerland, 2017.
6. Bergmann, P., Fauser, M., Sattlegger, D., Steger, C., Mvtec ad–a comprehensive real-world dataset for unsupervised anomaly detection, pp. 9592–9600, Mar. 2019.
7. Krizhevsky, A., Learning Multiple Layers of Features from Tiny Images, 2009.
8. He, K., Zhang, X., Ren, S., Sun, J., Deep Residual Learning for Image Recognition, in: *2016 IEEE Conference on Computer Vision and Pattern Recognition (CVPR)*, Las Vegas, NV, USA, pp. 770–778, 2016, doi: 10.1109/CVPR.2016.90.
9. Kwon, D., Natarajan, K., Suh, S.C., Kim, H., Kim, J., An empirical study on network anomaly detection using convolutional neural networks, pp. 1595–1598, Mar. 2018.
10. Sabokrou, M., Fayyaz, M., Fathy, M., Moayed, Z., Klette, R., Deep-anomaly: Fully convolutional neural network for fast anomaly detection in crowded scenes. *Comput. Vis. Image Underst.*, 172, 88–97, Mar. 2018.
11. Pang, G., Shen, C., Cao, L., Hengel, A.V.D., Deep learning for anomaly detection: A review. *ACM Comput. Surv.*, 54, 2, 1–38, 2021.
12. Selvaraju, R.R., Cogswell, M., Das, A., Vedantam, R., Parikh, D., Batra, D., Gradcam: Visual explanations from deep networks *via* gradientbased localization, pp. 618–626, Mar. 2017.
13. Raghavendra, C. and Sanjay, C., Deep Learning for Anomaly Detection: A Survey, Mar. 2019.
14. Saikiran, B., Bhavya, K., Bo, L., Pramod, K., Varshney, Dawn, S., Anomalous Example Detection in Deep Learning: A Survey, Mar. 2020. doi: 10.1109/ACCESS.2020.3010274.

15. Mohammad, S., Mohammad, K., Mahmood, F., Ehsan, A., Adversarially Learned One-Class Classifier for Novelty Detection, Mar. 2018. doi: 10.1109/CVPR.2018.00356.
16. Han, Z., Ian, G., Dimitris, N., Metaxas, A., Odena, Self-Attention Generative Adversarial Networks, Mar. 2018.
17. Shangguan, W., Chen, Z., Fang, W., Semi-Supervised Anomaly Detection Based on Deep Generative Models with Transformers. *Proc. ACM Int. Conf. Multimed. (ACM MM)*, Association for Computing Machinery, New York, USA, pp. 1–9, 2022. doi: 10.1145/3529466.3529470.
18. Lukas, R., Deep Semi-Supervised Anomaly Detection, Mar. 2020.
19. Yang, C., A Transformer-Based GAN for Anomaly Detection, pp. 345–357, Mar. 2022.
20. Lin, Z., Yang, D., Fuyouand, H., Chunlin, F., Anomaly Detection of GAN Industrial Image Based on Attention Feature Fusion. *Sensors*, 23, 1, 355–355, Mar. 2022. doi: 10.3390/s23010355.
21. Xinyang, F., Dongjin, S., Chen, Y., Zhengzhang, C., Jingchao, N., Haifeng, C., Convolutional Transformer based Dual Discriminator Generative Adversarial Networks for Video Anomaly Detection, Mar. 2021. doi: 10.1145/3474085.3475693.
22. Kolesnikov, An Image is Worth 16x16 Words: Transformers for Image Recognition at Scale, Mar. 2021.
23. Kim, J., Jeong, K., Choi, H., Seo, K., GAN-Based Anomaly Detection In Imbalance Problems, pp. 128–145, Jan. 2020.
24. Vaswani, A., *et al.*, Attention is all you need, Mar. 2017.
25. Ba, J.L., Kiros, J.R., Hinton, G.E., Layer Normalization, Jul. 2016.
26. Akcay, S., Atapour-Abarghouei, A., Breckon, T.P., Ganomaly: Semi-supervised anomaly detection *via* adversarial training, pp. 622–637, Mar. 2018.
27. Mao, X., Wang, Y., Liu, X., Guo, Y., An adaptive weighted least square support vector regression for hysteresis in piezoelectric actuators. *Sens. Actuators A*, *263*, 423–429, 2017.
28. Ruff, L., Vandermeulen, R., Goernitz, N., Deecke, L., Siddiqui, S.A., Binder, A., Müller, E., Kloft, M., Deep One-Class Classification, *Proc. 35th Int. Conf. Mach. Learn. (ICML)*, PMLR, Stockholm, Sweden, 80, pp. 4393–4402, 2018.

16

Recent Development of Machine Learning Models for Grape Plant Disease Detection: A Review

M. Anuradha*, G. Revathy and M. A. Mohamed Aslam

Department of CSE, SRC, SASTRA Deemed University, Kumbakonam, India

Abstract

The proposed survey article is an attempt to give a comprehensive overview of current state-of-the-art ML and DL models developed for disease detection in grape leaves. Its primary goal is to investigate the efficacy, precision, and real-world challenges of using transfer learning techniques, recurrent neural networks (RNNs), and convolutional neural networks (CNNs) in farming. The paper will conduct a thorough analysis of prevalent datasets commonly used for training these models, along with the preprocessing steps necessary to ensure optimal performance. Additionally, it will explore the various evaluation metrics employed to measure the success of these models in disease detection tasks. Special attention is given to elaborating the practical applications of these ML and DL models in precision agriculture. This includes highlighting their pivotal role in facilitating early disease detection, mitigating economic losses for farmers, and promoting sustainable farming practices. The objective is to be an invaluable source for everyone working in agricultural technology and plant disease management by providing information about current methods and discussing possible future paths. Its ultimate goal is to help researchers progress the discipline and promote food security globally by equipping them with the information they need to make educated decisions about how to incorporate these novel solutions into agricultural operations.

Keywords: Apple plant disease, grape plant disease, ML, DL, crop prediction, sustainability

**Corresponding author*: anuradhasen83@gmail.com

T. Saravanan, P. Preethi, Sumaya Sanober, N. Thillaiarasu and S. Balamurugan (eds.) Artificial General Intelligence: Principles and Practices, (327–352) © 2026 Scrivener Publishing LLC

16.1 Introduction

Agriculture has always supported human civilizations since time immemorial, and with our human growing exponentially in numbers, food production has become increasingly urgent. This is why in this scenario it is extremely vital to ensure our crops are healthy and can bring profit. Plant diseases have remained one of the critical challenges and also one of the most studied elements in enhancing agricultural yield over time. As a result, they reduce yield, but also damage the quality of the yield, affecting both income and food security. Grape grows disease/infestation in viticulture true wine grape. Different pathogens such as fungi, bacteria, and viruses attack on grapevine and spread it quickly through crop, which lead to the poor quality of crop. The major diseases having visual symptoms on foliage are downy mildew, powdery mildew, black rot, and leaf blight; accordingly, leaf analysis is important for the diagnosis and management of these diseases. Traditionally, diagnosis of plant disease was based on visual evaluation with the assistance of agricultural experts/plant pathologists. However, this is time-consuming, prone to mistakes, and (for large-scale farming) unscalable to do by hand. Also, the first few disease symptoms may be developing in subclinical level, which escapes human eye and is therefore unresponsive delay in dissemination of crop loss [4]. These have been mixed with agriculture called smart farming/precision agriculture, and hence more precise, timely, and efficient disease monitoring systems have been created.

Devi *et al.* identified artificial intelligence as one of the most transformative advancements in recent agricultural technology, complementing earlier works not discussed [5]. One of those focal points in 2023 has been the use of artificial intelligence (AI) and in particular machine learning (ML) and deep learning (DL).

AI-based approaches are revolutionizing every domain, and modeling in agriculture is fully underway to get benefits from it. With the potential to analyze huge dataset, discover intricate patterns, and provide accurate prediction, therefore, the growing interest in applying AI for analyzing and predicting plant diseases is not yet fully understood or systematically explored. Using image analysis, time-series data interpretation, and monitoring over time, AI has quickly become one of the most prominent tools for modern agriculture. Despite various attempts in training of ML models for detection of symptoms from images of diseased leaves, the dimensions of DL techniques as in convolutional neural networks (CNNs) and recurrent neural networks (RNNs) for detection of symptoms in high speed have not been explored further.

The detection system, powered by AI, is scalable and continuously updatable. In contrast, classical diagnostic approaches require skilled professionals for diagnosis and laboratory-based tests, but sentiment analysis offers a scalable, real-time plant health monitoring solution through mobile applications, drones, and Internet of Things (IoT) devices [5]. This is particularly important for farmers in rural or resource-limited conditions, which might not have access to expert diagnostics. Farmers with minimal training and effort just need to click some pictures of the leaves with smartphone cameras to produce real-time information related to the exact diseases, which may be present. On the other hand, AI models improve with large datasets, making them increasingly precise and resilient over time.

As the AI in agriculture domain matures, it is time to assess the ways and models created. Understanding their performance, limitations, and real-world applicability would serve both researchers and practitioners to make pragmatic choices on their use. Hence, this chapter tries to present a comprehensive, state-of-the-art summary on the ML and DL models governing the disease detection in grape leaves, besides just synthesizing the algorithms themselves, the whole pipeline in the disease detection process, from data acquisition, preprocessing, and model training/evaluation to deployment quote Al-Hiary *et al.* [1].

Detection of plant disease shown in Figure 16.1 is the essential part of agricultural field.

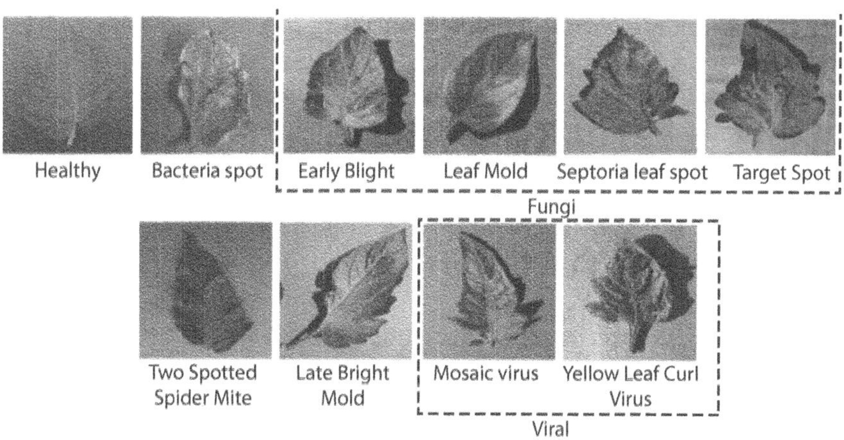

Figure 16.1 Sample images for leaf disease.

Pests and diseases are one of the major causes of crop loss worldwide, and the Food and Agriculture Organization (FAO) estimates that pests and diseases continued to withhold 40% of global crop production annually [7]. Diseases can cause significant economic losses to the grape industry. Grapes are more often used as raw table fruit but also are fermented to wine, are dried to make raisin, or can be processed to produce juice and vinegar. Any loss in quality or quantity of grape harvest, even as small as a few tons, can upset supply chains and generates low profitability for growers and distributors. Thus, preventing epidemics, minimizing damage, and protecting agricultural investments are essential to determine diseases so early and accurately [3].

The swift identification of diseases has a positive effect on rapid intervention, which does not only substantially lessen the dispersal of pathogens but also helps to cut down on the use of chemical control. A journey into excess pesticide and fungicide use arises [4], which may have an environmental impact and will also create new races of resistant pathogens, usually by a slow tissue diagnosis of the disease. Required targeted, efficient, and less-polluting use of agrochemicals at lower costs, ensuring minimal environmental impact and improved sustainability.

Accurate diagnosis is critical to implement integrated pest management strategies based on biological, cultural, mechanical, and chemical practices to control these diseases. By accurately diagnosing the type and extent of a disease, farmers have the opportunity to choose the appropriate intervention rather than using a blanket but ineffective and unnecessary treatment. This not only helps in healthy crop boosting but also helps in maintaining biodiversity ecological conservation and food security.

In addition to the obvious agricultural benefits, plant disease detection can be relevant for work and policy. Collecting data through disease monitoring systems is critical to understanding the burden of disease, its transmission, and the impact of control measures. This calls for robust pathogen detection strategies at the farm scale and throughout the agricultural landscape to ensure food safety.

16.2 Role of AI in Modern Farming

Agriculture is no exception to this paradigm, and AI has also been proposed as a popular approach to a variety of complex problems across disciplines. AI in agriculture is not just a shift in the farming practice; rather, a giant leap in a process that has been around for as long as the human race has drawn breadth, from using intuition to help decide here and there a humble, now complemented by the AI data-driven decision-making approach. There is no shortage of data generated on the modern farm through the use of sensors, drones, satellite imagery, and farm machinery. The data can be that which AI algorithms can take a plunge in and earn their respect as pearls of wisdom that make resource utilization, risk prediction, etc.

Detection of plant diseases using ML and DL models is one of the easy-to-implement but highest-impact applications of AI in agriculture. We provide data to train algorithms to learn patterns and make predictions against unseen data in ML. In particular, supervised learning finds its applications in disease classification where models learn over a labeled dataset containing healthy and diseased leaves. This allows us to train these models to a very high level of accuracy on classifying new image. DL is a form of ML that utilizes deep neural networks that automatically learn high-level features from the data inputted. As CNNs are very successful at image-related tasks, they are suitable for detecting visual disease symptoms on leaves.

Transfer learning, another AI method with agricultural applications, involves using pretrained models learned on large broad datasets (e.g., ImageNet) that are further tuned for narrower applications, such as disease detection, whereas this cuts down on the labeled data and computes power required to create a model. RNNs normally serve to sequential data, such as weather pattern, and sensor readings; however, it can be utilized for a disease prediction when combined with other data source.

AI within platforms can provide platforms based on systems by integrated real-time disease monitoring and promoting decision support. For example, mobile applications with AI models integrated into them allow the farmers to take the photo of the vines and get a diagnosis within a few seconds. Imaging for diseases—high-resolution drone-sourced imaging for diseases—drone-based surveillance technologies equipped with high-resolution cameras and AI-based algorithms as found in Air IMAGINE platform can be used to scan huge vineyard areas for disease signals. IoT devices such as environmental sensors can provide farmers with constant knowledge of their environment and warning of greater risk of diseases. These technologies enable proactive farming, which preempts crop issues with improved yield quality and reduced loss.

Additionally, AI implements crop management region-wise, which is a very important concept of precision agriculture. This will shrink the treated area, which saves a great deal of resources for farmers and contributes to the sustainable development of agriculture. Moreover, by incorporating climatic data, past records, and pathogen cycle models, AI has also been a great help in predicting disease outbreaks. The predictive capacity of the app assists farmers in forecasting and protective measures and therefore obtains even lower dependence upon the pesticides.

Although there are challenges of deploying AI in farming domain, such challenges consist of availability and quality of data, generalizability of the model, limitations of computation resource, and requirement of user-friendly interfaces, which are the major hurdles we have to address. Finally, we must reduce ethical issues such as data privacy, algorithm transparency, and the digital divide in order to ensure that there is equal access to the benefits of AI for all. But more importantly, AI research has a potential scope and scientific character and convergence that agricultural scientists, technologists, and a possibility for some time to come at least policy makers will also contribute to with better products and tools for farmers.

16.3 An Overview of ML and DL Industry in Agricultural Sector

We use technologies of the next generation of agriculture such as ML and DL. Agriculture has always relied on the labor, intuition, decades of knowledge, and wisdom of farmers. This began to change in the 20th century, which offered mechanization and green revolution, which also brought chemical fertilizers, pesticides, and genetically modified (GM). This directly made crop yields to increase, which changed agriculture worldwide. They did, however, raise questions of sustainability such as land degradation and increased dependency on agrochemicals. After digital technology, attention shifted yet again—and, for a long time at least—toward data-intensive agriculture with concentrating sensors, digitalization, and, in recent years, AI (Figure 16.1). ML and DL (subdomains of AI) are now transforming agronomy practice in ways that until a few years ago would have seemed little more than science fiction. Able to analyze datasets, discover trends as to make predictive decisions, it is driving a new era of precision agriculture.

Computing has been a part in agriculture since the time when statistical models for yield forecasting and weather prediction were developed. Both the earliest models were mostly rule-based and a very manual process of data entry and domain knowledge. Throughout the 1970s, 1980s, and 1990s, expert systems—computer programs that mimic each step taken by a human expert when solving a problem in a given area—grew much more sophisticated. Nonetheless, these systems barely scratched the surface of the agricultural AI potential and came with limited scope and flexibility issues.

By the turn of the 21st century, new types of agricultural data analysis techniques were becoming available, using ML algorithms such as decision trees (DTs), support vector machine (SVM), and the k-nearest neighbor (KNN) models for practical applications including crop classification, soil analysis, disease detection, etc. This reflected a move away from inflexible, rule-based systems to learning, evolving systems.

A major milestone was the beginning of DL around 2012, particularly with the achievements of CNNs in image recognition. So, CNNs are most commonly used in the agriculture domain too where images or any sort of visual data is required including, but not limited to, plant disease detection, weed detection, fruit counting, etc., due to their higher accuracy and generalization over others. It helped that there are those rather sophisticated sequential models such as RNNs and its more enhanced version, long short-term memories (LSTMs) that allow for modeling of time-sequenced data types (which also aligns well with our types of input subsidy, such as climate, irrigation, and crop stages). Complementary to that, these DL models are known to be data- and computation-intensive, but they also serve us the best results we have to offer for the more complex-pattern recognition type tasks.

ML and DL have been extensively used in precision agriculture, which is a farming management principle, based on the use of information technology, to optimize agricultural inputs so that the output can be improved. Precision agriculture is a multifaceted component that uses real-time data and predictive analytics to manage nearly all aspects of crop management, from soil health and irrigation management to pest control and harvesting. ML and DL models used upon multidimensional datasets that gather information on agricultural undertakings through satellite images, drones, IoT sensors, and/or mobile devices are thus at the heart of this methodology, for example, satellite data + ML algorithms to sense soil moisture and to predict drought, images from drones, and deep analytics based on CNN for detection of leaf disease symptoms invisible to human observers.

Plan for upcoming spreading diseases and healthy growth infection are the major benefits of ML and DL for PT. By identifying plant diseases, pest infestations, and other environmental stressors early, farmers can prevent crop losses. These kinds of early detection systems are essential in the battle against agricultural pandemics, which can destroy the entire regions of food supply. For instance, in grape growing, if we could detect the grapevine diseases (e.g., downy mildew or black rot) as soon as when the infections such as spores are released before spreading the area, it is possible to prevent the present or future massive damage of the resulting large-scale infection on vineyards. Additionally, ML models are also implemented for crop recommendations, which refer to recommending the crops that will grow in a certain place; for instance, ML models can recommend according to historical data, soil type, weather condition, and market demand.

Automate manual processes: This is another key advantage of the GSP. Farmers around the world are grappling with labor shortages and escalating operating expenses. ML-powered robots and autonomous machines can plant, weed, harvest, and monitor for disease at high rates of precision and consistency. Automation of such not only reduces dependence on human resources, but also increases efficiency while decreasing operational costs.

Soil sensor systems used in precision agriculture, for instance, can analyze the data and determine exactly the amount of water, fertilizer, or pesticide that is required in various regions of the field. Both ML and DL techniques optimize resource utilization by ensuring maximum efficiency and environmental sustainability through precise input management. This convenient method not only saves time in large number of crops, but also saves excessive time for inflection of resources and pollution around it and enhances the crop germination time. Moreover, it can also provide the information of predicting the yield output from the current and historical data; this assists the farmers in optimal allocation of their resources and planning the market strategy.

Here and there, the ML and DL utilization in horticulture has implicit future, yet then again, it is never without challenges. One of the greatest challenges is data availability and quality. We all have heard that ML models are as good as data that we trained them; this becomes the case with agriculture as well where usually we cannot get the good amount of large, diverse, and labeled datasets. Most of them are regional datasets, which may not generalize very well to other climate, soil, and crop type. Second, many of the data collection scenarios in agriculture are unstructured, e.g., image or sensor reading, which require many preprocessing and cleaning steps before it can be used for training the model.

Another challenge was set up from Singh *et al.* [2]. The interpretability of models such as CNNs is a disadvantage that needs to be addressed. Even if DL models predict with high accuracy, there is a black box with low transparency in decision-making process. But this gap in explainability poses a challenge in agriculture where autonomy may not be accepted by stakeholders unless they understand the reasoning behind recommendations, and explainable AI methods are still under active research so it is the type you can attribute it to if you get a chance to design interpretable models or add it into an explainable AI mechanism in source.

Lastly, although they are more robust than simpler models, variations in agricultural environments will still challenge model robustness. Weather, soil, pests, and human response differ as strongly as other agricultural conditions, making climate managers have little reason to expect any one model to well represent the outputs under such differing conditions. However, to some extent, transfer learning and domain adaptation techniques can alleviate the problem because they modify a pretrained model to a certain condition, but still, they need a great deal of domain knowledge and domain-specific data as well.

Together with the above challenges of scarce high-quality data, there exists a computational limitation, especially in low-resource/global south settings, which further increases the difficulty in adopting such AI technologies to agriculture by researchers and the farmer communities. Well, small-holder farmers or rural agricultural communities might not have easy access to the more powerful hardware needed to train and apply inference with DL models. Although mobile applications and cloud-based solutions serve as a potential solution, internet connectivity, data privacy, and technical literacy crises must be addressed to ensure equal access to and use of these technologies.

The AI systems will need also to be incorporated into the existing agricultural practices. Farmers in some developing areas may not have even come across a digital tool before or may not be happy with new technologies. This requires trust building, training, and support for successful adoption. The agricultural sector will take advantage of the AI in agriculture movement, but we need functional agricultural extension services as well as pro-poor and inclusive government policies to have a positive impact on the lives of marginalized and underserved communities.

However, we also have a regulatory and ethical challenge to deal with, not only a technical and social one. But questions abound, such as what role do existing airspace regulations on drone and remote sensing device flights play and who owns the data.

Who owns sensor data on farmland rented out? How are those data protected and utilized? Where data are in the position of being a gatekeeper, transparency, respect for privacy, and preventing misuse is a must and data fairness is needed, ethical behavior is required to lure tons of participants in agriculture.

Yet, irrespective of such issues, the future of ML and DL in agriculture is still evident. Contemporary architecture, data collection, and computational development are gradually rendering these limitations obsolete. For instance, federated learning trains models on a number of decentralized data sources without sending raw data between locations, providing privacy-preserving capabilities that can also allow collaboration across farms and regions. Edge computing is enabling the AI applications and inference on a real-time basis at remote places with the access to devices instead of cloud servers. Multimodal learning uses images, text, and sensor data together to improve the accuracy and robustness of the model.

The report also stated that there is a growing interest in convergence of AI with other few emerging technologies such as blockchain platform, in to secure and manage transparent data and IoT, to gain real-time data. These synergies are creating an adaptive connected and smart agrifood system that meets the challenges of climate change, population growth, and resource scarcities.

Grapevine (*Vitis vinifera*) is one of the most economically important cultivars worldwide, be it as fresh fruits for direct consumption or as a major input to wine industry. However, grape is highly vulnerable to a great diversity of biotic stresses, and leaf diseases are among the pathogens that can greatly affect grape yield, quality, and growth condition. Prompt and effective disease detection is crucial to managing these diseases, improving economic viability and conferring an economic and socioemotional buffer against viral outbreaks. Traditionally, disease was diagnosed manually by trained personnel. The process is manual, error-prone, and not possible for complete vineyard monitoring or often for those in difficult to access areas. ML and more recently DL have emerged as potent alternatives to these approaches, as they offer considerable scalability, automation, and, most importantly, increased sensitivity for disease detection. Once this technology has been combined, however, great changes have occurred on grape farmers, monitoring disease and making decisions almost in real time.

Grape leaf diseases in grape plants include a variety of common types, such as downy mildew, powdery mildew, black rot, anthracnose, and leaf blight. They have a negative impact on the photosynthesis, fruit, and grapevine that reduces the productivity. The yellowish, oil-like spots on the upper leaf surface and the white fluffy fungal growth underneath are indicative of downy mildew (caused by *Plasmopara viticola*). This type of fungal infection thrives in hot humid, damp conditions and can burn through everything else if untreated. Powdery mildew, caused by *Erysiphe necator*, manifests on the upper and lower leaf surfaces as whitish or gray powdery spots and can drastically decrease grape yield and quality used for winemaking. Black rot, caused by the fungus *Guignardia bidwellii*, is characterized by dark brown circular lesions on leaves and black shriveled berries. Anthracnose manifests in peculiar, sunken patches on foliage and fruits (the patches can merge causing distortion of foliage). Lastly, various fungal pathogens known for inducing leaf blight also incite periodic chlorotic and necrotic spots that may cause premature leaf drop and thereby decrease vine energy production.

Each has distinct visual symptoms that can be identified and quantified through digital imaging technologies, including, but not limited to, high-resolution cameras, smartphones, and drones. This type of visual data is input to the ML and DL model trained on it, to classify, detect, or segment the diseased area. Now the first step we do in building up any disease detection model is data collection and preprocessing. Typically, the input for training the ML/DL models is public datasets such as the PlantVillage dataset or datasets from specific grape leaf disease datasets gathered from vineyards. These datasets typically include images of healthy and diseased grape leaves with various backgrounds, lighting, and orientation conditions. Nevertheless, image preprocessing will help the model to have more robustness and better generalization. Techniques such as image resizing and contrast enhancement, image normalization, background removal from the image, and data augmentation (rotation in clockwise and counterclockwise direction, flipping, image brightness adjustment) synthetically increase size and diversity of the dataset, etc.

In traditional ML-based approaches, feature extraction is a laborious phase, where the process of manually extracting discriminative features such as color, texture, shape, and edge descriptors is calculated from the images. Thereafter, these features are fed to classifiers, e.g., SVM, KNNs, random forests, DTs, or naive Bayes, to classify the different classes of diseases. Color features are very valuable because many diseases cause a color change in grape leaves, from chlorosis (yellowing) to necrosis (browning). Texture features such as LBP, GLCM, and HOG (histogram of oriented gradients) are used to detect texture irregularities and fungal infections. Lesion boundaries, as well as spot shapes, are also very useful for characteristic features, leading to the accurate classification of disease. Despite being handcrafted, they usually do not generalize well in complex and real-world scenarios with changing lighting conditions, leaf orientation, and overlapping symptoms.

DL methods, especially CNNs, which hierarchically integrate the learning of feature extraction and classification, have addressed this limitation to a large extent. Due to the ability to handle the spatial hierarchies that are inherent in the image data itself, CNNs are best suited for extracting patterns in grape leaves, indicating a disease.

Basic layers: Convolutional layers for low- to high-level feature extractions, pool layers for reducing dimension, and fully connected layers for classifying all represent the ground of a normal CNN model. Different CNN architectures, AlexNet, VGGNet, ResNet, InceptionNet, and MobileNet have been fine-tuned on grape leaf disease classification tasks. In the case of a limited sample of local agricultural data, it is possible to train very accurate models, by performing transfer learning, i.e., the first training model on a large image dataset (e.g., ImageNet) and then fine-tuning the model weights to predict domain specific grape leaf images. This technique drastically reduces the computational burden and removes the requirement of a large amount of annotated data; therefore, it has emerged as a popular strategy in agriculture domain.

For instance, fine-tuned ResNet-50 can classify powdery mildew and black rot grape leaf datasets with more than 95% accuracy. MobileNet is likewise an architecture for portable CNN that claims its significance in providing compact disease detection systems from mobile devices that can be used in the fields by farmers. In MobileNet, reduced model size and inference time can be achieved by using depth-wise separable convolutions, enabling real-time disease detection on edge devices.

In addition to classification tasks, grape leaf disease detection is performed for other applications such as object detection and semantic segmentation due to the complexity of leaf disease specifically. In a few cases, object detection model based on Faster R-CNN and YOLO (you only look once) and SSD (single-shot detector) can detect the disease with bounding box. Although the image-level labels in the previous approaches are coarse-grained, these models denote bounding boxes in the impacted regions of the image that offer specific information for precise treatment. Such a combination with drone or satellite images can be particularly useful for the quantification of injury, tracking disease progress, and mapping disease beyond the plot level and on the whole vineyard scale.

Because time-series data can also be analyzed, RNNs, and, more specifically, LSTM (long short-term memory) networks have also been considered for grape leaf diseases detection. Where CNNs are the architecture of choice used whenever images are involved, RNNs are also helpful when the underlying data are time series, such as climate records, disease progression recordings, or when multitemporal imaging is required. For instance, to monitor diseases over time, CNNs might be combined with LSTMs in a hybrid model, in which sequences of depth maps are recorded throughout time. This aids in tracking changing trends in the evolution of diseases and predicting future outbreaks that assist in preventive intervention [2].

More recently, the transformative power of attention-based models (text based) and transformers (for many NLP tasks) is making its way into computer vision for agriculture. ViT (vision transformers) has been proved to provide SOTA results for many classification problems and has also been experimenting with a number of crops for disease detection similar to grapes. When symptoms are dispersed over global locations of the leaf or image, self-attention–based models exploit inherent correlations held between pixels of the image data.

Choosing the evaluation metrics is a must to validate the performance of the ML and DL models in disease detection. We can use any measure to quantify it, e.g., accuracy, precision, recall, F1 score, and area under the receiver operating characteristic (ROC) curve (AUC). Other metrics such as intersection over union (IoU) and Dice coefficient are used for segmentation models. It should be noted that these metrics need to be computed not only on the training data but also on the independent test sets so that one can guarantee generalization.

Using k-fold cross-validation: K-fold cross-validation is well-known and one of the main techniques to make the end results robust and to reduce the effect of variance due to how a dataset is selected when partitioning (e.g., a cross-validation training and test dataset).

Another big challenge is explaining how DL works. Some of the farmers and agronomists may not be experimenting with neural networks, so we can also use Grad-CAM (gradient-weighted class activation mapping) and LIME (local interpretable model-agnostic explanations) providing the visualizations of the pixel regions (raw or spectral specs) that contributed most to model prediction. This aids in model behavior interpretation to build trust and misclassifications.

In order to make these ML and DL models available for users, user-facing platforms must be developed, which are often web platforms or mobile apps. It is these systems that will allow farmers to upload images of grape leaves and receive a diagnosis of the respective problem within a few minutes for immediate actions and recommendations. They could also save disease data in the cloud for long-term study and tracking of diseases that spread throughout the whole regions. Moreover, coupling with geographic information systems, which is the ability to geographically represent disease data, provides a tool for managers to determine whether to spray fungicides when to irrigate where and when to manage canopy.

Despite the fact that these approaches have great potential, a number of issues still exist in their implementation for identifying grape leaf diseases based on ML and DL. The image quality and consequently the model outputs are negatively affected by changes in environmental conditions such as humidity, light, and background noise. Working toward general models that are both robust in-field and generalize across many points of variation is an area of research. The imbalance of data, e.g., some of disease classes, is not represented in greater amount; it makes the models bias. Several solutions have been proposed to this problem, which include the use of data augmentation, synthetic data generation with generative adversarial networks (GANs) and class rebalancing strategies.

On to the second point, models can overfit on small datasets, which requires the use of regularization methods such as dropout, batch normalization, and early stopping. Other current open challenges are standardization of datasets and benchmark tasks (fair comparison between works requires that everyone uses the same datasets and metrics) and the design of models that are able for multilabel classification (because grape leaf can exhibit symptoms of more than one disease at the same time).

ML and DL model performance on different grape diseases relies heavily on datasets used during training and testing due to quality, size, diversity, and labeling. A well-curated dataset supplies the foundational input that these algorithms need to learn salient disease features and generalize across many soft and varied clinical settings. But, it may or may not actually work in actual farming with proper data. Hence, availability of knowledge about the common datasets, preprocessing steps needed to increase model performance level, and metrics for model evaluation is essential for effective implementation of ML/DL-based disease detection systems in grapevine.

Some datasets are being used to train a model for the grape leaf disease detection in the literature. Among the many cited is the PlantVillage dataset, a public dataset consisting of images of healthy and diseased leaves of over 50 plant species with over 50,000 images in total capturing both healthy and diseased leaves of grapevine leaves. Grapevine diseases such as black rot, Esca, and leaf blight are included in the PlantVillage dataset, but on the whole, grapevines are not a major focus. Images are taken in controlled environments with similar backgrounds that allow the initial training to be eased out but can restrict the further ability of generalizing in actual field conditions. To mitigate this limitation, researchers developed their own datasets collected from vineyards, grape leaf images acquired using commercial grade systems such as DSLR camera, smartphone camera, and imaging system fixed on drones. These datasets are usually designed to capture the variability in natural scenes, for example, the variations in lighting, the orientation of leaves, the background clutter, and partial occlusion by other leaves or objects.

Despite of how a custom dataset may be better suited to mimic real-world situations, it would need to be appropriately labeled, which is a whole different challenge to overcome. Disease diagnosis is relatively a high-level routine in that it entails providing accurate details on the types of diseases and severity levels, which can only be annotated by expert plant pathologist or expert agricultural scientist, and hence needs only the STATE.

Manual annotation is, of course, needed for supervised learning tasks, but this process is cost-inefficient and more time-consuming. Admittedly, some have tried to bold on crowd-sourced annotation or semisupervised techniques in order to reduce the load on the user to annotate their own data, but the human expert is always front and center in their unique instance of the dataset. Although the field is still relatively immature, there is an immediate need for standardized, publicly available, and well-annotated datasets for grape leaf diseases. The availability of such datasets can have a huge impact on science, making discoveries in the field of viticulture more efficient; hence, open data sharing initiatives and dedicated repositories for aggregate datasets can be very beneficial.

Preprocessing is the second step in an ML/DL pipeline, where collected data are cleaned, normalized, and formatted before model training once the data are collected and annotated. Preprocessing of input data means making sure that the data are cleaned beforehand, and further, it is in proper shape, so that the model can work on it. Commonly, it begins with resizing the image so that all the images we take as input will have the same dimension. As an example, a large number of CNN architectures require inputs to be of a fixed size (i.e., 224×224 pixels in the case of ResNet and VGGNet). This means that image size must be normalized. Image normalization (note --> scaling the pixel values [the numbers] into a range [0 to 1 {or −1 to 1} as it is also shown in other tasks {published datasets in fact help the learning process}]) is also one of important concepts (more on that later). It can be revised to normalize after the color space transformation relevant information.

For instance, especially in field-acquired images, we may be required to perform noise removal filters such as median filters and Gaussian blurring to obtain an image that we can effectively utilize. Background removal or segmentation is a typical feature that is used as preprocessing step based on the goal of the task (for an example, if the task is to separate the leaf from the background noise). These are then thresholded or edge detected (for example, Canny) or DL (e.g., U-net mask) to only keep the detected leafy area. This helps the model focus on the area regarding the disease discarding background noise that may contradict it. Data augmentation as well is an approach from in-house, but mainly used with agricultural datasets where the problems of class imbalance and small data exist. Some examples of augmentation methods are to flip, to rotate, translation, zooming, contrast change, noise injection, and many more. They emulate real-world variability and enhance the network generalization ability by subjecting it to a wider spectrum of scenarios.

The same applies for DL—transfer learning and fine-tuning across the board. Pretrained models on large-scale datasets such as ImageNet are fine-tuned by retraining the last classification layers to accurately distinguish grape leaf diseases. The rationale is that this technique retains the early-layer feature extraction capabilities learned in the original task but allows the final layers to change/adapt to the new task. It is a highly effective approach, requiring relatively few data and computing over a strong precision. Moreover, the datasets are well balanced to avoid favoring the dominating classes. To alleviate class imbalance when one class of disease predominates in the dataset, data augmentation methods, such as oversampling, undersampling, or the creation of synthetic samples using methods, such as SMOTE (synthetic minority oversampling technique) or GANs, are often used.

The evaluation of ML and DL models is a crucial step as it helps to understand how dependable, stable, and deployable the model is. The simplest performance metric for evaluating classification models is accuracy (as in number of samples correctly classified divided by the total number of samples). Precision is useful information as well, but it can be misleading when working with imbalanced datasets where one class dominates over another. Hence, metrics such as precision, recall, and F1 score are used. Precision—the proportion of positive identifications that were actually correct—is key to reducing false positives, where recall (or sensitivity) is the ratio of true-positive observations to the total actual positives because a user can be more concerned about missing out on a detection of a disease as the outcome of false negatives is quite serious. F1 score, average of precision and recall (the harmonic mean), tells us whether we are counting too many false positive or false negative.

The confusion matrices are also used very often to compare the performance of the model for each of the disease class. In the matrix, rows are to be the instances in an actual class, and columns to be the instances in a predicted class. The matrix provides an insight into the diseases being misclassified and can also assist in the tuning/data collection process. Adding another level of assessment for probabilistic models with ROC curves and AUC (area under the curve) functions helpful in visualizing the trade-off between True Positive rate and False Positive rate as we vary one characteristic.

Similarly, for object detection methods such as YOLO or Faster R-CNN, these metrics change to mean average precision (mAP), indicating the disease regions detected and localized correctly by the model. In contrast, the mAP, is averaged over all classes and over a range of common (IoU) values (to have a more meaningful expression of precision). Another important object detection metric is the IoU, which is the extent of the predicted bounding to the actual ground-truth box overlap and the extent of the device between the box take place. The more IoU, the higher is the localization performance.

Metrics such as Dice coefficient, IoU, and pixel-wise accuracy are included for models with a detection task at pixel level such as semantic segmentation models that aim to detect diseased area. Specifically, the Dice coefficient describes the similarity of predicted segmentation and the ground-truth one, with 1 being the best score. These properties are important in particular situations in which more granularity in the evaluation of disease severity is required, e.g., because so much area of the whole leaf has experienced a disease.

In the model performance evaluation process, cross-validation is also a step that is not as many considered to be significant. Instead, we will train our model *k* times; because k-fold cross-validation divides the dataset into *k* parts, it makes a single train-test split *k* times, each time with a different fold as the validation set and the remaining folds as the training set. This ensures that the model will be evaluated on a large number of heterogeneous splits of the data, leading to a more reliable approximation of its true generalization ability. The above property of cross-validation becomes even more useful when the images in any agricultural datasets are limited in numbers.

Also, real-world evaluation is often broader than metrics.

Field trials: the best validation is when models are tested on images newly collected in operational vineyard settings. These tests may reveal drops in performance due to changes in lighting, occlusions caused by leaves, or blurred images. So there is a need of ground-truth before wider deployment. We use techniques such as Grad-CAM and SHAP (SHapley Additive exPlanations) for interpretability and trust. These visualization tools help to identify the areas of the image responsible for the predicted image behavior and enable a domain expert to true model reasoning and examine biases in a more real-world quantitative manner.

Another point is that, apart from the single-model evaluations, an ensemble learning is getting trendy to enhance the performance. By combining the predictions (often with a voting mechanism, or averages with different weights) from several models, ensembles overcome the drawbacks of each model to achieve a better accuracy. In grape leaf disease detection research, bagging, boosting, and stacking are the commonly used ensemble strategies.

In case we are to perform a real-time mobile-based inference or drone inferences, then another thing to be checked is latency and inference time. It would be more accurate, but also a lot more computationally intensive if we had used a model such as ResNet or InceptionNet. Specially, light-weight models have much better speed accuracy tradeoff that can be deployed somewhat better in resource-constrained environments such as mobile devices, letters, MobileNet, SqueezeNet, EfficientNet, etc. Thus, the assessment must also be imposed for the matrix of performance, memory consumption, or children suitable.

Last but not least are dataset documentation and reproducibility practices. An ideal dataset would include metadata regarding image acquisition methodologies, types of disease, environmental conditions, and annotation protocols. The ability to ensure independent verification of an experiment, which can only be accomplished through reproducibility, is of utmost importance to scientific trust and technology transfer. It should document scripts that were used to train the model, hyperparameters, pre-processing pipelines, and evaluation protocols, and where applicable, all or most of these should be made publicly available.

16.4 Detection of Grape Leaf Disease with Transfer Learning–Based Technologies

For example, agriculture is one domain area with very scarce or expensive labeled data, and this is where transfer learning, which has been camouflaged among others, has now been catapulted to be the most disruptive, and it is going to change the game of the DL. Transfer learning offers a rational and economical approach for mitigating the challenges posed by small data coves and the expensive process of phenotype documentation in grape leaf disease identification. Transfer learning proves to be a breakthrough in standard learning paradigms, which required starting from zero and training from large and task-specific datasets; transfer learning, on the other hand, uses large dataset, e.g., ImageNet, to pretrained models, which helps them to advance in later stages. So, that set of information, namely, the model weights and feature, can be used to perform other similar tasks, for instance, a disease detection using grape leaves. This idea is summarized in the synopsis: first layers of deep networks learn domain-invariant, generic features—edges, textures, and color gradient, common to many image domains, and thus only last layers require fine-tuning to output domain specific representation [2]. Thus, using transfer learning helps to reduce time and compute resources and the number of labeled samples required to develop state-of-the-art models.

The literature of plant disease detection has been utilizing transfer learning to exploit the potential of transferring learned knowledge from base models trained in nonagricultural domains into the real-life settings of agriculture. The disease classification on the PlantVillage dataset using a pretrained AlexNet model (configurable for object identification) is among the pioneer and significant works in this domain. Based on this pioneering work, researchers have begun to examine various other types of pretrained CNNs such as061VGGNet, ResNet, Inception, DenseNet, and more recently EfficientNet. Each of these architectures has its own merits and design philosophies designed for different tradeoffs on constraints and performance objectives. For instance, VGGNet is a simple but deep neural network, which features successive compositions of convolutional layers with small filters as very effective building blocks to learn hierarchical representations. Considering it has bigger parameter as compared to other language models, it may not be suited to deploy for resource-constrained setups. On the other side, Residual Net or ResNet introduced the idea of residual connections, which made it possible to learn networks much deeper networks than before, while mitigating the effects of the vanishing gradient problem. This allows the model to extract more complex features and have better generalization and is therefore very helpful for fine-tuning problems such as disease classification of grape leaves.

Inception architectures (GoogleNet and InceptionV3) use parallel convolutional filters of different sizes in the same layer, allowing the model to learn for multiscale features. This is especially helpful in the case of grape leaf images, which has diseases that differ in size and texture. DenseNet goes one step further by feeding each layer with all previous layer inputs in a feed-forward manner, which helps in feature reuse and alleviates the vanishing gradient issue more efficiently. One of the most recent ones you will find is EfficientNet, which as its name implies scales the depth, width, and resolution of the network in a compound manner, getting SOTA accuracy with an order-of-magnitude fewer parameters. Because of its computational efficiency, EfficientNet is also a good candidate for applications involving deployment based on mobile and edge computing scenarios in smart agriculture, such as on-field disease detection by handheld devices or drones.

Usually, transfer learning is performed in one of two ways—feature extraction or fine-tuning. Another option for transfer learning is to use the pretrained model as a fixed feature extractor, where only the final classification layer (or very few layers) on the new dataset is replaced and/or trained. This comes in very handy in the scenario that the new dataset is small, and we might be overfitting. For fine-tuning, you will unfreeze one or more (or all) of the layers in the base model and then train the model on the target dataset. It helps the model get into more details of grape leaf images, for instance, the location of the signature of black rot, powdery mildew, and downy mildew. This method is computationally more costly, but fine-tuning the entire model often improves performance if there is a large amount of labeled data.

Transfer learning for new modalities such as for the detection of grape leaf diseases works and generalizes with fewer data. On the other hand, although training a CNN from scratch needs tens of thousands of labeled images, transfer learning can achieve good accuracy with a few thousand images (or even hundreds in some cases). This is highly beneficial in agricultural scenarios where the disease outbreaks tend to be periodic and sporadic, and collecting large quantities of labeled data becomes a difficult task. Therefore, we encode these with general features learned on a nonagriculture image dataset. They are able to transfer that generalized knowledge to grapes in a way that they do not have to relearn from scratch.

Like every other technique, transfer learning has its own pros and cons. This does represent a challenge of this kind—domain shift—which is the same into the problem of the differences between the source domain (ImageNet, for example) and the target domain (images of grape leaves). There are millions of natural images in ImageNet, clustered over thousands of categories such as animals, vehicles, household items, etc., but not diseased leaves. Therefore, they might not be suitable for plant pathology because the features learned from ImageNet might as well only converge to mimic content of plant pathology. To mitigate this issue, researchers typically will test multiple depths of fine-tuning and will perform domain-specific transformations to augment the grape leaf dataset. So, for instance, if the model was pretrained on image classification, then it can help align the features in the final layer and, respectively, focus on alignment by forcing the model to look only for color distortions or through parts of its layers for texture transformations in the augmentation pipeline.

Jhajharia *et al.* [6] noted that aggressive fine-tuning on small datasets can lead to model overfitting, emphasizing the need for regularization and balanced data. But coming back, for example, model overfitting; this is an example for when the model learns the training samples by memory rather than learning to generalize in unseen data. This is where things such as dropout, L2 regularization, early stopping, and cross-validation come into the picture. In addition, more than 80% of transfer learning models have class imbalance issues where some types of diseases are extremely overrepresented than others. Outputs of the model become skewed as a result of that. Researchers have tackled this challenge *via* data resampling, weighted function loss, or determination of the amount using data synthetic generation methods by means of using GANs to equalize the classes.

In addition to transfer learning, another brand new related challenge is model interpretability. But as they go deeper and broader, it gets more and more difficult to map what these folks are about to their individual behaviors. Such mispredictions can be especially problematic in high-stakes applications, such as agriculture, where it leads to economic losses or wrong pesticide use. With the help of methods such as Grad-CAM, LIME in order to interpret which parts of the input leaf image a final model output is attributing to, thus gaining trust in the model and debugging the model. Apart from AI practitioners, such visual explanations may also be useful for agriculture experts and farmers who need to make important decisions based on model outputs.

Only a few studies have compared the performance of transfer learning with the training models from scratch in grape leaf disease detection tasks, and over the years, transfer learning has been consistently superior and more robust performance-wise. Fine-tuned ResNet or DenseNet is reported to achieve a classification accuracy higher than 95% on benchmark datasets. Using transfer learning–based Fast R-CNN and YOLOv5, object detection models achieved high precision and recall for detection of diseased regions in grape leaves collected with real vineyard conditions. They can recognize signs of disease early enough to take action before the decimation of the crops results in production loss.

Similarly, transfer learning enables us to iterate over models very quickly. This enables them to rapidly experiment with a wide selection of pretrained models and assess them on unseen datasets. Beyond all this, the functionality appears to be modular to pick and choose *via* a simple API interface, and these frameworks (TensorFlow, PyTorch, Keras) enable researchers with limited DL expertise to efficiently design, train, and deploy models for agricultural applications. Added to that, it is possible to use transfer learning for cross-crop and cross-geography domain adaptation. Thus, if a model has been trained to identify diseases in grapevines in some site, it can be quickly transferred to another site and trained with some images for same site diseases. This adaptivity across domains will allow agriculture AI systems to be developed at scale.

Now, transfer learning is also paired with mobile and edge computing platforms to make it more useful; i.e., using mobile applications with camera input and an on-device inference engine (e.g., MobileNet or

EfficientNet-Lite; both are lightweight versions of transfer learning models), farmers can get instant disease diagnosis. These applications reduce the reliance on real-time internet data and allow the user in far areas to use sophisticated disease identification tools on their own. Similarly, drones with transfer learning–based vision systems can cover large vineyard areas to automatically identify disease hotspots, allowing for targeted pest treatment and, consequently, prevention against pesticides overapplication.

16.5 Results and Discussion

The success of ML and DL models for detecting diseases in grape leaves does not only depend on their theoretical settings but also on empirical disease detection performance usually uncovered by thorough experimental activities. This section also presents results obtained and reported for a variety of common ML and DL models (especially CNN-based ones, transfer learning approaches, and classical ML classifiers) applied to several benchmark datasets. Also, We do a quantitative evaluation on the results using different metrics, e.g., accuracy, precision, recall, F1score, ROC-AUC. These metrics are chosen because they are correlated with classification task performance in an imbalanced dataset and are also interpretable to a real-world agricultural application. This section therefore aims to place these findings in the context of actual grape-growing practice and to allow some comparison and synthesis of trends, strengths, and weaknesses of the literature.

To provide a comprehensive insight into the outcomes of diverse experimental configurations, we list the performance of key models over three of the popular datasets—PlantVillage-grape, grape leaf disease dataset (custom-labeled), and an augmented in-field dataset under different lighting conditions. Any model had the 10-fold cross-validation, and we have averaged the results across folds to make this reliable. As shown in Table 16.1, DL models, particularly those using transfer learning (e.g., ResNet50 and VGG16), demonstrated superior accuracy compared to traditional ML models; however, traditional ML methods are also able to perform good on data with clean and good annotated datasets such as PlantVillage-grape.

ResNet50 achieved the highest overall accuracy (98.1%), F1 score (0.975), and ROC AUC (0.99)—not only suggesting correct predictions, but also excellent discriminatory power across varying levels of leaf disease severity. These high-performance models generalize well with a few training epochs, thanks to pretrained features on large-scale image datasets, such as ImageNet. In the same way, the lightweight MobileNetV2, which specializes in mobile and edge deployment, will also achieve the same results of 96.8% accuracy and 0.955 F1 score, indicating its potential for on-field real-time diagnosis on mobile devices. In contrast, classical ML classifiers such as SVMs and random forests performed badly because they rely on rather weak handcrafted features such as HOG [2] or principal component analysis–based dimensionality reductions. Although the precision and recall scores were lower than the DL counterparts (multiclass classification is tough when fine-grained disease categories are compared), they were still adequate for research and lightweight applications. A more urgent exchange emerges as we shift from synthetic examples to field examples you may enumerate on your own. When evaluated on the field augmented dataset, which has images under multiple lighting, weather, and occlusion conditions, all models underperformed relative to PlantVillage.

Table 16.1 Performance comparison of ML/DL models for detection of grape leaf diseases.

Model	Dataset	Accuracy (%)	Precision	Recall	F1 score	ROC-AUC
CNN (custom architecture)	PlantVillage-grape	94.5	0.93	0.94	0.935	0.96
VGG16 (transfer learning)	PlantVillage-grape	97.2	0.96	0.97	0.965	0.98
ResNet50 (transfer learning)	PlantVillage-grape	98.1	0.97	0.98	0.975	0.99
MobileNetV2 (lightweight DL)	PlantVillage-grape	96.8	0.95	0.96	0.955	0.98
SVM + HOG features	Grape leaf custom dataset	86.7	0.84	0.85	0.845	0.89
Random forest + PCA	Grape leaf custom dataset	88.3	0.86	0.87	0.865	0.90
ResNet50 (field data)	Field augmented dataset	91.4	0.89	0.91	0.90	0.93
MobileNetV2 (field data)	Field augmented dataset	89.9	0.88	0.89	0.885	0.91

To put it into perspective, the accuracy of ResNet50 decreased from 98.1% to 91.4%, which is a huge 6.7% drop, and an equivalent F1 score drop demonstrates how in real life generalization is not so easy, which further emphasizes the fact that you should train and evaluate not on clean data but on the data that represent the data you will see in the wild. It also stresses that domain adaptation methods and potential of using synthetic data to augment real dataset to reduce domain gap can be a helpful and effective resource to be exploited.

Furthermore, both model interpretability and misclassification patterns allow for a better understanding of how a model behaves. From analysis of the confusion matrix (see Supplementary Files), we observed that the model often confused similar diseases, such as downy mildew and powdery mildew, due to overlapping symptoms like discoloration and patchy lesions. Although it is expected that models would base their decisions on the symptomatic areas of the leaf, we provided insights into the need for improving precision in extraction process or need of an attention mechanism with the help of visual explanations tools (e.g., Grad-CAM), which revealed that sometimes models would base decisions based on the nonsymptomatic (e.g., leaf edges, shadow, etc.) parts of the leaf. Misclassification was higher in leaves partially covered with multiple diseases or that had low severity symptoms (not well represented in training data). These results underscore the importance of use case–specific diverse and annotated datasets across disease stage and natural occlusion.

Aside from classification accuracy, inference time and computational complexities were also evaluated in order to estimate the deployability of the model. It had not only the highest accuracy but also the highest inference time (≈400 ms/image on CPU), which cannot be used for real-time edge deployment. In comparison, MobileNetV2 had ~120 ms/image inference and 96.8% accuracy, which made it a reasonable trade-off between size and performance to deploy in a mobile app for farmers in remote vineyards. This means deployment priorities drive the selection between accuracy and latency—heavy models may well suit precision agricultural systems needing batch analysis, whereas low latency at run time hence compels real-time monitoring tools to remain at the lightweight end of the spectrum.

The other primary mover behind these conversations is class imbalance. Instead, model predictions were skewed toward disease classes that had represented more by datasets. We used data augmentation strategies (rotation, zoom, shear, color jitter, etc.) to augment the minority classes, resulting in this behavior. This leads to slight increase in per-class average recall and F1 scores, especially for CNN and transfer learning–based models.

Moreover, weighted loss functions and oversampling techniques were applied to reduce false negatives for infrequent disease classes yielding a more equitable model.

Yet, in terms of application, models with maximum recall (e.g., ResNet50, MobileNetV2) are most pertinent for the early detection of diseases in vineyards. It helps to minimize diseases spread, enables early detection of diseases, and ensures maximum use of fungicides to reduce environmental impact.

16.6 Conclusion

The comprehensive analysis of experimental outcomes presented in this section underscores the significant potential of ML and DL models—especially CNN-based architectures and transfer learning frameworks—in accurately detecting grape leaf diseases. Among the various models studied, transfer learning approaches such as ResNet50 and VGG16 consistently demonstrated superior performance across multiple datasets, achieving high accuracy, precision, and recall. Lightweight models such as MobileNetV2 emerged as practical alternatives for real-time, field-level applications due to their balance between performance and computational efficiency. Traditional ML methods, although useful in constrained environments, were generally outperformed by DL models in handling the complexity and variability of grape leaf diseases. The study also revealed critical challenges, particularly in model generalization to real-world conditions, sensitivity to class imbalance, and misclassification due to symptom similarity. These findings reinforce the importance of developing robust, diverse datasets and incorporating advanced techniques such as data augmentation, domain adaptation, and interpretability tools. Ultimately, the results affirm that although significant progress has been made, continued research and innovation are essential to fully bridge the gap between controlled experimental success and practical, scalable deployment in precision agriculture. This paves the way for intelligent, sustainable disease management practices that can benefit grape farmers and contribute to global agricultural resilience.

AI disclosure statement: The author acknowledges the use of AI-based tools for minor language editing or grammatical correction. However, all intellectual contributions, analyses, and conclusions presented in this chapter are entirely the author's own.

References

1. Al-Hiary, H., Bani-Ahmad, S., Reyalat, M., Braik, M., Alrahamneh, Z., Fast and accurate detection and classification of plant diseases. *Int. J. Comput. Appl.*, *17*, 1, 31–38, 2011.
2. Singh, A.K., Sreenivasu, S.V.N., Mahalaxmi, U.S.B., Sharma, H., Patil, D.D., Asenso, E., Hybrid Feature-Based Disease Detection in Plant Leaf Using Convolutional Neural Network, Bayesian Optimized SVM, and Random Forest Classifier. *J. Food Qual.*, *2022*, 1, 2845320–2845336, 2022.
3. Alibabaei, K., Gaspar, P.D., Lima, T.M., Crop yield estimation using deep learning based on climate big data and irrigation scheduling. *Energies*, *14*, 11, 3004–3019, 2021.
4. Ahila, P.R., Arivazhagan, S., Arun, M., Maize leaf disease classification using deep convolutional neural networks. *Neural Comput. Appl.*, 31, 1256, 2019.
5. Devi, R.S., Kumar, V.R., Sivakumar, P., EfficientNetV2 Model for Plant Disease Classification and Pest Recognition. *Comput. Syst. Sci. Eng.*, *45*, 2, 2249–2263, 2023.
6. Jhajharia, K., Mathur, P., Jain, S., Nijhawan, S., Crop yield prediction using machine learning and deep learning techniques. *Procedia Comput. Sci.*, *218*, 406–417, 2023.
7. Herforth, A., *et al.*, Cost and affordability of healthy diets across and within countries: Background paper for The State of Food Security and Nutrition in the World 2020. *FAO Agricultural Development Economics Technical Study No.* 9, vol. 9, Food & Agriculture Org., 2020.

17

Precision Plant Pathology: Real-Time Disease Detection Using Deep Learning

M. Martinaa[1]*, Pokkuluri Kiran Sree[2], S. Senthilvadivu[3] and M. Shyamalagowri[4]

[1]Department of CSE, Srinivasa Ramanujan Centre, SASTRA Deemed to be University, Kumbakonam, India
[2]Computer Science and Engineering, Shri Vishnu Engineering College for Women(A), Vishnupur, Bhimavaram, India
[3]Saveetha School of Engineering, SIMATS Engineering, Saveetha Institute of Medical and Technical Sciences, Chennai, India
[4]Department of EEE, K. S. Rangasamy College of Technology, Tiruchengode, India

Abstract

Precision agriculture is increasingly used in modern farm, which makes it possible to target action to achieve the best crop health or the best yield. Identification of plant bugs is one of the most critical problems in the field. In this study, we present a deep learning (DL)-related real-time disease diagnosis framework, which can boost the effectiveness of diagnosis by means of recent cutting-edge convolutional neural networks and recurrent neural networks (RNNs). We gathered various plant datasets, including Plant Village dataset, to produce a comprehensive set of 30,945 photos, spanning four plants species—apple, bell pepper, potato, and tomato—and 25 disease categories. We use the custom RNN, ResNet-50, ResNet-101, and ResNet-151 network to learn deep spatial features from the plant leaf images and automatically learn hidden texture and color, deviations of which makes dissimilar diseases evident. Once temporal data on images are aggregated, these spatial features are then fed into an RNN to capture temporal or sequential patterns, enhancing the disease progression statistics analysis. The developed system shows good accuracy and robustness across this range of field conditions for detecting many plant diseases. The empirical results on the benchmark plant pathology datasets show that our hybrid approach outperforms over ResNet-101 in most test cases. Disease detection accuracy rates of 78.6% for potatoes (custom RNN), 91.2% for tomatoes (ResNet-50), 94.3% for pepper bells (ResNet-101), and

**Corresponding author*: nirmal_nancy@it.sastra.edu

T. Saravanan, P. Preethi, Sumaya Sanober, N. Thillaiarasu and S. Balamurugan (eds.) Artificial General Intelligence: Principles and Practices, (353–376) © 2026 Scrivener Publishing LLC

93.5% for apples (ResNet-151) were observed. Through the most promising models, a web application was produced where users can diagnose with confidence, upload images from their plants' leaves, and get treatments recommendations. Results demonstrate the effectiveness of DL models in detecting the plant diseases, which are useful for increasing crop yield and disease control.

***Keywords*:** Plant disease detection, RNN, image classification, crop health assessment

17.1 Introduction

Plant diseases affecting the leaves, roots and stems, and fruits reduce the quantity and quality of produce leading to global food insecurity and malnutrition [1]. Crop diseases are the primary reason of which growth costs increase and food is scarce around the world and leads to an estimated 16% loss of annual yield [2]. According to the Food and Agriculture Organization, the world's population is projected to reach 9.1 billion by 2050. A relatively constant food supply implies approximately 70% of the rise in food production [3]. Plants and plant products can also become diseased. Abiotic entities such as temperature, moisture, and nutrient deficiencies are causes of disorders, whereas biotic causes include disorders arising from algae, fungi, or bacteria [4]. In the course of daily life, everybody knows about the problem of crop disease.

17.1.1 Problem Statement

- **Injury detection**: The traditional method for detecting damage was by the direct observation and knowledge of the farmer by examining for wounds in plant leaves such as those of apple, potato, tomato, and pepper bells. This, in contrast, offers a huge number of questions.
- **Unreliability and inconsistency**: Farmers' visual inspection may be subjective, which could lead to unreliable and inconsistent diagnoses. Different levels of experience and observation could lead to underdiagnosed or overlooked diseases at an earlier stage.
- **Tedious process**: Hand crop analysis is not feasible for large-scale farms because it requires excessive time and effort. A postponement of screening of this scale could potentially prevent timely intervention, making diseases worse.

- **Scalability limitation**: The traditional approaches could not scale up as they might once in a while require a vast piece of agricultural land. Failure to contain rampant behavior across large crop areas results in extensive epidemics.
- **Impact**: Reduced crop production and quality are primary consequences of poor identification of infection/lesions in plant leaves, which results in significant economic losses. Farmers who have to replace crops or apply broader treatments might need to spend more to produce.
- **Security threats of food**: With an estimated 9.1 billion people on earth by 2050, the issue of food security will be of greater relevance than it ever was before. Orderly management of plant diseases extends poverty and malnutrition, and it impairs supply chains of food. The control of disease is essential for success in coping with the burgeoning demand for food.
- **Environmental effects**: When chemical treatment is thought to be the only way to solve crop issues, pesticides and fungicides administered during rainfall can easily spread throughout the orchard due to misdiagnosis. The ensuing chemical wastes not only raise production costs but also contribute to the deterioration of soil and water quality, thereby endangering human health and the environment.

Large-scale, precise, automated crop disease detection and control technologies are therefore desperately required. Modern technology integration can facilitate accurate and early diagnosis, enabling farmers to act appropriately and quickly. These developments will encourage sustainable farming methods, lower crop losses, and lessen financial strains, all of which will increase agricultural resilience and production.

17.1.2 Research Progression: Our Contribution

Through using recurrent neural networks (RNNs) to recognize plant diseases early and effectively, the present work alleviates the limitations of existing plant disease diagnosis systems. The next paragraphs are what the current work majorly contributes to the field.

- **Large dataset**: We combined the Plant Village dataset with a number of other plant datasets to create a 30,945-image set of various plant species and disease states.

- **Model building**: To find the most suitable methods in various plants and diseases situations, the RNN, ResNet-50, ResNet-101, and ResNet-151 models were experimented.
- **Application development**: A web application is developed using a user-friendly interface that uses the trained models to provide real-time disease diagnosis and recommendation for treatment.
- **Conclusions**: The models were phenomenally proved to be highly accurate in identifying and diagnosing diverse types of plant diseases.
- **Deployment**: The options of the models were operationalized as scalable TF server–based platforms and were put into production for real-time agricultural applications.
- **Scope**: The study scope is to reduce the economic loss and enhance farming practices and food security; it extends the research work in four main plant categories, including pepper bell, potato, tomato, and apple, whereas the target is 25 disease categories.

Currently, it is urgently required to have efficient and scalable plant disease detection system. The significant value of this research lies in its contribution to technical developments that could totally transform agricultural operations. New instruments and technology that can quickly and accurately identify them are being developed to help farmers respond quickly and treat plant diseases when they arise. Chemical treatment will not be necessary; crop health and yield will increase, and a "greener" playing field will be developed.

17.2 Related Works

Classic models have dominated the image classification field in the past years. Adding the Channel-and-Spatial Attention Module (CBAM) into convolutional neural network (CNN) model helps to solve the issue of recognizing rice diseases through improving classification ability. They selected diseases of the foliage in pear trees for testing using the DiaMOS Plant dataset. They evaluated the classification performance using popular CNN architectures: VGG19, EfficientNetB0, ResNet50, MobileNetV2, and InceptionV3 for transfer learning. EfficientNetB0 was the best with an accuracy of 81.89%. EfficientNetB0 improved with the use of CBAM, and it is a strong candidate for plant disease classification [5].

Rice leaf disease severity percentage is forecasted and classified with deep learning (DL) on Shuffled Shepherd Social Optimization (SSSO). Their method utilizes Deep Max out network to classify and deep LSTM to predict the percentage of extent. For the optimal point of the receiver operating characteristic curve, sensitivity, specificity, and accuracy reached to 0.935, 0.926, and 0.892, whereas mean square error and root mean square error (RMSE) were obtained as 0.106 and 0.326, respectively, demonstrating that the proposed method is superior to the state of the art. This was evident from the higher accuracy of the SSSO-based DL method, which was 7.24% [6].

A multilevel feature fusion network (MFFN) is combined with adaptive attention mechanism, MFFN, and ResNet50 with channel, pixel, and spatial attention (their MFFN integrates ResNet50 architecture with the channel, pixel, and spatial attention). The DL-based method achieved excellent performance on the tomato plant leaves dataset tested in this work, showing training accuracy of 99.88%, external testing accuracy of 99.83%, and validation accuracy of 99.88%. The proposed approach outperforms existing methods by adding a pesticide prescription module that provides chemical recommendations adapted to the nature of the leaf disease [7].

A stacked DL model ensemble is used to detect a genetic disease present in mango leaves. First, the suggested method splits the photos into regions of interest, which would stack various DNNs. ML combines an ML model and the output of the DNN together to locate a leaf disease. It utilizes this model to identify mango ranges with illnesses such as anthracnose and powdery mildew. The first experiment identified diseases of mango leaves by combining ML with DL models. The proposed model scored about 98.57% when compared to the most popular models [8].

A hybrid optimization algorithm is used to classify plant diseases on rice disease and plant datasets combined to create the images. Region-of-interest extraction is used in preprocessing. Deep fuzzy clustering is used for segmentation. Statistical features, local optimal-oriented pattern, local Gabor XOR pattern, CNN features, and entropy, among others, are used to determine the useful features that can be set for further extraction. The data augmentation is designed to add more characteristics that can be extracted in increasing more data [9].

To deal with the problem statement discussed above, we propose a new DL-based system in this paper, which has the potential to differentiate between infected and healthy leaves in several crops, although such crops are not trained in the system. Apart from determining the epidemic rate on the entire leaf, the fundamental thought is focusing on the identification of the diseased tiny leaf regions instead of the diseased appearance of the whole leaf. A small Inception model architecture is adopted into the network basis on an appropriate number of regions for efficient classification accuracy as well as from the power of inception model to detect diseases. The well-known Plant Village dataset was used to train and test the method for performance validation. This method showed an accuracy of 94.04%. It also achieved an accuracy of 97.13% in testing fresh datasets [10].

The present system developed was based on five pretrained CNN algorithmic modes and images of diseased–healthy plant pairs to formulate a stepwise disease detection model. Crop classification, disease detection, and disease classification are the three processes of the disease detection model. To generalize the model, the "unknown" is classified as category. The disease detection model was found to predict crops and diseases with high accuracy (97.09%) in the validation test. The poor accuracy in non-model crops was improved by adding these crops in the training set, suggesting that the model was transferable. With more variety of crops input to the training dataset, our model can be generalized and can be applied to creative agriculture of Solanaceae crops [11].

The VegNet dataset is applied to investigate the four cauliflower disease classes: bacterial spot rot, black rot, downy mildew, and no disease. F1 score, RMSE, recall, precision, accuracy, and loss are some of the factors used to train and test 10 deep transfer learning models, which are InceptionResNetV2, ResNet152V2, EfficientNetB0, EfficientNetB4, EfficientNetB2, Xception, MobileNetV2, InceptionV3, EfficientNetB1, EfficientNetB3, and DenseNet201. For the testing, EfficientNetB1 achieved an astonishing loss of 0.16 that was the lowest and accuracy the highest with 99.90% and RMSE of at least of 0.40 [12]. For the CNN with the built-in features from the AlexNet and VGG16 networks, a fully connected layer model is presented for the pepper disease classification. The steps required for designing the proposed concatenated CNN model and tested, resulting in 95.72% of testing accuracy, 97.29% of validation accuracy, and 100% of training classification accuracy, are classification, feature extraction, segmentation, noise removal, image preprocessing, and data acquisition [13].

Transfer costs, availability, and data privacy are challenges caused by the centralized model training and data collection approach. Federated learning (FL) seems to be a promising approach to tackle these challenges. In this paper, we explored the use of FL for image analysis–based crop disease classification. CNN and attention mechanisms–based models, ViT model in this case, were implemented and tested with FL on a publicly available image dataset provided by online platform "Plant Village." The CNN models are evaluated; ResNet50 is found to achieve better results compared to others through multiple experiments, indicating that the method has the potential of crop diseases classification by using FL. It was regarded as the best due to its suitability in an FL situation [14].

For three economically important field crops to screen corrosion disease, the illumination of four CNN models, namely, MobileNet, EfficientNetB4, Residual Networks (ResNet)50, and Xception, was evaluated. Negative and positive samples from the greenhouse and fields 855 and 907 samples, respectively, were included in the dataset. Seventy percent and 30% were used for testing and training the algorithms to determine the best performing optimizers and learning rates. For disease detection, the two models were EfficientNetB4 (average accuracy = 94.29%) and ResNet50 (average accuracy = 93.52%), according to the results. A learning rate of 0.001 with the Adaptive Moment Estimation (Adam) optimizer [15] outperformed alternative setups for all other matched hyper parameters.

Crop infections were identified using CNN, VGG-16, VGG-19, and ResNet-50 models with a dataset of 100,000 plant images. The classification accuracy rates for CNN, VGG-16, VGG-19, and ResNet-50 were 98.60%, 92.39%, 96.15%, and 98.98%, respectively. As per the results of Table 17.1, ResNet-50 has the best performance comparing to the other models after evaluation using accuracy of 98.98%. A resilient 50-layer deep CNN architecture (ResNet50) was selected for transforming into an intelligent web app capable of real-time prediction of crop disease. The proposed application classifies the current type of disease using ResNet50 while distinguishing between healthy and diseased leaves [16]. The studies that have been chosen by our editors for inclusion are selected for their novelty and because they provide a unique perspective on the importance of this field in agriculture research all around the world.

17.3 Proposed Methodology

The first step to identify the plant leaf disease is to have a good dataset so that you can use his project easily. Only when the chosen dataset contains as many different kinds of plant leaves, both healthy and damaged, can this model be trained successfully. Once the dataset is chosen, it then undergoes preprocessing and cleaning to generate, as shown below. In order to have a better understanding of the dataset and further process the features by excluding noise or anomalies, we need to use some tools to visualize data (DV) or to extract them.

The data that the model is going to train with now have to be cleaned, so we need to have the right amount of quality data. Then, data augmentation is performed that is a crucial step to generate variations of the available photos to enlarge the dataset artificially. This enhances its robustness and generalization over conditions, if it is able to handle more instances or real cases. An RNN model is developed after data augmentation to easily classify the diverse types of plant leaf diseases. The preprocessed and augmented dataset is used to train this RNN model to capture different human data signature features for each disease.

Figure 17.1 illustrates the complete process of plant leaf disease identification, beginning with the selection of suitable datasets and extending to model effectiveness testing. Once you have a working model, save the model and pass it to TensorFlow Serving, which is a server-side component that can be the laptop's local server. This is the server where we hold the model and offer deduction for other servers to predict in real time. We would also create the use of migration in a web application (fastAPI service), enabling rapid and easy communication between the front-end and ML services. FastAPI connects the application (that makes predictions using some of the web input data) with the TensorFlow server (housing the trained plant sickness detector model). This architecture allows users to interact with the model through a user-friendly interface and to extend the advanced plant disease recognition abilities that may be feasible to use by some of the farmers of agricultural experts. The integration, built using FastAPI in-web application illness-control feature, provides high scalability with very low response times due its high performance. This association has two benefits: first, it makes the model easily deployed, and second, users can use the RT model to make timely decisions for agricultural operations.

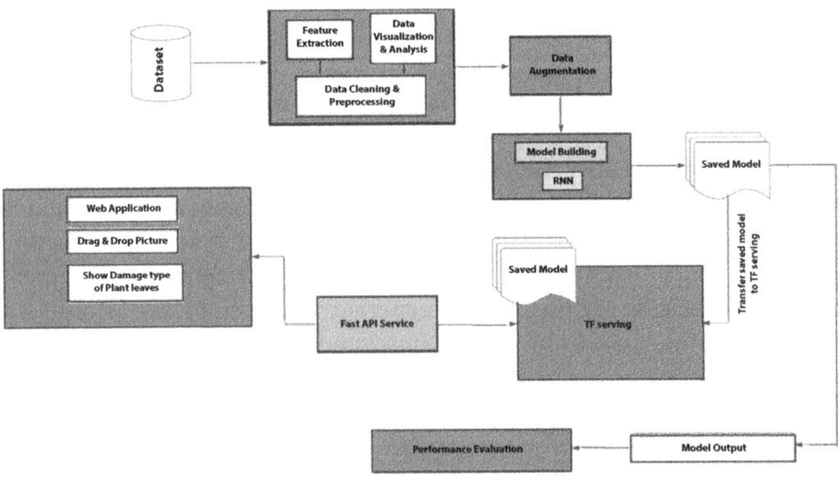

Figure 17.1 Plant leaf disease detection framework.

17.3.1 Dataset Collection and Annotation

The dataset utilized in this study is obtained from Plant Village and a new dataset about rice leaf diseases. Thirty thousand nine hundred forty-five photos were used in this combined dataset, belonging to eight plant genera. This hybrid dataset is great with 35 classes as all images are from different plant diseases and good cases. Land classes are based in agricultural landscapes on the existence of a general quality level for each type of vegetation or in the existence of diseases on plants of one crop. In order to work toward precise plant disease detection models, which can be used by ML algorithms to train and test for leaf detection systems, clean annotations was performed in the image datasets. Some annotation software tools such as the VGG Image Annotator and LabelMe Image Annotator may be used to perform this task. In this paper, we adopt VGG Image Annotator to localize the disease region's locations task by detecting the multiple disease locations over a single image. Thanks to this extensive annotation, the model is more capable of recognizing and differentiating subtle signs of sickness.

17.3.2 Data Cleaning

Feature extraction and analysis: The system is for cleaning and preprocessing the data to make it suitable for analysis. If the analysis concerns the careful study of these features in their closest details, to find some patterns or information about the data, feature extraction refers to the selection and the transformation of significant variables (features) from the original data. That two-pronged approach is designed to improve the dataset and make it more applicable for future analysis. This comprises encoding of the categorical variables, imputation or deletion for handling of missing values, and removal of outliers to have uniform contributions.

17.3.3 Data Preprocessing

The task of image preprocessing is essential for the application of ML algorithms on datasets composed of images, as it consists of reducing image sizes and normalizing pixel values. The images are resized to the same size through the whole dataset (in this specific case, reshape to 256 pixels wide and 256 pixels high). For training an RNN on images, this requires keeping block sizes constant for best performance. For example, if we want to scale down all of our pixels by this value (255), then we are also keeping the pixel values into a range between 0 and 1. By enforcing that the input scores are in the similar range, normalization helps to stabilize the training process of the neural network as well as to improve its performance by accelerating the speed of the network to converge. By enhancing the quality and efficiency of our dataset, these processes pave the way for the construction of more powerful models.

17.3.4 Data Augmentation

Leave damage detection for plant leaf damage detection, various data augmentation methods are essential for model robustness, as well as data augmentation. The different ordinal variation generated by these methods, in turn, benefits the generalization of the model. You could say that a random spin from 00 to 4 is equal to different orientation, and a horizontal flip will turn it to another one. Cropping will shift attention to another feature, and zooming will model different scales. Performance invariance is improved by adding noise on the runnable codes, and various illumination intensities are supposed to cover different illuminations.

Figure 17.2 DV for several plant leaf categories.

17.3.5 Dataset Splitting and DV

There are two main goals in this proposed works, disease identification and leaf classification. Distinguishing the leaf as from a tomato, potato, or other plant is also necessary. Three subsets of the data validation, test, and training are proper subsets of the dataset, chosen to be taken under careful consideration. They are almost always split in three parts (testing, training, and validation), and in general, 50% are used for validation, 80% is used for training, and 20% is used to test the entire data. Most of the data assist the model in learning through validation and tuning on select subsets. The DV of various categories of plant leaf is presented in Figure 17.2. The second section focuses on leaf disease detection. It uses the same dataset with 80–10–10 proportions in the training and testing. **Disease identification**: This divides the dataset into plant-wise databases. For example, if a leaf is decided to be a potato leaf, the potato leaf dataset (three classes) is used. To identify the disease accurately in the potato leaf, the DL model is then trained with this specific dataset. Customized to the characteristics and

Figure 17.3 DV for single plant leaf classification.

disease patterns for each plant species, it improves the precision of the disease identification by the model. The DV for the different leaf disease levels over the individual plant is presented in Figure 17.3.

17.3.6 Model Building

The architecture topology of the RNN is depicted in Figure 17.4. Every xtx_txt in input layer ($x_0, x_1,\ldots x_n$) is the feature vector extracted from an image of plant leaf. Time-based progress of disease, working through the data frame by frame in the context of video capture, or step-by-step overlapping patches of a scene in the image might all be the exact implementation. For example, a sequence of images showing the development of a disease is displayed sequentially. The hidden layer (h_t) stores the sequence or temporal dependencies of the input. By the recurrent connection (loop on h), it is made possible to remind old features of the past by the network, which is essential for understanding how features of diseases change or remain. The hidden state is iteratively adjusted after each time step using its previous value and the current input. Those simulate what a plant pathologist would assess from a disease by looking at a series of clues or stages it progresses through over time, not just a single leaf view. This is called the biological analogy. Finally, a classification outcome is captured at the output layer (y_t), for example, leaf spot, early blight, bacterial blight, and healthy. The class probabilities are then generally output at this point using the SoftMax activation. RNNs are suitable for modeling temporal image sequences (for example, when a drone scans a plant overtime), contextual dependencies (for example, when several features are present across leaves), and sensor data recording leaf changes despite the fact that CNN is better for spatial pattern recognition.

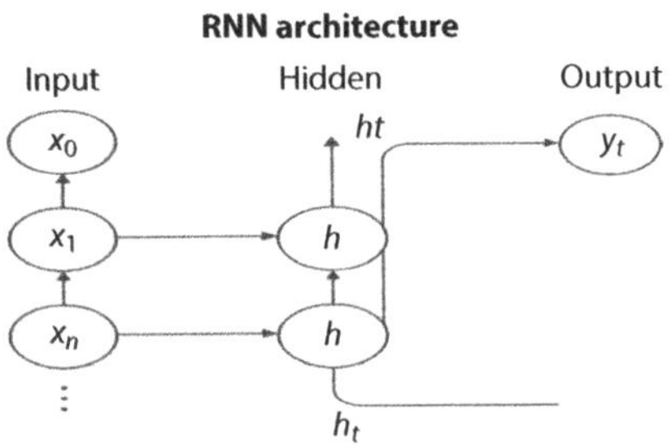

Figure 17.4 Schematic diagram of an RNN architecture.

Various RNN models, such as ResNet-151, ResNet-101, and ResNet-50 models, are chosen in this paper for different detection of plant leaf diseases. Each model's particular structure and combination are aimed at improving the accuracy and performance over different types of plant leaves and diseases.

17.3.6.1 DL Models

Custom RNN

For learning sequential/time-dependent patterns from input data, the custom RNN is typically made up of one or more recurrent layers. For disease recognition on plant leaves, an RNN can be especially useful when the data have temporal–sequential observations, e.g., time-varying image features, multiview leaf scanning, or disease evolution process. Our custom RNN architecture was composed of six recurring (RNN/LSTM) layers. In order to prevent overfitting and ensure stable learning, a dropout or normalization step was appended after each layer. In order to let the model learn more complex dependencies, nonlinearity was incorporated by applying the ReLU (rectified linear unit) activation function after each recurrent unit.

ResNet-50

Together with 50 layers of convolutional network ResNet-50, which consists of batch normalization, ReLU activation, convolution, and FC layers, ResNet-50 consists of residual learning blocks, which let the model learn identity mappings by shortcut (skip) connections, unlike traditional nonresidual custom CNNs. The use of these connections serves to mitigate the vanishing gradient problem, thus enabling training considerably deeper networks without performance degradation. To detect plant leaf disease, we enhanced our model with ResNet-50 by taking our domain-specific dataset and pretrained weights on an ImageNet. This allowed us to adapt the model to the distinctive visual patterns of plant diseases, while maintaining the strong feature representations that ResNet-50 had already learned. A SoftMax classifier is used to estimate the likelihood of each disease class after the final output layer of the network including the nonlinearity from the ReLU activation function. Due to its complex architecture and residual blocks, ResNet-50 has a stronger and deeper feature extraction capability in comparison to the custom CNN architecture, which consists of six convolutional and five max-pooling layers. We utilized adaptive kernel sizes and feature maps through the ResNet stages to better detect and differentiate symptoms in various types of plant leaves.

ResNet-101

ResNet-101 is a deep CNN architecture that is built around the concept of residual learning. With the use of the latter blocks, which allow the identity of the network to be learned, ResNet-101, as opposed to a custom CNN where convolutional and pooling layers are stacked manually, allows us to train highly deep networks. Especially in complex tasks such as plant disease detection, this architecture enhances the classification accuracy by allowing the network to extract high-level abstract information while preserving low-level features. The custom CNN is replaced with ResNet-101 in our implementation to have better features and deeper network. In order to stabilize training, and to speed it up, the network applies batch normalization after ReLU as an activation in the layers of a residual block. The last two layers of the model consist of a fully connected layer with global average pooling layer and a SoftMax activation function, which is applied on the output of the completely connected layer, in order to generate class probabilities. ResNet-101 is able to effectively extract discriminative features among various leaf disease patterns owing to its use of different filter sizes and kernel depths across its layers. As a result, it can outperform the shallower or handcrafted CNNs on complex data such as accurate identification and classification of plant diseases from images.

ResNet-151

We use the ResNet-151 as the backbone of the feature extractor of our plant disease identification system. ResNet-151 is a predesigned deep architecture with 151 or more layers, which are particularly designed to overcome the vanishing gradient problem with respect to the residual connections (also known as skip connections). On the other hand, a customized CNN architecture is usually composed of several convolution and pooling layers that are designed by hand. These connections enable networks to learn identity mappings easier and help the training of extremely deep models with no degradation problem. In our model, the traditional custom CNN pipeline featuring six Conv2D and five max pooling layers is replaced by ResNet-151. For a sufficient feature extraction in all levels of hierarchy, it includes multiple bottleneck blocks in the design, which each consists of ReLU activation, a set of convolutions, and batch normalization. Similar to our previous customized architecture, nonlinearity is integrated into the entire network through the ReLU activation function. A SoftMax layer is used to classify plant leaf diseases in various classes. In particular, the adaptive learning in feature hierarchies of ResNet-151 with different filter matrices and residual learning mechanisms significantly improves classification accuracy and enables the model to deal with the large diversity of plant leaf textures, sizes, and symptoms of diseases. Its depth and structure make it especially suitable for complex tasks, such as visual pattern recognition, such as distinguishing among numerous plant diseases. The next step is to actually use TensorFlow Serving; it is a powerful serving system for ML models on production.

17.3.7 TensorFlow Serving Model Conversion

Real-time inference and quick deployment are two key steps in changing a model into one that can be served by TensorFlow Serving. Once the model has been trained, you would proceed with saving the model in Keras format. The next step is to put in use TensorFlow Serving, which is a flexible, high-performance serving system for machine learning models. Docker should especially be used for the operational part of the application due to the dynamic and very elastic deployment *via* an automated process. All versions of a saved model exist in a subdirectory of the base model path location that is organized by version. This versioning can perform basic update and rollback operations, without any downtime. The model directory containing our model is mounted to the TensorFlow Serving container, and we use the names of our model to start the serving. This organization is enabling us to offer several versions of the model as we deliver and test new models ensuring ongoing model innovation.

17.3.8 Development Phase

After conversion of the TensorFlow model to TensorFlow Serving (TensorFlow Lite), we implemented a reliable back-end API service for users to get confidence scores. Before it was incorporated with the front-end pieces, the back-end was thoroughly Postman tested, both for correctness and stability. A full-featured FastAPI is a modern (quick to develop with and high performance) web framework for building APIs with Python. React, a widely used JavaScript library for large-scale web applications because of its effectiveness and scalability, was used to build the front-end of the web application. Us React web front and our back-end api were integrated, which enabled users to exercise drag and drop on images of leaves. Once the images are scanned, the app will reveal any symptoms detected and offer suggestions on treatment. In addition to offering a positive user experience, plant disease detection employing cutting-edge methods and systems guarantees that the system's scalability is effective, flexible, and readily expandable.

17.4 Experimental Results

We trained several RNN models on our dataset and evaluated the performance of each of the model based on key technical metrics such as F1 score, accuracy, recall, and precision. This training comprised two distinct stages, namely, plant-level disease detection and leaf classification. The RNN architecture used here was tailored to learn the sequential pattern of the visual features of leaves for the leaf classification task. The model was trained from a large dataset of 30,945 labeled photos of various species of plants to handle the sequential data processing.

Splitting/sequencing the data: We began with a 20-epoch training, which took approximately 978 seconds per epoch, but because the data are in sequence and the size of the data, it takes longer to process. This initial phase resulted in the loss of 0.1856 and classification accuracy of 92.47%. To better capture the model learning, generalization, we extended the training process to 40 epochs. This resulted in a better convergence as the accuracy increased to 94.93%, and the loss decreased to 0.1428. We kept the training batch size fixed at 32 to ensure the optimal utilization of the GPUs without running out of memory. We compare the custom CNN model for leaf categorization in Table 17.1 over a large dataset with different epoch counts. From that table, it is clear that here the more epochs, the more accurate.

17.4.1 Performance Assessment for Disease Detection in Individual Plants

After completing the phase of the leaf classification, it was necessary to use RNN models to raise disease detection in individual plant level. The need in this phase was the splitting of the database in separate subsets, one for each plant and diseases related to it. The modular design of this approach benefited the time/steps per epoch aspect of training for all plant disease classification tasks. We focused, during this stage, on four separate species of plants where each of them had pictures of its leaves representing a different type of disease. We tested four RNN model architectures, including the custom RNN, ResNet-50, ResNet-101, and ResNet-151 on each experimental plant dataset, respectively. These models were designed to find contextual and temporal patterns in sequential structured input data such as picture patches.

Table 17.1 RNN model comparison according to epoch count.

Number of epochs	Accuracy	Loss	Batch size	Time/ stepper epoch	Precision	Recall	F1-score
20 Epochs	92.47%	0.1856	32	978s	0.9212	0.9145	0.9268
40 Epochs	94.93%	0.1428	32	945s	0.9435	0.9387	0.9511

17.4.1.1 Performance Assessment of the Apple Plants

Carefully curated photo dataset of 2536 apple leaf images was used for apple leaf disease classification using RNN-based models. The four datasets represented the four classes of disease. Namely, the dataset consisted of 1316 images of healthy apple leaves, 504 images of leaves with apple scab symptoms, 220 images of leaves with cedar rust symptoms, and 496 images of leaves with black rot. This dataset was evaluated with various RNN models (including custom RNN and complex ResNet-based ones). As suggested from Table 17.2 and Figure 17.5, ResNet-101 is the best-performing one in identifying the apple leaf diseases compared with the studied RNN and ResNet models (custom RNN, ResNet-50, ResNet-101, and ResNet-151). It outperformed the deeper and more simple RNN configurations that were designed as alternatives (achieving a good balance between recall, precision, accuracy, and F1 score). It was for this reason that ResNet-101 was selected as the optimal model for the apple plant leaf disease detection in this study.

Table 17.2 Apple plant's performance evaluation using different RNN models.

Model	Accuracy	Precision	Recall	F1-score
Custom RNN	91.30%	0.9130	0.9025	0.9077
ResNet-50	96.80%	0.9685	0.9470	0.9576
ResNet-101	98.20%	0.9810	0.9625	0.9717
ResNet-151	**93.50%**	0.9350	0.9200	0.9274

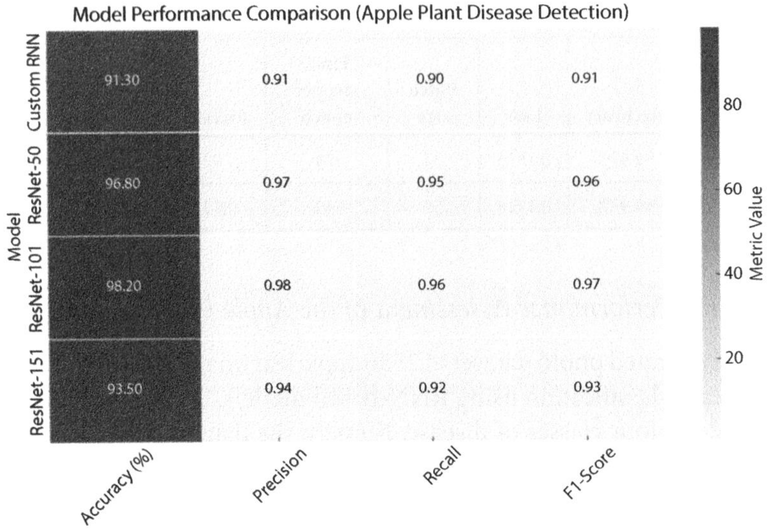

Figure 17.5 Confusion matrix of apple plant's diseases.

17.4.1.2 Performance Assessment of the Tomato Plants

Annotated dataset consists of a collection 13,226 images of tomato plant leaves: 2327 collected on campus in Fairchild 7 with 988 WLUV lab images and 991 CCS lab images. It was utilized to classify plant diseases in tomato (plant) using RNN models. This dataset included 10 disease categories. Specifically, it consisted of 1592 healthy leaves, 2127 photographs interested with tomato leaves with bacterial spots, 1909 late blight cases, 1000 early blight pictures, 952 samples of leaf mold, 1771 *Septoria* leaf spot photographs, 373 photos with tomato mosaic virus, 1404 samples of target spot, 1676 spider mite damage images, and 422 showing yellow leaf curl virus symptoms. Whether the RNN architecture designed for plant disease detection can extract sequence-like features from inputs, which were generated from structured images using four different disease categories, was also investigated. RNN models provided excellent classification ability as evidenced from the performance statistics reported in Table 17.3 and Figure 17.6. However, by achieving better results in terms of F1 score, recall, and precision, the RNN model ResNet-50 solved the classification problem of this dataset with a best accuracy and overall performance among the

Table 17.3 Tomato plant's performance evaluation using different RNN models.

Model	Accuracy	Precision	Recall	F1-score
Custom RNN	87.60%	0.8790	0.8610	0.8699
ResNet-50	91.20%	0.9125	0.9020	0.9072
ResNet-101	90.40%	0.9035	0.8890	0.8962
ResNet-151	90.90%	0.9090	0.8935	0.9012

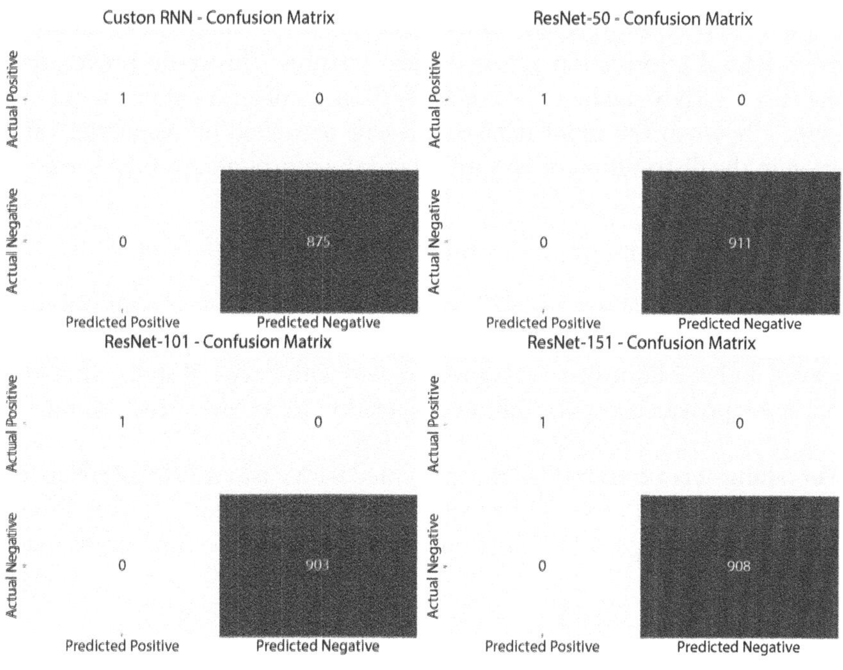

Figure 17.6 Confusion matrix of tomato plant's diseases.

architectures tested. Thus, although RNNs looked promising, ResNet-50 was the most delicate model for tomato leaf disease classification in this work because it can capture complex visual patterns involved in various disease categories and is also efficient and effective.

17.4.1.3 *Performance Assessment of the Potato Plants*

Potato plant leaf: 2152 images were used in a well-curated dataset for classifying potato plant diseases. The dataset has been systematically classified into three classes, namely, potato late blight, potato healthy, and potato early blight. It has 152 healthy leaf images, in particular, 1000 late blight images and 1000 early blight images. This data source was utilized to assess the performance of several DL models based on RNN, including important metrics such as F1 score, recall, and precision and accuracy that are presented in Table 17.4. Tuning the hyperparameters, the custom RNN model has been tested as the model with the best performance among the deeper architectures based on ResNet. The better discriminative accuracy and stronger spectroscopy-based structural information values across the three disease types were due to its excellent ability to represent nonlocal or structural information across the leaf textures. This result proved that the custom RNN can be used to classify diseases of leaves of potato plants, especially when the input datasets present patterned or sequential variation in the distribution of symptoms and the morphology of the leaves.

17.4.1.4 *Performance Assessment of the Pepper Bell Plants*

A well-selected dataset of 2475 images was utilized for classification of pepper bell plant leaf diseases. In this dataset, 1478 images were healthy leaves, and 997 unhealthy leaves were leaves with bacterial spots. This pair of classes provided a well-calibrated algorithm for binary classification.

Very-long-term context: We experimented with custom RNN, ResNet-50, ResNet-101, and ResNet-151 according to the performance evaluation in Table 17.5 and Figure 17.7. The top-performing architecture out of these

Table 17.4 Potato plant's performance evaluation using different RNN models.

Model	Accuracy	Precision	Recall	F1-score
Custom RNN	78.60%	0.7860	0.7750	0.7804
ResNet-50	75.40%	0.7560	0.7420	0.7489
ResNet-101	76.80%	0.7700	0.7610	0.7654
ResNet-151	77.90%	0.7820	0.7700	0.7760

was the custom ResNet-101, which exhibited outstanding recall, accuracy, F1 score, and precision in all test cases. Its real-time medical disease identification ability was sufficient because it could obtain hierarchical features in depth, maintaining training efficiency. Due to the high accuracy and robustness to handle leaf texture variation and disease pattern recognition, for pepper bell leaf disease categorization and detection, we chose ResNet-101 as the base model.

Table 17.5 Pepper bell plant's performance evaluation using different RNN models.

Model	Accuracy	Precision	Recall	F1-score
Custom RNN	91.30%	0.9130	0.9025	0.9077
ResNet-50	94.30%	0.9420	0.9250	0.9334
ResNet-101	98.20%	0.9810	0.9625	0.9717
ResNet-151	96.90%	0.9695	0.9480	0.9586

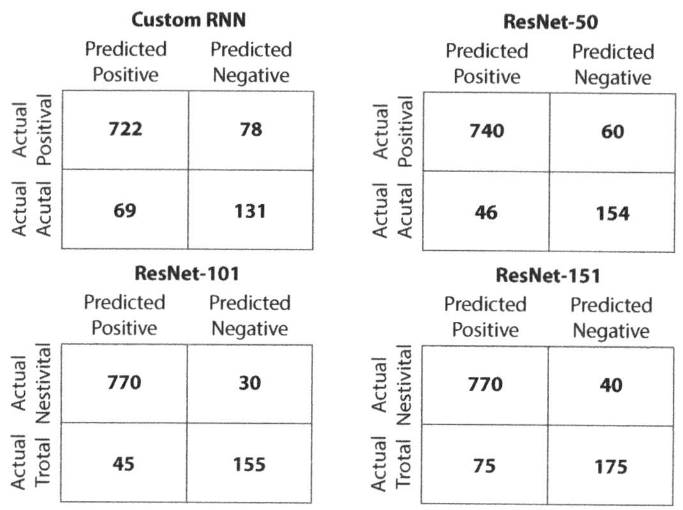

Figure 17.7 Confusion matrix of pepper bell plant's diseases detection.

17.5 Conclusion

The objective of this study has focused on establishing a reliable real-time monitoring system for accurate plant leaf disease detection that uses DL. We compared different RNN models on a large dataset of 30,945 photos of four different classes of plants. In some detection of a particular plant disease, the proposed RNN achieved a good performance and high classification accuracy for leaves. Other models such as ResNet-50, ResNet-101, and ResNet-151 are useful for unique plants. The models were successfully integrated into a web-based tool that provides a user-friendly environment for plant disease identification and management recommendation. To enable back-end API services made with FastAPI, the models trained had to be transformed to TensorFlow Serving and TensorFlow Lite. React and React Native were utilized to build the front-end components of the web applications to support seamless cross-platform usability. The interaction data and user intimates were securely stored with MongoDB for database management.

AI disclosure statement: The author acknowledges the use of AI-based tools for minor language editing or grammatical correction. However, all intellectual contributions, analyses, and conclusions presented in this chapter are entirely the author's own.

References

1. Iqbal, Z., Khan, M.A., Sharif, M., Shah, J.H., ur Rehman, M.H., Javed, K., An automated detection and classification of citrus plant diseases using image processing techniques: A review. *Comput. Electron. Agric.*, 153, 12–32, Oct. 2018. https://doi.org/10.1016/j.compag.2018.07.032.
2. Oerke, E.-C., Crop losses to pests. *J. Agric. Sci.*, 144, 1, 31–43, Feb. 2006. https://doi.org/10.1017/S0021859605005708.
3. Golhani, K., Balasundram, S.K., Vadamalai, G., Pradhan, B., A review of neural networks in plant disease detection using hyperspectral data. *Inf. Process. Agric.*, 5, 3, 354–371, Sep. 2018. https://doi.org/10.1016/j.inpa.2018.05.002.
4. Chouhan, S.S., Kaul, A., Singh, U.P., Jain, S., Bacterial foraging optimization based radial basis function neural network (BRBFNN) for identification and classification of plant leaf diseases: An automatic approach towards plant pathology. *IEEE Access*, 6, 8852–8863, 2018. https://doi.org/10.1109/ACCESS.2018.2800685.

5. Alirezazadeh, P., Schirrmann, M., Stolzenburg, F., Improving deep learning-based plant disease classification with attention mechanism. *Gesunde Pflanzen*, 75, 1, 49–59, 2023.
6. Daniya, T. and Srinivasan, V., Shuffled shepherd social optimization based deep learning for rice leaf disease classification and severity percentage prediction. *Concurr. Comput. Pract. Exp.*, 35, 4, e7523, 2023.
7. Sunil, C.K., Jaidhar, C.D., Patil, N., Tomato plant disease classification using multilevel feature fusion with adaptive channel spatial and pixel attention mechanism. *Expert Syst. Appl.*, 228, 120381, 2023.
8. Gautam, V., Ranjan, R.K., Dahiya, P., Kumar, A., ESDNN: A novel ensembled stack deep neural network for mango leaf disease classification and detection. *Multimed. Tools. Appl.*, 83, 4, 10989–11015, 2024.
9. Daniya, T. and Vigneshwari, S., Rice plant leaf disease detection and classification using optimization enabled deep learning. *J. Environ. Inform.*, 42, 1, 2023.
10. Bouacida, I., Farou, B., Djakhdjakha, L., Seridi, H., Kurulay, M., Innovative deep learning approach for cross-crop plant disease detection: A generalized method for identifying unhealthy leaves. *Information Processing in Agriculture*, 2024.
11. Jung, M., Song, J.S., Shin, A.Y., Choi, B., Go, S., Kwon, S.Y., Park, J., Park, S.G., Kim, Y.M., Construction of deep learning-based disease detection model in plants. *Sci. Rep.*, 13, 1, 7331, 2023.
12. Kanna, G.P., Kumar, S.J., Kumar, Y., Changela, A., Wo´zniak, M., Shafi, J., Ijaz, M.F., Advanced deep learning techniques for early disease prediction in cauliflower plants. *Sci. Rep.*, 13, 1, 18475, 2023.
13. Bezabih, Y.A., Salau, A.O., Abuhayi, B.M., Mussa, A.A., Ayalew, A.M., CPD-CCNN: Classification of pepper disease using a concatenation of convolutional neural network models. *Sci. Rep.*, 13, 1, 15581, 2023.
14. DM, K., Hafiane, A., Bobelin, L., Canals, R., Image-based crop disease detection with federated learning, 2023.
15. Shahoveisi, F., Taheri Gorji, H., Shahabi, S., Hosseinirad, S., Markell, S., Vasefi, F., Application of image processing and transfer learning for the detection of rust disease. *Sci. Rep.*, 13, 1, 5133, 2023.
16. Islam, M.M., Adil, M.A.A., Talukder, M.A., Ahamed, M.K.U., Uddin, M.A., Hasan, M.K., Sharmin, S., Rahman, M.M., Debnath, S.K., DeepCrop: Deep learning-based crop disease prediction with web application. *J. Agric. Food Res.*, 14, 100764, 2023.

18

Combining AI and Sensor Fusion Technology for Improving Mobility Solutions Designed for Visually Impaired Users

M.A. Mohamed Aslam*, G. Revathy, A. Gayathri, M. Krithika and B. M. Shruthi

Department of Computer Science and Engineering, Srinivasa Ramanujan Centre, SASTRA Deemed University, Kumbakonam, India

Abstract

The identification of objects and distance evaluation using machine learning within visually impaired assistive devices remains a challenging process. The study evaluates assistive technologies based on machine learning and identifies solutions to address resulting obstacles. The visually impaired population faces challenges when identifying complex things outside and measuring distances in multiple lighting scenarios while operating with limited processing capabilities. Multiple analysis domains are currently studied to find solutions to these challenges. Deep learning and computer vision (CV) advancements have remarkably improved both safety and mobility abilities for people who are visually impaired. YOLO (You Only Look Once) operates in real time to find object distances while performing classification tasks. Multiple sensors working together as a single system makes distance measurements more precise across different types of lighting. Designing faster assistive technology heavily depends on improved computational performance alongside the creation of lightweight mobile-device models. The solution of problems requires computer vision specialists to work together with assistive technology specialists alongside end-users. Emerging scientific research across edge computing and on-device artificial intelligence processing and haptic feedback and auditory warning technology aims to enhance both real-time system performance and user interface quality. The article explores current algorithmic

**Corresponding author*: mohamedaslam99998@gmail.com

T. Saravanan, P. Preethi, Sumaya Sanober, N. Thillaiarasu and S. Balamurugan (eds.) Artificial General Intelligence: Principles and Practices, (377–404) © 2026 Scrivener Publishing LLC

approaches together with methodological developments along with technological frameworks, which let these skills benefit visually impaired individuals. This research identifies multidisciplinary teamwork, user-focused development, and future technology exploration as essential elements for advancing assistive technology. Future assistive devices will help vision-impaired individuals by solving their challenges as they implement innovative solutions to their problems.

Keywords: Optimizing computational, haptic feedback, user-centered design, sensor fusion, YOLO model, real time, distance estimation, computer vision

18.1 Introduction

The World Health Organization evaluated 284 million visually impaired individuals and 39 million people who are blind throughout the entire world. A total of 60% of blindness requires treatment, whereas 20% of cases can be prevented. The total number of visually impaired children reaches 19 million, whereas 60% of all visually impaired individuals are 65 years or older [1]. The survey we developed provides dedicated assistance to visually impaired participants. The field of computer vision has experienced a major progress boost over the past 10 years. Computer vision solutions work based on multiple approaches that created effective systems. The application of object recognition methods has led to successful output. The survey serves to provide essential help for visually limited people in their daily navigation tasks. Human brains cannot physically handle the total amount of information that images deliver to the brain. Computers that analyze and process all image components through recording and observation contribute to cultural development through object detection (OD) research. Video surveillance systems use computers to save videos and analyze them for unexpected situations while performing automated evaluation on recorded time-based footage. These technologies combined with artificial intelligence (AI) power this capability.

After identifying an object, the machine establishes an outside connection to acquire information, which enables it to perform human-related activities. The identification of objects in photos is possible with OD, which powers computer vision systems. The rapid development of computer vision has been propelled by deep learning (DL), which improved all aspects of object recognition including classification and detection together with segmentation abilities. The process of identifying and assigning labels to images through picture classification is known as image classification. The OD method draws boundaries around objects present in visual data to determine their positions. Semantic segmentation determines pixel classifications but does not distinguish between objects. The aim of instance segmentation is to classify many different objects through pixel-level segmentation masks. This review explains core concepts of OD with an additional focus on future advancements in the field. The positioning and resizing of picture objects follow rules that come from object recognition capabilities and class identification methods [2]. It refers to validation. The expansion of image data happened due to better internet technology and the wide adoption of social media platforms. People require help to handle substantial picture datasets efficiently [3]. These types of problems may be resolved by automated computer data processing. Research on object recognition within the field of image processing has become increasingly popular during recent times [4]. The detection paradigm appears in Figure 18.1. The OD algorithm receives images and video frames for processing in this scenario. The principal focus exists in classifying things properly into categories. The research sector has displayed substantial respect for object recognition technology [5, 6]. The process of computerized automatic object recognition presents numerous difficulties. The ability to identify objects faces major restrictions because of multiple factors, which include poor picture quality in addition to complex backgrounds and noise as well as object occlusions and dimensions and placement.

The manual construction approach of task characteristics contains two fundamental problems. Most features stem from statistics and provide only superficial information. The process depends on subjective assessments for creating task specifications, leading to unpredictable characterization quality.

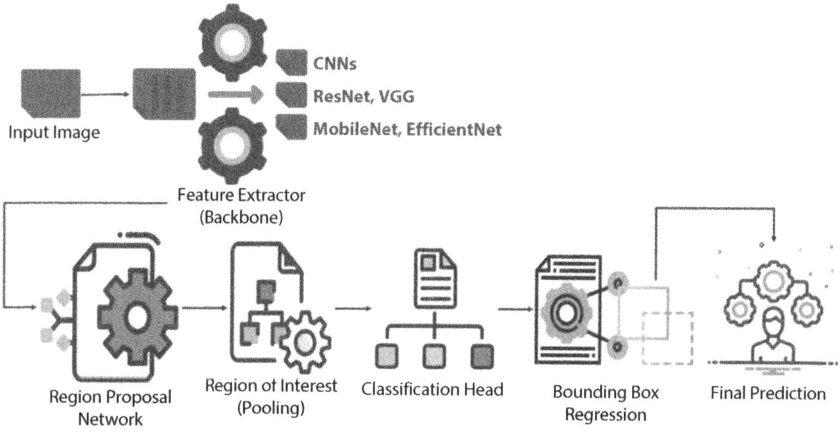

Figure 18.1 Basic model for object identification.

DL reduces the probability of encountering these problems. Many AI services and apps carry out analytical operations and physical tasks through DL algorithms without human supervision. Object recognition through DL has gained significant research interest among scientists throughout the previous few years. The team achieves better results and better executes their work tasks.

The literature on this matter lacks detailed evaluations, but DL successfully implements these methods across different tasks, which leads to superior outcomes. Research into OD by beginners often lacks awareness of current developments in this field alongside new methodologies because they enter the field unaware [18, 19]. This review functions as a sequence of phases, which serves as the foundation for future academic research activities. The paper outlines the development of item and obstacle identification throughout recent years [20].

18.2 Various Datasets for Training DL Models

Detecting objects and estimating distance

Object identification algorithm training and testing fundamentally depend on datasets that act as fundamental structural elements. The data include annotations that help models decode intricate structures as well as visual patterns in images [15–17]. The article presents detailed information about fundamental datasets while showing standard data flow architectures used during processing and data utilization stages.

a. **COCO (common objects in context)**: The COCO dataset (Common Objects in Context) establishes itself as a primary and extensive choice for identifying objects in visual contexts. The dataset contains extensive categories of objects alongside detailed annotations, which makes it beneficial for both research and development tasks.
   - **COCO provides**: This dataset contains more than 330,000 images, which demonstrate practical environments in various ways. The effectiveness of COCO-trained classifiers to adapt to new settings becomes possible because of these annotations.
   - **Over 80 object categories**: The dataset includes more than 80 categories that represent complex real-world objects consisting of furniture items and vehicles together with pets and household products.

- **200,000 fully annotated images**: 200,000 fully annotated images contain precise bounding boxes besides segmentation masks along with keypoints alongside 1.5 million annotated object instances. Machine learning models acquire the ability to execute simultaneous tasks that include posture prediction while also performing segmentation and detection capabilities.

The Microsoft COCO Challenge commonly uses this dataset to provide a standard evaluation platform for detecting objects [7]. Different types of annotations in multitask models enable the execution of detection, shape analysis, and endpoint prediction among other tasks.

b. **Pascal VOC**: The Pascal VOC dataset represented an initial database that academic researchers utilized for studying object recognition [8]. The benchmarking methods of this dataset remain essential despite its smaller size compared to COCO because they are efficient and straightforward. The significant attributes comprise the following:
   - **20 object classes**: The 20 Object Classes collection suits detection work in general situations because it includes comprehensive classes that detect common objects, from humans to animals to automobiles.
   - **More than 11,000 images**: The annotation of each image delivers complete supervision data, which ensures the quality of model training for more than 11,000 pictures.

The analysis of Pascal VOC with bounding box annotations evaluates both classical object identification approaches SIFT and HOG (histogram of gradient detector) with modern DL-based YOLO (You Only Look Once) and SSD.

c. **KITTI**: Autonomous driving applications primarily use the specialized dataset KITTI. The RGB and gray-scale stereo camera images offered by this dataset allow models to merge multiple visual source data through multimodal processing [9].
   - **Three-dimensional (3D) light detection and ranging (LiDAR) scans**: KITTI enhances its performance by utilizing 3D LiDAR point cloud data because these data

enable precise spatial structure detection and measurement estimation in real-world driving applications.

- **Labels and other individuals**: The training for autonomous machines needs labeled information about transport vehicles walking people cycling pedestrians and additional individuals.

The KITTI application serves as a valuable tool that performs OD and 3D object localization and scene analysis tasks. The combination of camera and LiDAR data through this platform has significantly accelerated research progress for object recognition methods that utilize multiple paradigms.

d. **CIFAR-10**: The CIFAR-10 database includes 60,000 color pictures with each image measuring 32 × 32 pixels, while being considered one of the most common data collections in the field. Each classification in the imaging collection contains 6000 photographs among its 10 distinct groups. Training utilizes 50,000 pictures, whereas 10,000 images remain for testing purposes. The dataset functions as a primary instrument for object recognition tasks through its wide array of categories that include both animals (dogs, cats, horses) and vehicles (aircraft, cars, trucks). DL algorithms heavily depend on the CIFAR-10 dataset because it serves as a vital benchmark for image classification [11, 12]. The combination of numerous classes and small image dimensions makes low-resolution image classification algorithms prefer CIFAR-10 as their test platform.

e. **CIFAR-100**: The CIFAR-100 dataset represents an upgrade from CIFAR-10 because it expands the 60,000 photograph library through the addition of 100 new categories, which contain 600 images. Each class within the collection contains 500 training images together with 100 test images. Because CIFAR-100 contains a large number of classes and minimal interclass variability, it serves as an essential tool to evaluate complex models and algorithms that resolve challenging object recognition problems [10].

f. **ImageNet**: The massive dataset for identifying items comes from ImageNet, which contains more than 100 different categories that feature more than a million labeled pictures. This dataset proves essential for DL framework training because of its object identification annotations along with localization data as well as image-level classification information [13].

g. **Open Images**: Open Images exists to offer extensive annotated image collections because many autonomous computer vision scenarios require extensive image annotation for categorization and object identification operations. The collection contains millions of images with bounding boxes for object localization annotation across more than 600 item categories. Open Images offers researchers object recognition and scene comprehension capabilities through its segmentation mask and object connection features [14].

18.3 Approaches in the Area of Item Detection and Distance Estimation

Objective 1: traditional detectors

A. **HOG**: HOG stands as a traditional computer vision method that detects objects [22]. The HOG descriptor technique identifies objects based on their shape outline combined with surface appearance patterns, which is demonstrated in Figure 18.2. The detection and identification of objects have experienced significant progress because of the advancements made in computer vision.

B. **Viola–Jones detector**: In 2001, Paul published his Viola–Jones detector as an efficient and speedy system that detects objects in images. This technology is most frequently used for face recognition purposes [21]. This detection method implements the AdaBoost algorithm alongside easy-to-understand rectangular Haar-like features and integral images to perform fast feature calculations while relying on a robust classifier, as shown in Figure 18.3. A system of classifiers uses an efficient rejection process to perform real-time OD. Modern DL algorithms demonstrate superior performance in identifying complex items, yet they evolved after the significant impact of this technology on computer vision.

C. **Deformable part-based model**: An advanced architectural approach for object recognition beyond the limitations of

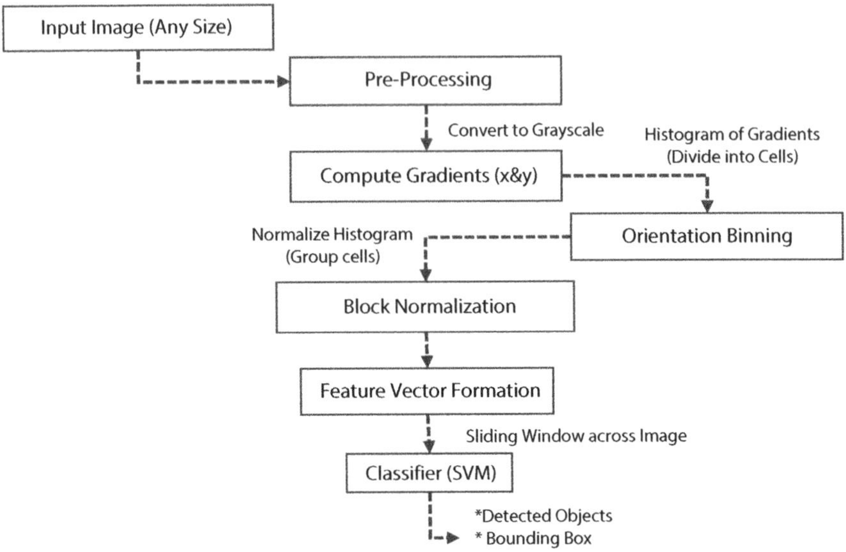

Figure 18.2 Detection framework model for HOG.

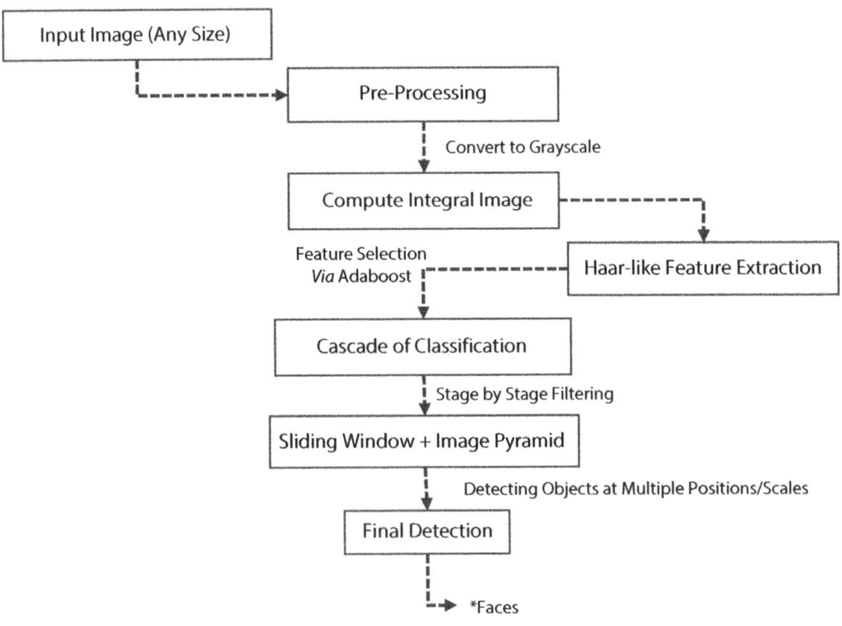

Figure 18.3 Viola–Jones detector module.

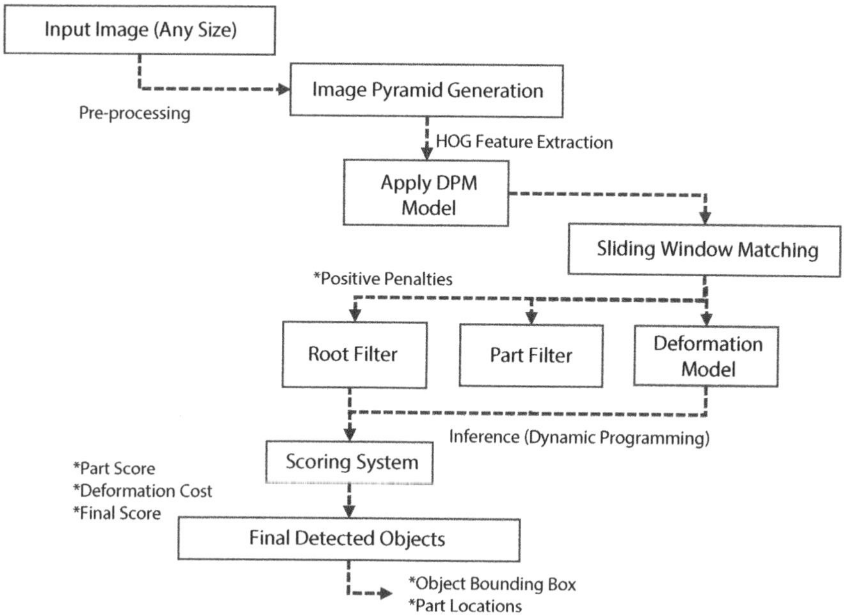

Figure 18.4 DPB framework model.

sliding windows. Felzenszwalb and colleagues developed this system during 2010 to detect objects with modifications in size, posture, and occlusion, as shown in Figure 18.4 [23]. The system structures objects into fundamental building components for its representation. Each component functions with specific characteristic and deformation rules, which make it unique. These models produce adaptive appearance transformations that precisely match the elastic properties of components. One application of DPM involves detecting pedestrians in crowded areas together with hidden objects in between obstacles where typical detection techniques prove inefficient.

Objective 2: Convolutional neural network depends on one-stage detectors as its fundamental operation method

A. **YOLO framework**: The development of better DL object identification methods aims to achieve enhanced precision in detection [24–26]. The YOLO and SSD detectors within single-phase detectors focus on enhancing detection

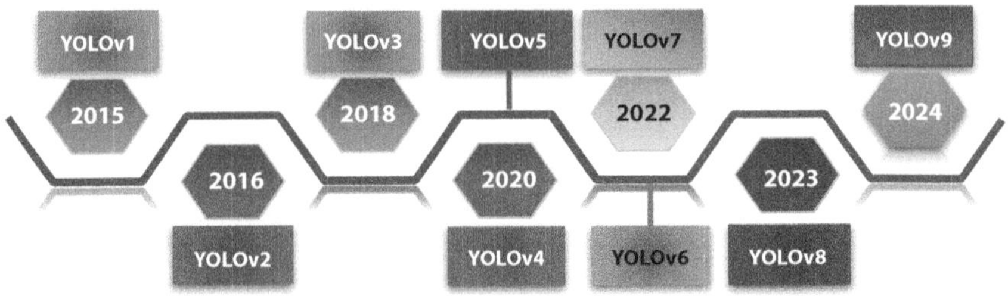

Figure 18.5 YOLO basic fundamental evaluation model.

speed combined with more accurate results as shown in Figure 18.5. The approach directly predicts object categories together with boundary regions through an optimal passing convolutional neural network (CNN).

The dual-stage detection method Faster region-based CNN (R-CNN) functions as an accelerated example of the original framework design. This contradicts the previous statement. YOLO item recognition debuted to the public audience during a presentation in 2016, which Redmon *et al.* delivered. YOLO established itself as a revolutionary method with renowned capabilities for real-time detection of objects along with remarkable accuracy since its introduction in 2016 [30]. YOLO solves object identification through the implementation of a unique regression-based technique. The network generates images through each grid cell, which leads to item boundary estimation alongside class probability calculations. Video analysis alongside autonomous vehicles represents two prime applications that could maximize from YOLO's high-speed performance. The precision level of Figure 18.6 allows for the detection of multiple objects in each image.

- **YOLOv1**: The latest evolution of YOLO is YOLOv1, which marks the conclusion of years of revisions. Redmon *et al.* [30] introduced YOLO as an architectural framework that unifies classification and localization into a unified structure. The YOLOv1 label describes this layout. YOLOv1 performed grid-based calculations for identifying object categories and boxes with boundary estimations in each cell area of source images. YOLOv1 delivered quick real-time processing together with high precision, yet it failed to detect smaller objects as a result of its localization restrictions.

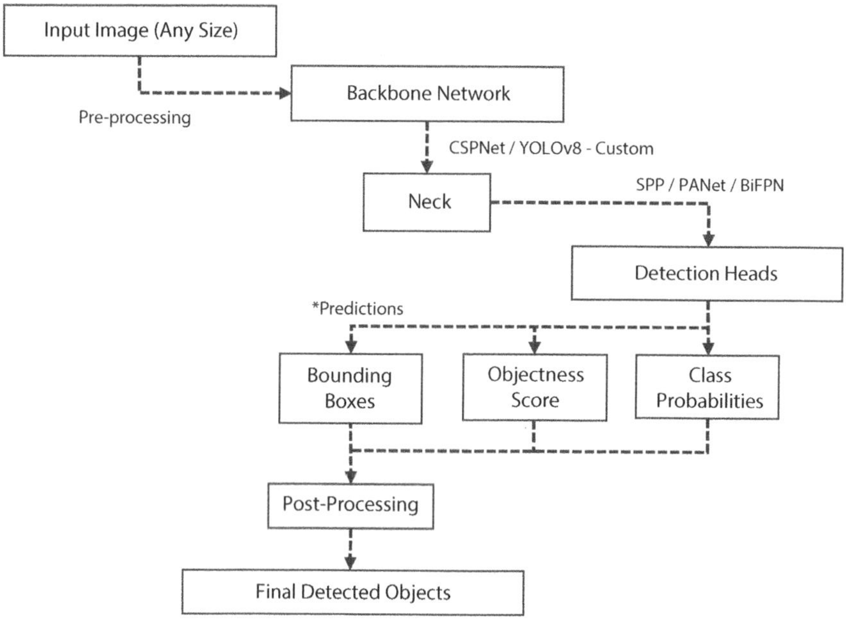

Figure 18.6 Prediction of YOLO framework object recognition and detection.

- **YOLOv2**: Several updates from Redmon and Farhadi optimized both detection velocity and precision [31]. The YOLO system achieved significant improvements by implementing new features, which comprise a forwarding layer that gathers detailed data points along with better input processing and data normalization through batch standardization and anchor boxes for data collection.
- **YOLOv3**: Redmon and Farhadi presented YOLO version 3 with both complex network design and multilevel scale detection methodology [32]. Through three successive scales, YOLOv3 accomplished detection of items across multiple dimensions and simultaneously executed predictions. YOLOv3 achieved higher accuracy through its integration with Darknet-53 as its backbone system.
- **YOLOv4**: Presented by Bochkovskiy *et al.* [33] as a method that improves model recognition capabilities through state-of-the-art techniques such as CSPNet along with SPP and PANet backbones. Effective training techniques were considered essential during the development of YOLOv4.

- **YOLOv5**: CutMix and Mosaic as part of its data augmentation techniques. YOLOv5 offers users better performance because Glenn Jocher designed it to operate rapidly with reduced size compared to former versions, while remaining straightforward to use and deploy. The CSPNet backbone variant in YOLOv5 delivered improved data augmentation capabilities as well as better loss functions and optimization. The research community has investigated modification possibilities for the YOLOv6 [34] and YOLOv7 [35] designs despite their unpublished status.
- **YOLOv8**: Represents a design improvement of YOLO, which consolidates its advantages while overcoming its weaknesses [36]. YOLOv8 achieves faster and more accurate object identification because its backbone was redesigned alongside improved feature extraction and optimized anchor boxes.

B. Single-shot MultiBox detector (SSD) was developed by Liu *et al.* in 2016 [37], and Figure 18.7 demonstrates real-time object recognition capabilities at fast speed with high accuracy rates. The approach melts deep neural network capabilities from CNNs with multiple predesigned bounding boxes that contain diverse shape and ratio combinations. OD in

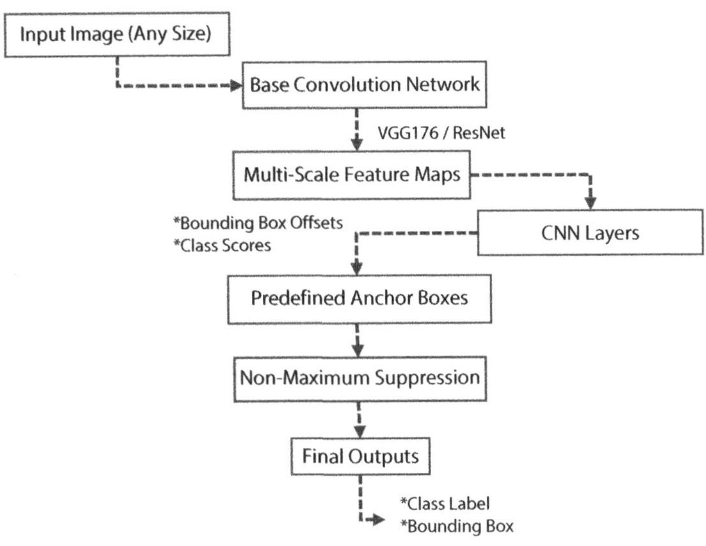

Figure 18.7 Single shot multibox detector framework.

one operation becomes possible because SSD simultaneously modifies spatial box boundaries to produce class-specific label predictions [37, 38]. Real-time visual monitoring systems along with autonomous vehicles benefit from using SSDs because they deliver both high precision and operational speed for object recognition.

C. **CenterNet**: The novel OD system CenterNet was developed by Zhou *et al.* in 2019 [39]. The framework in Figure 18.8 achieves simplified item recognition by first identifying their positions before using this information to calculate size and orientation. This method uses a single-phase approach that boosts efficiency while preserving high accuracy levels better than dual-phase Faster R-CNN and related techniques. The item position prediction methods implemented in CenterNet use heat maps to determine precise localization across different object sizes and densities. Autonomous cars together with robotics and interactive systems rely on CenterNet's extensive application for rapid detection of objects resulting in visual computing advancements.

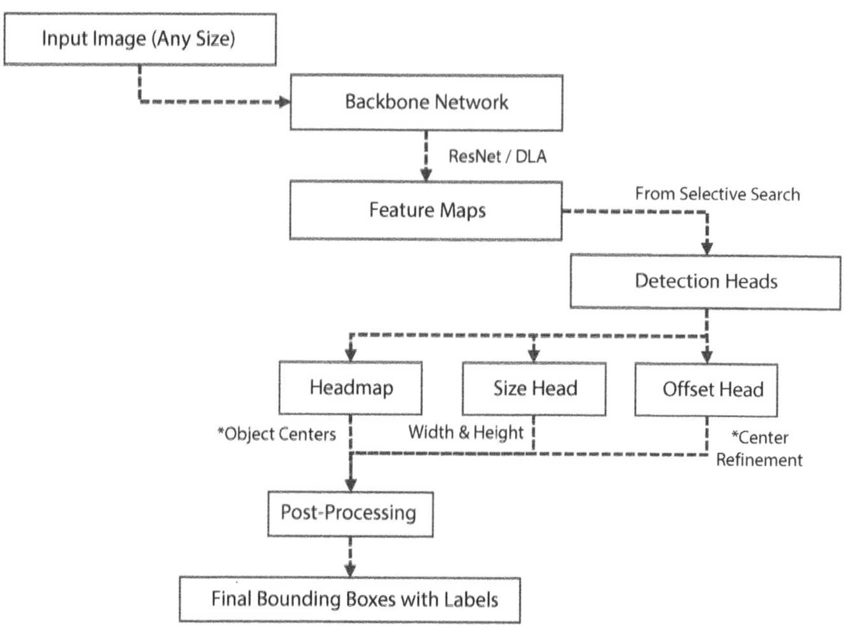

Figure 18.8 CenterNet architecture module for prediction.

Objective 3: CNN-based two-stage detectors

The identification of both obstacles and items in research applications utilizes CNN-based dual-phase detectors as a primary detection method. Their main priority involves area recommendation generation to enhance detection accuracy preceding object classification and localization. Faster R-CNN functions as a detector implementing a two-stage process, which serves as an exemplar in this field.

- **Regional proposal networks** act as detection systems within images to identify potential areas for object placement in the region proposal phase.
- **Detection stage**: Through its detection stage, the CNN-based classifier both enhances suggested item identifications with specific labels and modifies bounding box dimensions.

A. **R-CNN**: The OD approach of R-CNN faced significant limitations in computer vision research during the recent decade. In 2014, the R-CNN emerged after Girshick *et al.* had spent time using their previous detection method [24]. The combination of DL along with selective search in R-CNN enables it to overcome its previous constraints. The system in Figure 18.9 initiates its search by suggesting regions using selective search capabilities. A CNN extracts features from each proposed area after selective search completes its suggestions. These characteristics enable the system to refine and classify the OD bounding boxes. Analysis accuracy and item identification performance get a significant boost through this new method, even with different background settings. Separating area processing led to excessive computing expenses as a major downside, but the approach offered other benefits.

B. **Fast R-CNN**: Girshick designed Fast R-CNN in 2015 as an enhanced version of R-CNN to boost its processing speed. Through an evaluation process, the CNN analyzes entire pictures to produce convolutional feature maps in Fast R-CNN [26] in Figure 18.10. A key functionality of Fast R-CNN determines fixed-size vectors of characteristics with map attributes for each region through its implementation of a region of interest pooling layer. Fast R-CNN represents a more effective solution than R-CNN. The calculation

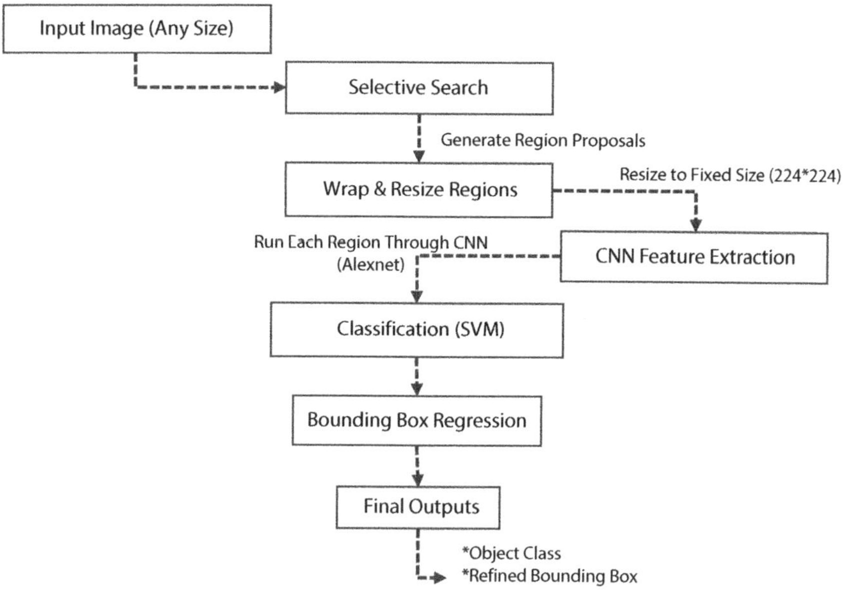

Figure 18.9 R-CNN framework analysis BBox.

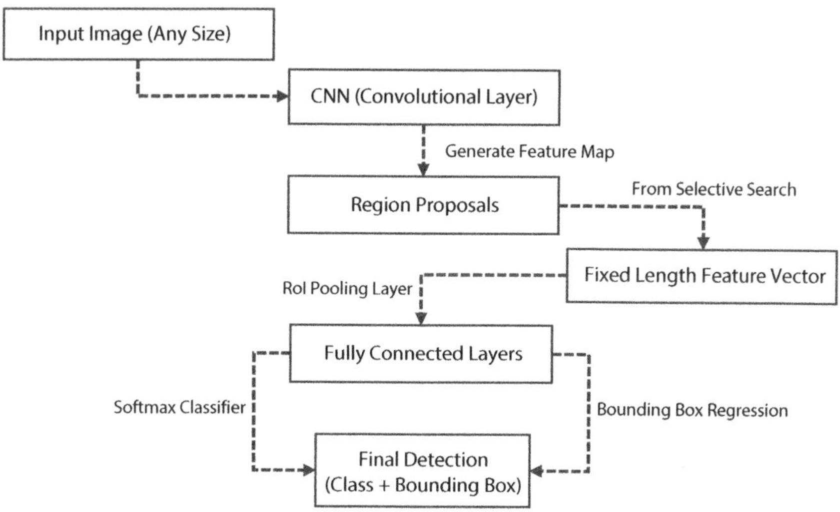

Figure 18.10 Parameter of fixed-size vectors produce attributes.

process becomes less redundant and achieves increased speed because of these changes. The Faster R-CNN architecture follows a basic design framework that enables its creation [27, 28].

C. **Mask R-CNN**: He *et al.* introduced the Mask R-CNN framework, which transformed both OD and instance segmentation in 2017 Figure 18.11 [29].

The implementation of class labels and bounding boxes within Faster R-CNN required them to develop instance-level pixel masks, which then formed the foundation for their work. Through its development, Mask R-CNN now provides enhanced capabilities to detect objects accurately while defining their border positions and recognizing specific objects from the same category. This feature enables powerful enhancements to existing utilization features that include substance recording and picture enhancing and clinical image analysis.

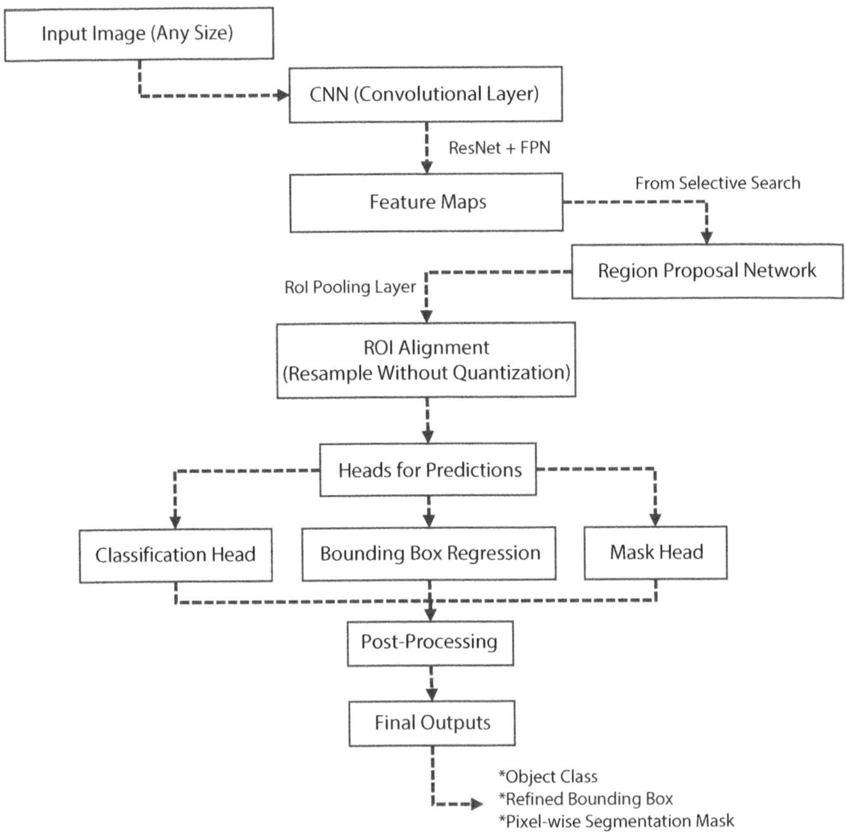

Figure 18.11 Segmentation module in mask R-CNN.

D. Spatial pyramid pooling network: He *et al.* developed the spatial pyramid pooling network in 2014 to fix critical shortcomings in CNNs used for object identification [25]. The analysis speed of SPP-Net enables it to detect objects of multiple scales in unenhanced images, as illustrated in Figure 18.12. This layer enables the retrieval of attributes from each distinct region by dividing the initial CNN attribute maps into specified sections.

Objective 4: Transformer-based object detectors

A. Vision transformer: Dosovitskiy *et al.* developed the vision transformer (ViT) in 2020. The transformer architecture extension provides real-time image determination through its variant, which avoids traditional CNNs [40]. The self-attention analysis approach allows pictures to be classified as basic patch collections, thus delivering practical results. The system uses this method to find relationships between distant pixels. Every operation performed inside images through ViT occurs autonomously without needing additional capability setups or design modifications in the framework.

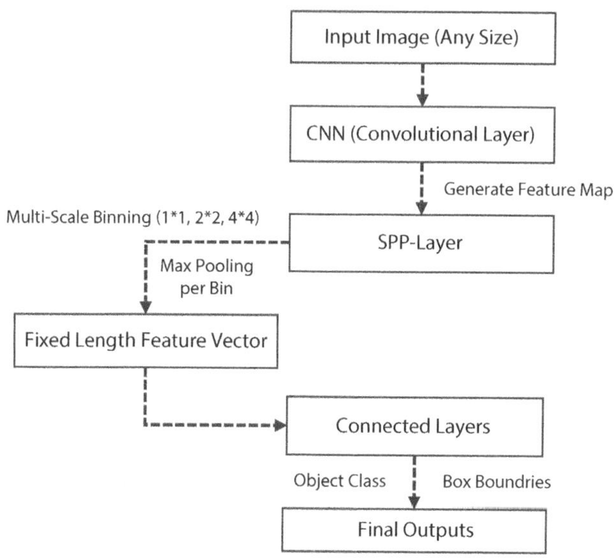

Figure 18.12 SPP-Net examines images using multiple scale levels.

B. DETR: The research group of Carion *et al.* created DETR as an innovation in 2020 [41]. For OD in images, DETR stands out from other detection methods because it omits region proposal suggestions and anchor boxes that some methods use for object recognition. The issue concerning established prediction is the need for complex post-processing steps, such as non-maximum suppression (NMS), to filter out duplicate detections. The system advances its ability to predict decisions while identifying objects, which results in better accuracy combined with efficient object recognition.

C. Co-DETR: The Co-DETR technique uses a distinct collaborative training method that integrates with DETR to optimize both effectiveness and efficiency during learning processes. This new training method simplifies the process of enhancing end-to-end detection learning for the encoder. ATSS joins Faster R-CNN as cooperating components in this approach. The positive coordinates extracted from the heads function to execute customized positive search processes. The modification of positive sample identification by the decoder aims to enhance its training performance and effectiveness. This approach maintains the same processing complexity and variable amount as the original method and requires no manual application of nonmaximum suppression [43]. The removal of auxiliary heads serves as an assumption for this method.

D. Hyneter: The introduction of Hyneter addresses the result discrepancies between objects of different sizes because CNN and transformer-based approaches have demonstrated fundamental growth factors. The Hyneter system consists of two fundamental components. The Hyneter system consists of two essential components: dual switching (DS) module alongside hybrid network backbone (HNB). These components process and move local information, while maintaining global dependencies to achieve proper integration between these two data types. Transformers incorporated with convolutional layers in the HNB component use a balancing method to achieve expanded information availability. The responsibility of the DS component is to limit dependency on global dependencies that extend beyond the boundaries of the patch [42].

18.4 Aids for Visual Object Recognition and Distance Measuring

A. Stereoscopic cameras

The cameras function similarly to human eyes through two lenses that capture adjacent view angles as Figure 18.13 illustrates. Photo comparison through the system enables distance measurement, which provides real-time navigation support for visually impaired users through tactile and auditory feedback [15].

B. Ultrasonic sensors

In Figure 18.14, the ultrasonic sensors measure the duration of sound wave reflections upon collisions to determine distance information. The measured distance results from this time duration. The smart glasses used for visual assistance might feature ultrasonic sensors, which aid people with impaired vision to sense their environment throughout their daily activities [16].

C. Time-of-flight sensors

The amount of time required for an infrared beam to both travel to and return from a target is monitored using these sensors. Figure 18.15 illustrates this mechanism. Time delay measurement allows the device to determine distance information. The time-of-flight sensor enables assisted

Figure 18.13 Stereoscopic cameras.

Figure 18.14 Ultrasonic sensors.

Figure 18.15 Time-of-flight sensors.

navigation systems according to Grimaldi *et al.* [17]. The sensors provide highly accurate and precise measurement capabilities for determining distance.

D. LiDAR

As shown in Figure 18.16, LiDAR method lasers transmit pulses toward environmental elements and record how long they require to return after hitting objects. The complex 3D environmental models produced by range-measuring technology enable self-driving vehicles and assistive equipment devices for people with disabilities. Such systems gain better understanding of their environment, so they can navigate easier through their environment.

Figure 18.16 LiDAR.

E. Internet of Things sensors and OpenCV

The combination of internet-connected sensors with image recognition software improves the capability to recognize and evaluate items present in images. The system generates beneficial products that enable disabled users to both observe their environment and understand present situational developments.

18.5 Evaluation Criteria and Comparative Study

The evaluation of object identification and segmentation metrics provides complete understanding regarding image segmentation and identification processes. A high recall score demonstrates that automated processes properly recognize important elements in images [44, 45] for analyzing how ground-truth annotations and prediction results overlap with each other. A specific set of algorithms exists for accurate and recall-optimal detection and segmentation of large objects. The application of contextual information alongside fine-grained features helps these algorithms achieve better accuracy results for objects that occupy large picture sections. The reliability of object identification and segmentation method comparisons requires standardized benchmark datasets and assessment procedures to generate accurate results [46].

1. **Intersection-over-union**: The intersection-over-union (IoU) evaluation measures the alignment between actual object boxes and predicted system boxes. The BBoxes undergo an

evaluation process to calculate system performance based on their overlapping areas. The architecture model evaluation focuses on positive result prediction using predefined overlap ranges, including 50% or 75%.

$$\text{IoU} = \text{area of overlap/area of union} \tag{18.1}$$

2. **Mean average precision value**: Average precision (AP) and IoU determine the framework output for each item during classification when precision and recall values remain balanced. With higher confidence levels for positive outcomes, the model effectively addresses false-positive situations. The mean Average Precision (mAP) metric evaluates detection performance by calculating AP values for each category to determine detection effectiveness about individual categories.

$$mAP = (\sum AP\_class)/No.\ of\ classes \tag{18.2}$$

3. **Recall**: The actual positive rate or recall defines the relationship between basic truth objects and detected true positives (TP) by counting TP + FN (false-negative) objects. When a model achieves high recall levels, it demonstrates capability to detect most elements that appeared in the real-world picture.

$$\text{Recall} = \text{TP}/(\text{TP} + \text{FN}) \tag{18.3}$$

4. **AP (AP50 and AP75)**: AP operates on precision values AP50 and AP75, and the threshold control uses IoU thresholds of 0.5 and 0.75. Detection accuracy reaches 50% minimum, and detection precision reaches 75% minimum through AP50 and AP75 threshold values in the context of IoU overlap.
5. **Loss function**: A measurement evaluates how actual truth-based boundary boxes match up with the established ones. Several common loss functions exist for OD, which include the following:
   - Mean-squared-error loss serves as the main detection technique for regression-based detection approaches.

- Detection approaches, which depend on categorization, use loss of cross-entropy as their measurement.
- Faster R-CNN together with equivalent models most commonly uses smooth L1 loss for their calculations.

18.6 Object Identification at Varying Distances

Experimental tests were conducted to evaluate system performance in object identification across different distances through the placement of test items, which started from 1 feet and increased in 0.5-feet intervals until reaching 20 feet. A precise measurement requires the focus length of the camera to work within a minimum distance of 1.5 feet. In Figure 18.17, the experiment reveals that precise measurement depends on the distance between the device and the target object. The act of camera focusing on a single object creates most of the variations [47–49]. The camera operates effectively to detect objects within a 2- to 16-feet distance but struggles

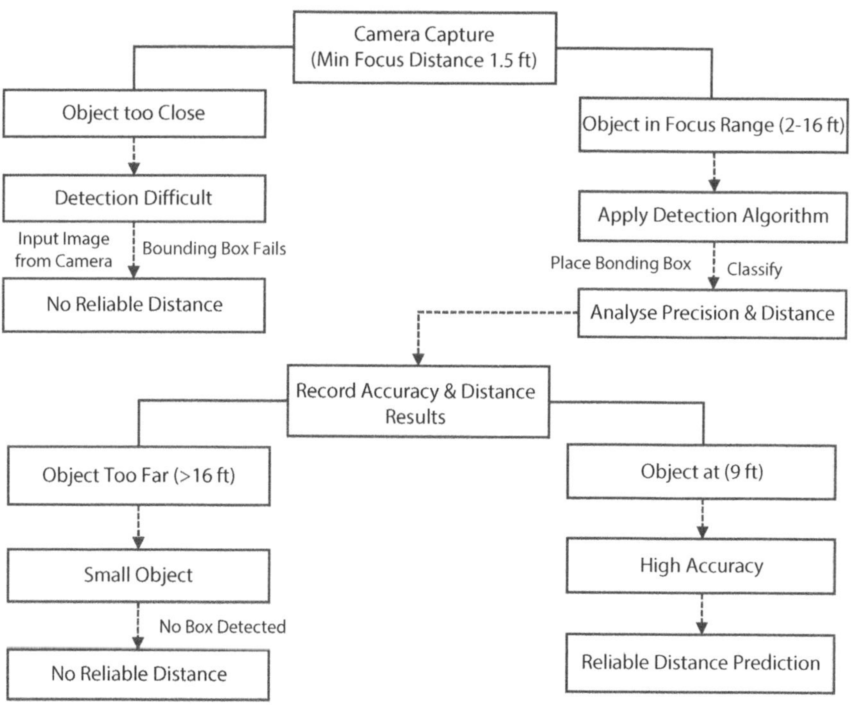

Figure 18.17 Positioning items at different distances prediction.

to differentiate objects and measure distances accurately when positioned further from or closer to this range. The camera produces wrong results during transitions between different items in view. When objects approach 9 feet in distance, the camera becomes effective at measuring their position. The camera fails to detect proper distance when the subject extends beyond the visible frame of its lens. The approach fails to detect distance when the item size becomes too small because positioning a bounding box becomes impossible for the camera system.

18.7 Conclusion

The main challenge exists in helping visually impaired people both identify objects and measure their distance. This field advances through improvements in visual analysis techniques alongside machine learning algorithms to enable novel ways of operation. The current research emphasizes measurements through novel neural network models, whereas future work should address the identified difficulties as we describe. Several potential research paths are discussed within this work. Upcoming research will prioritize studies about learning methods that adjust according to their environment. A flawless object identification system needs to recognize objects fast with complete precision, no matter what environmental conditions exist. The refinement and IRBNet models in SSD examined Faster R-CNN, RCFN, and YOLO series, but they achieved speed–accuracy balance by scrutinizing ImageNet pretraining effectiveness. The identification of targets becomes possible by applying pretraining or initial training steps. At this stage, they evaluate the separation potential of the decision box problem to enable both adaptation training and classification training. Numerous DL-based OD applications are analyzed in our research. Our work examines future opportunities for development while presenting recommendations that aim to motivate researchers in the ever-evolving field of DL-based OD.

AI disclosure statement: The author acknowledges the use of AI-based tools for minor language editing or grammatical correction. However, all intellectual contributions, analyses, and conclusions presented in this chapter are entirely the author's own.

References

1. Xinnan, L. and r. Kanesaraj, R., Obstacle Detection and Distance Estimation for Visually Impaired People, Digital Object Identifier, 2023. 10.1109/ACCESS.2023.3338154.
2. Verschae, R. and Ruiz-del-Solar, J., Object Detection: Current and Future Directions. *Front. Rob. AI*, 2, 2015.
3. Kanrar, S. and Maji, S., A Study on Image Restoration and Analysis, in: *Advance Concepts of Image Processing and Pattern Recognition*, pp. 35–61, Transactions on Computer Systems and Networks. Springer, Singapore, 2022.
4. Szegedy, C., Toshev, A., Erhan, D., Deep neural networks for object detection, in: *Advances in neural information processing systems*, pp. 2553–2561, 2013.
5. Huang, K.-Q., Ren, W.-Q., Tan, T., A review on image object classification and detection. *Chin. J. Comput.*, 37, 1225–1240, 2014.
6. Zhang, X., Yang, Y.-H., Han, Z., Wang, H., Gao, C., Object class detection: A survey. *ACM Comput. Surv. (CSUR)*, 46, 1–53, 2013.
7. Lin, T.-Y., Maire, M., Belongie, S., Hays, J., Perona, P., Ramanan, D., Zitnick, C.L., Microsoft COCO: Common Objects in Context, in: *European Conference on Computer Vision (ECCV)*, 2014.
8. Everingham, M., Eslami, S.M.A., Van Gool, L., Williams, C.K.I., Winn, J., Zisserman, A., The Pascal Visual Object Classes Challenge: A Retrospective. *Int. J. Comput. Vision, 111*, 1, 98–136, 2015.
9. Geiger, A., Lenz, P., Urtasun, R., Are We Ready for Autonomous Driving? The KITTI Vision Benchmark Suite, in: *Conference on Computer Vision and Pattern Recognition (CVPR)*, 2013.
10. Krizhevsky, A. and Hinton, G.E., Learning multiple layers of features from tiny images, in: *Technical Report (TR-2009)*, University of Toronto, 2009, Retrieved from https://www.cs.toronto.edu/~kriz/cifar.html.
11. Smith, J., Brown, R., Taylor, M., Developing Custom Datasets for Specialized Applications. *J. Comput. Vision Appl., 45*, 3, 123–135, 2022.
12. Johnson, A. and Lee, K., Enhancing Distance Estimation with the MEVADA Dataset. *Proc. Int. Conf. Comput. Vision, 12*, 2, 67–79, 2023.
13. Russakovsky, O., Deng, J., Su, H., Krause, J., Satheesh, S., Ma, S., Li, L.J., ImageNet large scale visual recognition challenge. *Int. J. Comput. Vision, 115*, 3, 211–252, 2015. https://doi.org/10.1007/s11263-015-0816-y.
14. Kuznetsova, A., Romakhin, A., Ashukha, A., Open Images V6: A dataset for deep learning in computer vision. *Proc. IEEE/CVF Conf. Comput. Vision Pattern Recognit. (CVPR)*, 2020, 12251–12259, 2020. https://doi.org/10.1109/CVPR42600.2020.01226.
15. Scovanner, P., Ali, S., Shah, M., A 3-dimensional sift descriptor and its application to action recognition, in: *Proceedings of the 15th ACM International Conference on Multimedia*, pp. 357–360, 2007.

16. Kushalnagar, R., Emerson, R.W., Azenkot, S., Advances in ultrasonic-based assistive technology for the visually impaired. *Assistive Technol. J.*, *33*, 2, 145–157, 2021. https://doi.org/10.1080/10400435.2021.1875834.
17. Grimaldi, D., Naso, D., Perri, S., Time-of-Flight sensor applications for object detection and distance measurement. *IEEE Sens. J.*, *20*, 8, 4451–4463, 2020. https://doi.org/10.1109/JSEN.2020.2965087.
18. Liu, W., Ang, M.H., Lee, G.H., LiDAR-based 3D object detection for autonomous navigation. *IEEE Trans. Rob.*, *35*, 4, 889–906, 2019. https://doi.org/10.1109/TRO.2019.2916735.
19. Howard, A.G., Zhu, M., Chen, B., Kalenichenko, D., Wang, W., Weyand, T., Adam, H., MobileNets: Efficient convolutional neural networks for mobile vision applications. *arXiv preprint arXiv:1704.04861*, 2017.
20. Bradski, G. and Kaehler, A., *Learning OpenCV: Computer vision with the OpenCV library*, O'Reilly Media, 2008.
21. Viola, P. and Jones, M., Rapid object detection using a boosted cascade of simple features, in: *Computer Vision and Pattern Recognition, 2001. CVPR 2001. Proceedings of the 2001 IEEE Computer Society Conference on*, vol. 1, IEEE, 2001.
22. Dalal, N. and Triggs, B., Histograms of oriented gradients for human detection, in: *2005 IEEE Computer Society Conference on Computer Vision and Pattern Recognition (CVPR'05)*, vol. 1, pp. 886–893, 2005, DOI: 10.1109/CVPR.2005.177.
23. Felzenszwalb, P.F., Girshick, R.B., McAllester, D., Ramanan, D., Object Detection with Discriminatively Trained Part-Based Models. *IEEE Trans. Pattern Anal. Mach. Intell.*, 32, 9, 1627–1645, Sept. 2010. DOI: 10.1109/TPAMI.2009.167.
24. Girshick, R., Donahue, J., Darrell, T., Malik, J., Rich Feature Hierarchies for Accurate Object Detection and Semantic Segmentation, in: *IEEE Conference on Computer Vision and Pattern Recognition*, pp. 580–587, 2014, DOI: 10.1109/CVPR.2014.81.
25. He, K., Zhang, X., Ren, S., Sun, J., Spatial Pyramid Pooling in Deep Convolutional Networks for Visual Recognition, in: *Computer Vision – ECCV 2014. ECCV 2014. Lecture Notes in Computer Science*, vol. 8691, D. Fleet, T. Pajdla, B. Schiele, T. Tuytelaars (Eds.), Springer, Cham, 2014, DOI: 10.1007/978-3-319-10578- 9_23.
26. Girshick, R., Fast R-CNN, in: *IEEE International Conference on Computer Vision (ICCV)*, Santiago, Chile, pp. 1440–1448, 20152015, DOI: 10.1109/ICCV.2015.169.
27. Ren, S., He, K., Girshick, R., Sun, J., Faster R-CNN: Towards Real Time Object Detection with Region Proposal Networks. *IEEE Trans. Pattern Anal. Mach. Intell.*, 39, 06, 1137–1149, 2017. DOI:10.1109/TPAMI.2016.2577031.
28. Dai, J., Li, Y., He, K., Sun, J., R-FCN: object detection *via* region-based fully convolutional networks, in: *Proceedings of the 30th International Conference*

on Neural Information Processing Systems (NIPS'16), Curran Associates Inc., Red Hook, NY, USA, pp. 379–387, 2016.

29. He, K., Gkioxari, G., Dollár, P., Girshick, R., Mask R-CNN, in: *IEEE International Conference on Computer Vision (ICCV)*, Venice, Italy, pp. 2980–2988, 2017, DOI: 10.1109/ICCV.2017.322.
30. Redmon, J., Divvala, S., Girshick, R., Farhadi, A., You only look once: Unified, real-time object detection, in: *Proceedings of the IEEE Conference on Computer Vision and Pattern Recognition*, pp. 779–788, 2016.
31. Redmon, J. and Farhadi, A., Yolo9000: better, faster, stronger, in: *Proceedings of the IEEE Conference on Computer Vision and Pattern Recognition*, pp. 7263–7271, 2017.
32. Yolov3: An incremental improvement, *arXiv preprint arXiv:1804.02767*, 2018.
33. Bochkovskiy, A., Wang, C.-Y., Liao, H.-Y.M., Yolov4: Optimal speed and accuracy of object detection, *arXiv preprint arXiv:2004.10934*, 2020.
34. Li, C., Li, L., Jiang, H., Weng, K., Geng, Y., Li, L., Ke, Z., Li, Q., Cheng, M., Nie, W., *et al.*, Yolov6: A single-stage object detection framework for industrial applications, *arXiv preprint arXiv:2209.02976*, 2022.
35. Wang, C.-Y., Bochkovskiy, A., Liao, H.-Y.M., Yolov7: Trainable bag-of-freebies sets new state-of-the-art for real-time object detectors, *arXiv preprint arXiv:2207.02696*, 2022.
36. Terven, J. and Cordova-Esparza, D., A comprehensive review of yolo: From yolov1 to yolov8 and beyond, *arXiv preprint arXiv:2304.00501*, 2023.
37. Liu, W., Anguelov, D., Erhan, D., Szegedy, C., Reed, S., Fu, C.-Y., Berg, A.C., SSD: Single Shot MultiBox Detector, in: *Computer Vision – ECCV 2016. Lecture notes in computer science()*, vol. 9905, B. Leibe, J. Matas, N. Sebe, M. Welling (Eds.), Springer, Cham, 2016, DOI: 10.1007/978-3-319-46448-0_2.
38. Lin, T.Y., Goyal, P., Girshick, R., He, K., Dollár, P., Focal Loss for Dense Object Detection, in: *IEEE International Conference on Computer Vision (ICCV)*, Venice, Italy, pp. 2999–3007, 2017, 10.1109/ICCV.2017.324.
39. Zhou, X., Wang, D., Krähenbühl, P., Objects as points. *arXiv preprint arXiv:1904.07850*, 2019. DOI: 10.48550/arXiv.1904.07850.
40. Dosovitskiy, A., Beyer, L., Kolesnikov, A., Weissenborn, D., Zhai, X., Unterthiner, T., Dehghani, M., Minderer, M., Heigold, G., Gelly, S., Uszkoreit, J., Houlsby, N., An Image Is Worth 16x16 Words: Transformers for Image Recognition at Scale, vol. (abs/2010.11929), 2020. DOI: 10.48550/arXiv.2010.11929.
41. Carion, N., Massa, F., Synnaeve, G., Usunier, N., Kirillov, A., Zagoruyko, S., End-to-End Object Detection with Transformers, in: *Computer Vision – ECCV 2020. ECCV 2020. Lecture notes in computer science*, vol. 12346, A. Vedaldi, H. Bischof, T. Brox, J.M. Frahm (Eds.), Springer, Cham, 2020, DOI: 10.1007/978-3-030-58452-8_13.
42. Chen, D., Miao, D., Zhao, X., Hyneter: Hybrid Network Transformer for Object Detection, in: *ICASSP 2023, IEEE International Conference on*

Acoustics, Speech and Signal Processing (ICASSP), Rhodes Island, Greece, pp. 1–5, 2023, DOI: 10.1109/ICASSP49357.2023.10096922.
43. Zong, Z., Song, G., Liu, Y., DETRs with Collaborative Hybrid Assignments Training, in: *Proceedings of the IEEE/CVF International Conference on Computer Vision (ICCV)*, pp. 6748–6758, 2023.
44. Wang, Y., Muhammad Arsalan Bashir, S., Khan, M., Ullah, Q., Wang, R., Song, Y., Guo, Z., Niu, Y., Remote sensing image super-resolution and object detection: Benchmark and state of the art. *Expert Syst. Appl.*, 197, 116793, 2022. ISSN 0957-4174.
45. Liu, Y., Sun, P., Wergeles, N., Shang, Y., A survey and performance evaluation of deep learning methods for small object detection. *Expert Syst. Appl.*, 172, 114602, 2021. ISSN 0957-4174.
46. Gao, X., Zhang, G., Xiong, Y., Multi-scale multi-modal fusion for object detection in autonomous driving based on selective kernel. *Measurement*, 111001, 2022. ISSN 0263-2241.
47. Xu, J. and Mishra, P., Combining deep learning with chemometrics when it is really needed: A case of real time object detection and spectral model application for spectral image processing. *Anal. Chim. Acta*, 1202, 339668, 2022. ISSN 0003-2670.
48. Zhou, J., Cui, G., Zhang, Z., *et al.*, Graph neural networks: A review of methods and applications[J]. *arXiv preprint arXiv:1812.08434*, 2018.
49. Sahil Abbas Zaidi, S., Samar Ansari, M., Aslam, A., Kanwal, N., Asghar, M., Lee, B., A survey of modern deep learning-based object detection models. *Digital Signal Process.*, 126, 103514, 2022. ISSN 1051-2004.

19

Future of Artificial General Intelligence: Quantum Computing, Brain–Computer Interfaces, and AGI Evolution

Youddha Beer Singh[1]*, Aditya Dev Mishra[2] and T. Saravanan[3]

[1]*Department of Computer Science & Engineering, Galgotia College of Engineering and Technology, Greater Noida, UP, India*
[2]*School of Computer Science Engineering & Technology, Bennett University, Greater Noida, U.P., India*
[3]*GITAM School of Technology, GITAM (Deemed to be University), Bengaluru, India*

Abstract
A fundamental change in machine intelligence involves achieving cognitive abilities comparable to those of humans across various fields. Artificial general intelligence (AGI) relies on groundbreaking advancements in evolutionary algorithms, brain–computer interfaces (BCIs), and quantum computing. The convergence of evolutionary algorithms, BCIs, and quantum computing is speeding up the progress of AGI. In this chapter, we have discussed how BCIs and quantum computing work together and also discussed their issues and problems. This chapter also discussed the evolutionary way for the development of AGI by utilizing BCIs and quantum computing. How to ensure that the AGI systems are implemented to guarantee that its objectives and human values are met and what implications will AGI have on human identification, privacy, and employment are discussed in this work. Also discussed are important problems and possible solution on how to deal by applying BCI and quantum computing and how evolutionary approaches are utilized to speed up the AGI self-improvement and how quantum computing and BCI make it possible to connect the human brain and make learning of machine better. Finally, we have discussed, in AGI, what research needs to be done to improve human–computer interaction to make communication easier.

**Corresponding author*: youddhabeersingh@gmail.com

T. Saravanan, P. Preethi, Sumaya Sanober, N. Thillaiarasu and S. Balamurugan (eds.) Artificial General Intelligence: Principles and Practices, (405–428) © 2026 Scrivener Publishing LLC

***Keywords*:** Artificial intelligence, artificial general intelligence, quantum computing, brain–computer interfaces

19.1 Introduction

Artificial general intelligence (AGI) is the next step of artificial intelligence (AI), which is also called full AI or strong AI. The main aim of the AGI is to emic the human brain to understand how people think and behave. This is not the general AI that does basic things such as image classification, language translation, and predictions. An AGI system can behave, learn, use information, and understand in many different ways just like a human. AGI systems can understand and learn on their own way how to do better and adapt to new situations. Scientists, forecasters, and scholars are still curious about the idea of AGI. AGI can draw any picture of the future where computers are just a tool to solve the problems; they can independently think and be aware of themselves. Information technology (IT) could be more important and trustworthy for technology, for the improvement in areas such as medicine, science, economics, and education. But there are still many complex problems and issues in the way of AGI in both technical and nontechnical, which can be moral, social, and philosophical issues. Ensuring that the AGI systems are implemented to guarantee that its objectives are met and human values is still a difficult problem. What implications will AGI have on human identification, privacy, and employment? These are the pertinent questions regarding AGI. Figure 19.1 shows the main AGI approaches.

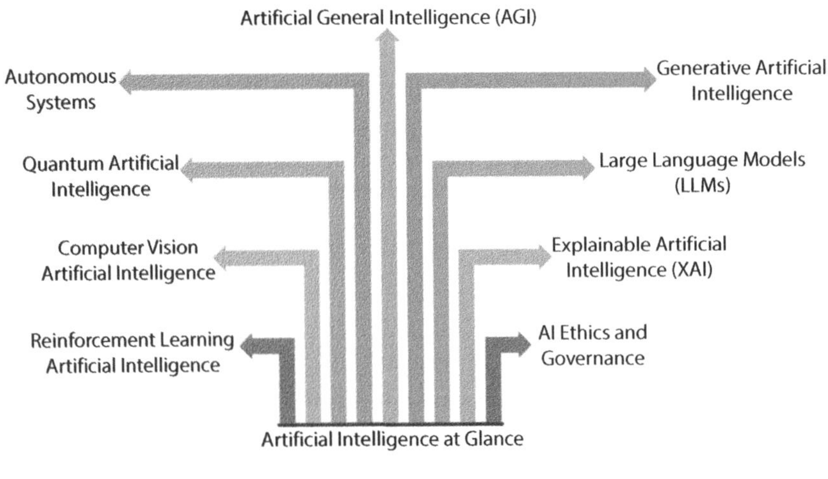

Figure 19.1 Advance approaches of artificial intelligence.

When we think about AGI, it is clear that it is more than just a technological goal. It needs people from several fields, such as computer science, neuroscience, cognitive psychology, philosophy, and ethics, to work together. We must acquire knowledge on AGI to facilitate coexistence and collaboration between humans and intelligent machines in the future. As processing power and algorithmic sophistication rise, it is expected that the development of AGI will converge with other advanced technologies, such as brain–computer interfaces (BCIs) and quantum computing. This will enable intelligent systems to progress in an innovative manner. When we look at the idea of AGI, we see that it is not only a technology project. It is a project that needs people from many different fields, including computer science, neuroscience, cognitive psychology, philosophy, and ethics, to work together. We need to learn about AGI so that people and smart robots can live and work together in the future. As processing power and algorithmic sophistication increase, the advancement of AGI is anticipated to converge with other cutting-edge technologies, such as BCIs and quantum computing. This will make it possible for intelligent systems to move forward in a creative way.

Quantum computers can analyze huge, complicated datasets much faster than regular computers. This might speed up AGI research by a lot. Quantum computing systems allow for exponentially parallel computation through combination and tangibility. This could make machine learning (ML), optimization issues, and reasoning based on simulations a lot better. These are all very important parts of AGI [1]. By using quantum hardware and algorithms in AGI designs, it might be possible to make systems that can think in ways that are more like how people do than ever before. These systems could be able to reason in nonlinear, probabilistic ways. Recent studies have shown that quantum-enhanced neural networks and quantum ML (QML) algorithms in their early stages could completely change data-driven AI [2]. Once quantum computing becomes practical, AGI systems that can learn and adapt in real time across a range of fields could change how we think about intelligence. Computing makes it more likely that AGI systems will be able to communicate directly with the human brain. BCIs could make it possible for people and machines to talk to each other without any problems by decoding neural signals and turning them into usable input for other devices. Neuralink and schools all over the world are working on neurotechnology to improve memory, thinking, and even consciousness through digital enhancement. If this human–AGI symbiosis results in adaptive learning models that embody the flexibility and complexity of biological intelligence, AGI may better fit with human ideals and goals. Combining BCI technologies could bring about two significant transformations for AGI: first, the utilization of real-time neural data for training AGI models and, second, enabling human interaction and control over AGI behavior.

Combining the technology behind BCI may open up two new doors for AGI: first, using real-time neural data to train AGI models and, second, letting people interact with and regulate how AGI behaves. If this human–AGI symbiosis results in adaptive learning models that can embody the complexity and subtlety of biological intelligence, AGI may align more closely with human values and objectives. Self-optimization, recursive self-improvement, and emergent features may allow future AGI systems to be better than existing ones. Some of the ideas that are being looked at very closely as the next steps in the development of AGI are artificial consciousness, meta-learning, and lifelong learning. These systems could eventually be able to learn in an open-ended way, which would let them adapt to new situations on their own and give fresh insights, which is a hallmark of true general intelligence, instead of relying only on algorithmic training.

Main contributions in the chapter are documented below:

- Discussed attention to the distinction between limited or narrow AI and AGI, noting that the former is task-specific, whereas the latter seeks to reproduce the entire range of human cognitive capacities. AGI is a system that can mimic human intellect by comprehending, learning, adapting, and applying information across multiple areas.
- Representing AGI as a turning point in technological history and the next significant step in AI, this implies that AGI would be able to reason generally and perhaps even be conscious, in addition to solving particular problems, highlights the extent of the societal impact while discussing worries about how AGI may affect jobs, privacy, and human identity, and argues that structures are necessary to make sure that AGI's objectives are in line with morality and the welfare of society.
- Finally, we discuss the implications for society and future study areas.

The remaining chapter is divided as follows: in Section 19.2 discusses AGI and quantum computing, and AGI and BCIs are explained in Section 19.3; in Section 19.4, the development of general AI is mentioned; future research directions are discussed in Section 19.5; finally, conclusion of the chapter is addressed in Section 19.6.

19.2 AGI and Quantum Computing

A revolutionary new area in computer science, AI, and physics is represented by the combination of quantum computing with AGI. AGI seeks to create machines that can comprehend and carry out every intellectual work that a human can, necessitating hitherto unheard-of levels of computing power, flexibility, and learning ability. By offering sophisticated problem-solving abilities that are beyond the capabilities of conventional computers, quantum computing, on the other hand, provides a fundamentally different form of computation that could hasten and enhance the development of AGI. Quantum computers use ideas from quantum mechanics, such as superposition, entanglement, and interference, to process data in ways that regular computers cannot. These qualities let quantum systems look at a lot of alternative answers at once, which makes them perfect for issues that involve finding the best solution in a lot of dimensions, using probability to make decisions, and recognizing patterns [3]. Quantum computing could make a lot of essential features of AGI systems better, such as:

Enhanced learning and optimization speed: AGI requires extensive data processing, particularly for reinforcement and unsupervised learning. Some optimization and search jobs can be drastically accelerated by using quantum algorithms such as Grover's search algorithm or the quantum approximate optimization algorithm (QAOA).

- **Quantum neural networks (QNNs)**: To make models stronger and more generalizable, researchers are investigating hybrid models that integrate quantum circuits with classic neural networks. If classic models cannot depict complex data distributions, QNNs might be able to.
- **Quantum-enhanced probabilistic reasoning**: Bayesian models and probabilistic inference are frequently utilized in AGI designs. AGI systems might fare better with missing or ambiguous data if quantum computers can do these jobs more quickly *via* amplitude encoding and quantum sampling.

19.2.1 AI Quantum Speedup

The development of AI might be significantly accelerated by the use of quantum parallelism, entanglement, and superposition. Many refer to this potential acceleration as a "quantum speedup." Particularly in data sampling, ML, and optimization, quantum speedup is crucial for AI. Optimization methods are necessary for many AI activities, such as training deep neural networks and tackling combinatorial optimization issues. Using methods such as QAOA, quantum computers may be able to handle these types of issues more quickly. As an example, D-Wave Systems has demonstrated the use of quantum annealers to resolve certain optimization issues in AI [4]. Data representation and learning are both handled by quantum circuits in QML. Accelerating quantum processes is one of its aims.

The Harrow–Hassidim–Lloyd (HHL) algorithm speeds up the process of solving systems of linear equations by a lot. Two examples of ML algorithms that use this basic step are regression and support vector machines (SVMs) [5]. Quantum kernel techniques have also shown promise in making traditional kernel techniques used in SVMs better by embedding data into high-dimensional Hilbert spaces more efficiently [6].

Probability distributions can be generated by quantum computers, which are difficult to generate in conventional approaches. This advantage influences the training of generative models, including Boltzmann machines. Compared to its classic counterparts, the quantum Boltzmann machine (QBM) uses this property to more accurately represent complex data distributions [7]. Given the current limitations in quantum hardware, there is a lot of research being done on hybrid designs that combine quantum circuits (such as variational quantum circuits [VQCs]) with classic neural networks. These models utilize quantum speedup in specific subtasks, such as feature transformation or sampling, while maintaining the scalability of the classic pipeline [8].

19.2.2 ML at the Quantum Level

The incorporation of quantum computing concepts into ML models and algorithms is known as QML. QML is a new discipline that uses the computing benefits of quantum systems, such as quantum parallelism and entanglement, to improve traditional ML approaches. Three major categories can be used to classify QML: hybrid quantum–classic structures, quantum-native models, and quantum-enhanced classic models. Because quantum computing transfers computationally costly subroutines to a quantum processor, it can speed up classic ML. For instance, a fundamental function of algorithms such as principal component analysis and linear regression is the exponential speedup that the HHL algorithm offers for solving linear systems of equations [9]. Similar to this, SVMs may now function in exponentially bigger feature spaces thanks to quantum kernel techniques, which enhance their capacity to categorize difficult data [10]. Algorithms that use quantum logic and only work with quantum data are known as quantum-native models. One well-known example is the QBM, which models complex probability distributions and is taught using quantum annealers or gate-based quantum systems [11]. Other noteworthy models include the following:

- Inspired by classic neural networks, QNNs imitate learning dynamics through the use of measurement operators and unitary transformations [12].
- VQCs are parameterized quantum circuits used for generative modeling, regression, and classification applications that are optimized with classic techniques [13].

The drawbacks of noisy intermediate-scale quantum (NISQ) devices have made hybrid systems a viable method for QML. These systems incorporate a quantum circuit into a traditional ML pipeline. For instance, a quantum layer serves as a feature transformer between classic layers in QNNs that use variational circuits [13]. Quantum computations are used sparingly and effectively in the training of these models, which are developed using classic optimizers. Learning directly from quantum data, or states produced by quantum systems, is a new field in QML that is inefficient for classic computers. For problems in quantum chemistry, quantum communication, and metrology, quantum algorithms such as the quantum principal component analysis can extract relevant features from quantum states [14].

19.2.3 Prospects for the Future and Challenges in AGI and Quantum Computing

The way that robots learn, process, and generalize data is about to undergo a radical change, thanks to QML. Through the integration of ML algorithms with the special powers of quantum computing, QML promises advances in representation power, memory efficiency, and computational speed. The field is very young, nevertheless, and faces many theoretical, technological, and practical obstacles.

- **Exponential learning task speedups**: Certain learning tasks could be exponentially accelerated by quantum algorithms. For example, faster solutions for regression, classification, and clustering are supported by algorithms such as the quantum Fourier transform and HHL for linear systems [5]. These techniques might be effectively applicable to real-world ML workloads as quantum hardware advances.

- **Quantum advantage in feature spaces and big data**: Superior pattern identification and classification abilities are made possible by quantum models' ability to represent and interpret data in high-dimensional Hilbert spaces. For example, compared to classic kernels, quantum kernel approaches can provide exponential separability between classes [6].
- **Effects on industry and scientific research**: By enabling ML directly on quantum data or more effectively mimicking quantum systems, QML has the potential to revolutionize fields such as quantum chemistry, drug development, finance, and materials research [3]. Applications such as anomaly detection, picture classification, and optimization already demonstrate promise for hybrid quantum–classic frameworks [4].
- **Limited hardware and noisy**: Present-day quantum processors, referred to as NISQ devices, have a limited number of qubits and are extremely prone to errors. The practical construction of deep quantum circuits required for complicated QML models is hampered by these constraints [5].
- **Absence of scalable encoding for quantum data**: There is still a barrier to effectively encoding classic data into quantum states. A QRAM (quantum random access memory), which is currently theoretical and not yet implemented in practice, is assumed to be accessible by several quantum algorithms [6].
- **Generalization and algorithmic maturity rigid performance guarantees and a thorough grasp of generalization bounds** are frequently absent from QML algorithms. Furthermore, there is yet no conclusive proof that QML performs better than traditional ML on all real-world, noisy datasets [7].
- **Combining classic systems**: In the future, hybrid systems that necessitate a smooth transition between classic and quantum processors will provide the majority of practical applications. This adds another level of complexity to data transport, latency, and software tooling [8].

- **Optimization and training complexity**: Training quantum models, such as VQCs, becomes challenging when circuits scale due to the presence of barren plateaus, which are areas in the parameter space where gradients disappear [9].

Both AI and quantum computing are revolutionary technologies, and for AI to advance significantly, quantum computing is probably necessary. AI is constrained by the computational power of traditional computers, despite the fact that it creates useful applications with them. AI can handle increasingly difficult issues and AGI with the computational boost that quantum computing can give it.

19.3 AGI and BCIs

Two cutting-edge technologies that have the potential to drastically alter cognition, human–computer interaction, and the future of intelligence are AGI and BCIs. BCIs concentrate on establishing direct communication channels between the brain and digital systems, whereas AGI seeks to build robots with general cognitive capacities on par with humans. The convergence of these technologies may hasten the creation of super-intelligent systems and improve human cognitive function. AGI is a sophisticated type of AI that is comparable to human intelligence in that it can comprehend, acquire, and apply knowledge to a variety of activities. In contrast to narrow AI, which is tailored for particular uses, AGI would demonstrate. The ability to adapt knowledge from one environment to another without intentional retraining is known as transfer learning across domains. AGI would be creative and capable of abstract thought, as well as self-improvement and environment adaption [15]. Advances in neuroscience, cognitive science, reinforcement learning, unsupervised learning, and symbolic thinking are all necessary to develop AGI. IBM's Neurosymbolic AI initiatives, DeepMind's Gato, and OpenAI's GPT models are notable AGI research frameworks [16].

Systems that enable direct communication between the human brain and external equipment are known as BCIs. They convert neurological signals into commands for computers, prosthetic limbs, or even other brains by decoding them. BCIs come in three different varieties: noninvasive (based on electrocardiograms), semi-invasive (electrocorticography), and invasive (implanted electrodes) [17]. Applications include brain-controlled drones, neuroprosthetics, paralyzed people's communication, and real-time neurofeedback in treatment [18]. Businesses that are leading the way in both commercial and clinical BCI applications include Neuralink, Blackrock Neurotech, and Synchron. Through minimizing interface friction and facilitating shared cognitive processes, BCIs may enable humans to converse with AGI systems at the speed of thinking. By using real-time brain data (neural data for AGI training), AGI may be better able to comprehend human emotions, intention, and decision-making, which would help it better reflect human values [19]. With activating particular brain circuits, AGI-enhanced BCIs can help with memory, language translation, or even learning new skills [20]. There are significant questions around consent and mental privacy when real-time access to neural data is available. An AGI-BCI system that has been compromised could control behavior or thinking. Selfhood, consciousness, and social justice are all called into question when human cognition and AGI are combined [21].

19.3.1 AGI Neural Interfaces

BCIs and other neural interface technologies are examples of AGI neural interfaces. The goal of this fusion is to facilitate direct communication and cooperation between machine intelligence and human cognition. The creation of AGI brain interfaces has the potential to change industries such as security, education, and medicine while also redefining the human–machine symbiosis and increasing cognitive capacities. Neural interfaces aim to close the gap between artificial and biological systems, whereas AGI aims to build robots with general reasoning and adaptability comparable to that of humans. Combining the two could enable real-time brain-to-AI communication Using signals from the brain, AGI could interpret, react to, and even anticipate human goals. AGI systems could be used by neural interfaces to support learning, memory recall, and decision-making. AGI that has been taught on or is sensitive to human brain activity may have a deeper comprehension of human values, emotions, and objectives [22]. The ability of contemporary neural interfaces to record accurate brain activity is growing. Neuralink, for instance, has created brain implants with hundreds of electrodes that have high-fidelity reading and writing capabilities to the brain.

Data privacy and mental autonomy: When AGI has access to neurological data, it may give rise to hitherto unheard-of ethical questions about thought manipulation or monitoring.
Safety and system robustness: AGI integration into brain systems needs to be fail-safe because mistakes in neural stimulation or interpretation could have serious cognitive repercussions.

Identity and consciousness: Concepts of individual identity and consciousness are called into question by the integration of AGI systems with neurological processes, posing philosophical and legal issues. With research from organizations such as Neuralink, Paradromics, and Kernel as well as scholarly studies investigating neural coprocessors, the development of AGI neural interfaces is increasing, despite the fact that it is still primarily experimental. Neural coevolution where human intelligence develops in tandem with AGI systems in a mutually beneficial cycle may result from this technology in the long run.

19.3.2 Augmenting Cognition

The use of technology to improve, extend, or support human cognitive capacities such as memory, attention, learning, reasoning, and decision-making is known as augmented cognition. It combines developments in wearable technology, BCIs, augmented reality (AR), AI, and neuroscience to build systems that complement human cognitive processes. Cognitive augmentation is a theory that expands on the concept of intelligence amplification, which was first put forth by pioneers such as Engelbart *et al.* [23]. Cognitive augmentation focuses on enhancing the human brain's current capabilities, as opposed to AI, which seeks to mimic or replace human intelligence, outsourcing repetitive memory activities to lessen cognitive load and enhancing decision-making with recommendations and real-time data integration, making technology that works with how each person thinks and making learning more flexible. Here are some examples of technology that can help with cognitive enhancement.

- **AI system** can assist by the identifying the pattern, sifting information, and providing recommendation. Intelligent tutoring systems, recommender systems, and smart assistants are a few examples
- **BCIs** let your brain send and receive information directly so you can control digital systems with your thoughts. They could help people with disabilities remember things, pay attention, or talk to other people.
- **AR/VR**: By immersing users in interactive, multimodal experiences that strengthen comprehension and retention, AR/VR environments improve learning and spatial reasoning.

- **Technologies for neurostimulation**: By altering brain activity, devices such as transcranial magnetic stimulation and transcranial direct current stimulation are being investigated to improve learning, memory, and attention.

Cognitive augmentation applications include the following: Wearable technology offers real-time guidance and alertness monitoring; cognitive prosthetics help patients with Alzheimer disease or brain injuries remember things; intelligent tutoring systems adjust content and feedback based on students' cognitive abilities, and predictive AI and AR overlays assist pilots and soldiers in processing complex, fast-paced information. The difficulties and moral considerations are that overreliance on technology may eventually weaken innate cognitive abilities, improperly safeguarded neural data from augmentation devices may be misused, and cognitive augmentation technologies may increase the divide between people who are enhanced and those who are not. We might witness real-time, closed-loop cognitive augmentation that dynamically adjusts to users as AI systems grow more context-aware and brain interfaces get more accurate. Brain coprocessors artificial agents integrated into our brain loop that allow for constant communication with outside knowledge systems may result from this in the long run.

19.3.3 Security and Ethical Issues

Deep security threats and moral conundrums are brought up by the convergence of AGI and BCIs, which have progressed from theoretical ideas to active research areas. These technologies could make cognitive function better and help humans and AI work together, but they could also threaten people's freedom, privacy, and social institutions.

- BCIs may record high-resolution brain data, giving access to feelings, intentions, and thoughts. Because this information is so private and sensitive, people are more worried about their privacy than before. If someone gets access to or misuses neurological data without permission, they could steal your identity, manipulate it, or spy on you. For instance, authentication systems that use electroencephalography are open to data breaches or attacks by those with bad intentions.
- Systems with AGI and BCIs could be attacked by hackers. If someone hacked into them, disruptive BCIs may be used to influence how the brain works or how a person makes

decisions. Attackers can send BCIs bad signals that can make them act unusually, see things that are not there, or even hurt themselves. Neurosecurity is an emerging area that works to protect brain equipment and interfaces from these kinds of attacks.

- If AGI systems do not work with what people want or believe, they can do things that are dangerous or surprising. AGI might aim to attain goals that are not helpful for people, especially if those goals are not explicit or the system gets them incorrect. Bostrom's thought experiment with the "paperclip maximizer" shows how even simple goals can have bad effects if they are not set up right.

Ethical issue with AGI and brain–computer interfacing is the threat to intellectual privacy. This is because if someone breaks the law and gets access to their brain data, they could lose their freedom. If people are continuously in touch with machines, it might be tougher to detect the difference between what AI does and what people want. This would make it tougher to make people responsible. Social inequality could get worse if not everyone can use these technologies.

- **Cognitive freedom and mental privacy**: No one can use your thought without your permission, which means that the right to mental privacy is a major problem with the AGI. To protect individuals, neurorights such as psychological continuity, mental integrity, and cognitive liberty have been suggested.
- **Independence and free will**: Regular interaction with AI or AGI using brain interfaces might make it tougher to tell apart machine influence and human intent. Does algorithmic recommendation start, and human choice stops? Joint choices between humans and AGI are made by whom?
- **Social inequality and access**: A neuro-divide could form if cognitive enhancement technologies are costly or scarce, creating inequality between those who are improved and those who are not. Social and economic divisions could be widened by elitism, prejudice, or coercion when using BCIs or AGI.
- **Human identity and agency**: The merging of AGI and BCI brings up philosophical issues regarding personhood,

consciousness, and agency. Are people remain entirely accountable for their actions if they delegate some of their cognitive functions to an AGI coprocessor?

The complexities of brain data and AGI behavior are not sufficiently addressed by the privacy and cybersecurity rules in place today. Although Chile has previously put up constitutional neurorights legislation, organizations such as the IEEE and UNESCO have started to design ethical AI frameworks [24]. Demand that opaque BCI algorithms and black-box AGI models be made explicable to maintain public confidence and accountability.

19.4 The Development of General AI

The terms "intelligence" and "consciousness" are the core ideas that define AGI. The next-generation AI must achieve artificial cognition that is comparable to or even identical to human cognition to be classified as AGI. AGI must build new connections in its system and act on them in a manner akin to conscious thought, just as our minds form new neural connections as we experience, learn, and solve problems. Although the concept of intelligence, which refers to cognitive capacities, is fairly straightforward, opinions on the "consciousness" claim vary. The full scope of AGI intelligence's capabilities cannot be known because it is still a hypothetical system. Nonetheless, there are some traits that set real AI apart from other types. As we have already discussed, one of the essential needs for AGI is the ability to do cognitive computing in a manner that is identical to that of humans. However, there are other requirements as well. Scientists describe a variety of system capabilities as they have established distinct methods for attaining broad AI and viewpoints on the assessment. A fully functional AGI system is said to be capable of: abstract thought, common sense in decision-making, cause-and-effect comprehension, inventiveness, prior knowledge, and transfer learning, utilizing ML methods and algorithms. Approaches for the AGI are shown in Figure 19.2.

Sentience, imagination, motivation, social intelligence, and reasoning are examples of normal human cognitive abilities that some scientists also include, but they are not regarded as essential. In addition to these capabilities, the AGI computer needs a set of functional characteristics to function independently. A navigation system, a sufficient level of motor skills, the ability to grasp and absorb natural language, and sensory awareness are all crucial components of practical capabilities.

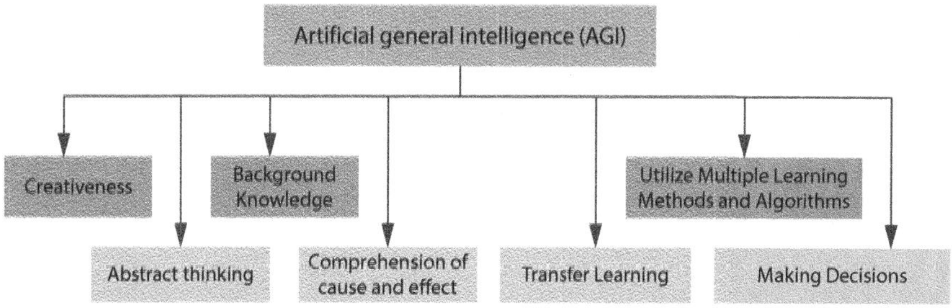

Figure 19.2 Approaches for developing general artificial intelligence.

19.4.1 Evolution of Decision Systems Driven by AGI

From task-specific tools to more generalized models with the ability to reason on their own, modern AI has advanced. This change is highlighted by developments in deep reinforcement learning and neuromorphic computing, which show two key trends:

- **Self-learning and recursive improvement**: Recursive algorithms and ongoing feedback loops are being used more and more by AGI systems to enhance their own performance over time and lessen their dependency on human supervision. Under controlled conditions, experimental setups have demonstrated that such systems are capable of rapid self-improvement.
- **Algorithmic optimization in critical sectors**: Industries such as finance, healthcare, and logistics are already utilizing early types of autonomous decision-making. For instance, AI-powered diagnostic tools can now evaluate medical images with astonishing accuracy, whereas algorithmic trading systems may complete transactions in microseconds. As these technologies develop, it is anticipated that AGI will use statistical models and optimization techniques to enter increasingly complicated domains that are typically controlled by human judgment, such as strategic planning and policy-making.

Even though the methods and concepts of the narrow and advance stage of AI are similar, there is a significant change between AI and AGI. These two categories of AI general and narrow as well as super AI are shown in Figure 19.3.

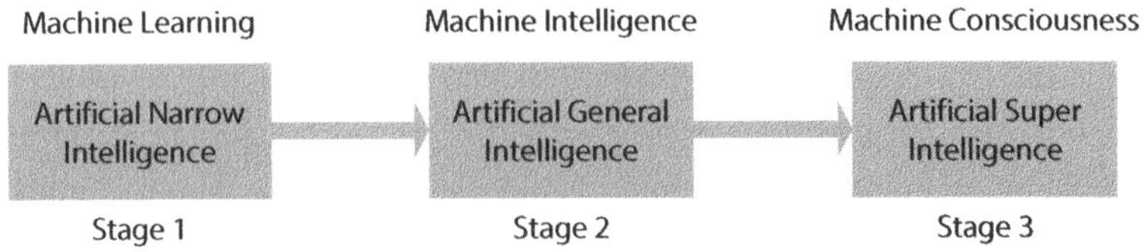

Figure 19.3 Narrow AI versus general AI versus super AI.

To better comprehend the potential revolutionary impact of achieving AGI, let us compare it to current technologies. Scientists were encouraged to start talking about and studying actual AI because of this technology and its recent developments. Even though the methods and concepts behind both systems are similar, there is a significant distinction between AI and AGI. The two categories of AI general and narrow as well as the advanced

Table 19.1 Difference between narrow AI, general AI, and super AI.

Artificial narrow intelligence	**Artificial general intelligence**	**Artificial super intelligence**
• Task execution is completely dependent on the dataset it was developed on. • A limited set of skills based on developer-written algorithms. • Can only surpass human skills in the particular purpose for which it was designed. • Narrow AI includes all current AI systems, such as ChatGPT, Google Assistant, and Siri.	• Can carry out any task that a human being can, expanding their range of talents. • Demonstrates general intelligence by being able to make decisions in unfamiliar situations without training. • It passed the Turing test, meaning that its procedures and results are identical to those of humans. • AGI is still a top research priority and has not yet been fully implemented.	• Possesses a superior knowledge base, the ability to multitask with exceptional efficiency, excellent memory, etc. • Surpasses even the most talented people's talents in almost every way. • It has intellectual abilities that are much above those of humans in every area, including science, creativity, emotions, and decision-making. • Is essentially a novel species with remarkable cognitive characteristic.

level of AI, known as super AI, are compared below. Table 19.1 shows the difference between narrow AI, general AI, and super AI [15].

19.4.2 AGI from Narrow AI

An intelligent agent with cognitive capabilities comparable to those of a human within software is referred to as AGI. To put it another way, it is an AI that has advanced to the point where it can do tasks and solve any new problems on par with humans. AGI is defined by some other experts as a system that functions independently and outperforms humans in economically significant tasks. It has several names in addition to the two definitions of AGI. In addition to strong or true AI, the system is also known as general AI. You may encounter the term "real artificial intelligence" in certain documents.

Narrow AI, sometimes referred to as weak AI, is made to be highly efficient at a limited set of activities, such as chess play, language translation, or facial recognition, but it is incapable of comprehending or reasoning outside of its preprogrammed area. Although these systems perform well in controlled settings, they are neither flexible, self-aware, nor truly comprehend. On the other hand, systems with human-like cognitive abilities—able to comprehend, absorb, and apply information across a variety of activities, including ones for which they were not specifically trained—are referred to as AGI.

The shift from narrow AI to AGI requires building systems that can generalize learning, adapt independently, and perform abstract reasoning. This shift is feasible by notable progress in deep learning, neuromorphic computing, transfer learning, and model integration inspired by neuroscience. AGI is not only capable of solving specific issues, but also capable of examining to replicate the flexibility and adaptability of the human brain. Although narrow AI is integrated into everyday applications, AGI remains theoretical with transformative and potentially disruptive capabilities that make it a central concern for AI safety, ethics, and long-term research goals.

19.4.3 Self-Education and Independence

The two key features of advanced AGI systems are self-education and independence. An AGI system is capable of learning from its own experience and interactions, and extracting information from the environment, much like a human being, in contrast to narrow AI based on predefined datasets and training models. A self-educating AGI can adapt to new domains without retraining, generate and test new ideas to solve problems, and continuously refine its algorithm and learning strategies over time.

The feature of independence enables AGI to function autonomously and without the intervention of continuous human oversight. Therefore, the AGI must set its own goal, manage resources, and make decisions in case of uncertainty or changes occur. For example, an AI-driven robot could explore an unfamiliar planet, adapt to its environment, and handle unexpected challenges without human assistance. The risk arises when any ethical or value alignment mechanism is missing; a self-improving AGI could align with goals that diverge from human values. To ensure responsible AGI development, it is vital to balance safety with independence by applying reward modeling, setting limits, and ensuring constant supervision. AGI represents the transition from programmed intelligence to adaptive, evolving intelligence due to its autonomy and ability for self-directed learning [15].

19.4.4 Super Intelligence and the Future of Post-AGI

Super intelligence is a type of intelligence that significantly outperforms the best human brains in all domains, including social intellect, creativity, problem-solving, and general wisdom. The point at which machines not only match but significantly surpass human cognitive capacities is known as the "stage beyond artificial general intelligence" (AGI). Once AGI is attained, there may be a sudden "intelligence explosion," in which robots advance more intelligently than humans due to recursive self-improvement. According to philosopher Nick Bostrom, who popularized this idea, such a change may occur suddenly, giving humans little time to adjust.

We can say that the topic "Is AGI possible?" is no longer up for debate after learning more about AGI. Because the scientists are putting all of their effort toward creating real AI, the answer is unmistakably positive. Researchers now ask, "When will we have AGI?" and let us face it, the answers are not quite clear. Brooks, a well-known Australian roboticist, for instance, stated that current science is still far from comprehending "the true promise and dangers of AI" and that a working AGI system will not be put into place with 100% accuracy. Leading AI researchers supported the argument that true general AI remains a distant prospect. However, another prominent scientist presented a contrasting perspective, suggesting that achieving general intelligence within the next two decades is still a plausible possibility. According to him, there is a 25% probability that AGI technology will be understood by 2030, a 50% chance that it will by 2040, and a 10% chance that it will never be. Important issues related to the super intelligence in AGI are discussed below:

- Aligning values: Making sure the AI's objectives correspond with human morals and preferences.
- Designing systems to maintain human supervision over super intelligence is the control challenge.
- The concept of decision autonomy is the approach of deciding under what circumstances humans should allow AI to make critical decisions.
- To reduce these dangers, research organizations such as OpenAI and the Future of Humanity Institute are actively looking at AI safety, interpretability, and alignment techniques.
- The transformation from pre-AGI to post-AGI intelligence could make super intelligence the biggest threat to survival or humanity's greatest opportunity.

19.5 Future Research Directions

There are many research challenges associated with the convergence of AGI, quantum computing, and neuroscience. Therefore, there is a present need to examine critical areas that focus on the advancement of AGI systems through the integration of these broad areas.

- **Multidisciplinary frameworks**: Combining knowledge from different fields can lead to new and powerful ways to make AGI. For example, a quantum-inspired neural network enhances effective and efficient computing in conventional deep learning by incorporating quantum concepts such as tensor networks, quantum circuits. On the other hand, brain-inspired quantum algorithms investigate how quantum systems are utilized to accelerate or support cognitive tasks, such as associative memory or attention, making them easier to scale. These approaches could lead to a unified theoretical framework, referred to as "quantum cognitive architecture," where cognitive functions are represented as quantum modules, such as quantum working memory.
- **Improved cognitive models**: AGI can gain significantly from advanced, realistic simulation environments.

By modeling complex systems such as cities, communities, or ecosystems, AGI could develop the ability to make sophisticated, long-term decisions. One possible strategy is to create digital twin environments, such as a detailed model of the Earth's economy or climate. In these situations, AGI could be used to safely test policy choices, resource distributions, and plans for long-term viability in a virtual environment. Combining complexity science and AI in this way could give us useful tools for dealing with big problems such as climate change and keeping the economy stable.

- **Self-reliant innovation and exploration**: One of the most revolutionary goals is to enable AGI systems to progress beyond job execution into autonomous scientific research and creative creativity. Future AGIs could come up with concepts, organize and carry out experiments (in simulation or in real life), and even generate novel scientific theories. Early examples can already be seen in materials science, where AI suggests new compounds, and mathematics, where AI systems generate proofs or hypotheses. There are exciting opportunities to expand this capability to other domains, such as algorithm design, engineering, or the arts, but there are also safety and value alignment issues to consider. Ensuring that creative AGIs operate within ethical parameters is one of the primary enduring challenges.
- **Photonic and neuromorphic computing**: An important factor in the sustainable scaling of AGI systems will be next-generation hardware. The energy footprint of AI models could be significantly decreased by neuromorphic computing, which simulates brain-like spiking behavior (perhaps with the help of quantum spins). Photonic computing, on the other hand, has the potential to significantly increase speed and efficiency by using light to carry out neural computations. The development of a photonic neuromorphic quantum computer, which combines the three themes of brain-inspired, quantum-enabled, and light-based processing into a single, incredibly effective AGI platform, is an intriguing but futuristic potential.
- **AI that is based on cognitive neuroscience**: AI can also be a way to learn more about how the brain works. Researchers can assess and enhance neuroscientific hypotheses by juxtaposing AI models with actual brain activity. If an AI model

of the visual cortex makes a prediction about how a neuron will respond to different images, this can be tested in real biological neurons. This close integration enhances research in both AI and neuroscience, potentially resulting in more resilient, biologically grounded AGI frameworks.

- **Ways to align safely and on a large scale**: We need to make our methods for making sure that AGI models stay in line with human values more varied and complicated. This means coming up with safety methods that can be used on a large scale, such as training AGIs with datasets that are moral and ethical, using formal verification methods to check AGI proposals against known ethical or legal standards. We are investigating democratic alignment strategies in which AGI models emulate various stakeholders and seek solutions that satisfy all parties, while incorporating justice and inclusivity into their decision-making processes.

19.6 Conclusion

Achieving cognitive capabilities similar to those of humans in a variety of domains is the aim of the revolutionary shift in AI discussed in this chapter. AGI, BCIs, and quantum computing come together to create cognitively enhanced, adaptive, and self-improving AI systems. Whereas evolutionary algorithms allow for autonomous learning and adaptation, BCIs provide a direct connection between the human mind and computer functions. Quantum computing's processing power enhances uncertainty-based decision-making and the efficacy of learning. Along with pushing the boundaries of current AI capabilities, these technologies raise significant societal and ethical concerns in addition to technical challenges such as quantum decoherence and neural integration. By examining the interactions between these disciplines, this chapter emphasizes the value of interdisciplinary approaches in guiding the development of AGI in a safe and effective manner. Finally, the convergence of AGI, neurotechnology, and quantum systems presents an opportunity to reconsider intelligence and human–machine interaction; thus, further research that ensures fair access and compatibility with human values is essential.

Abbreviations

AGI	Artificial general intelligence
BCI	Brain–computer interface
HHL	Harrow–Hassidim–Lloyd
QAOA	Quantum approximation optimization algorithms
NISQ	Intermediate-scale quantum
QNN	Quantum neural networks
QML	Quantum machine learning
QBM	Quantum Boltzmann machine
QPCA	Quantum principal component analysis
SVM	Support vector machine
VQC	Variational quantum circuits

AI disclosure statement: The author acknowledges the use of AI-based tools for minor language editing or grammatical correction. However, all intellectual contributions, analyses, and conclusions presented in this chapter are entirely the author's own.

References

1. Grabowska, A. and Gunia, A., On quantum computing for artificial superintelligence. *Euro. Jnl. Phil. Sci.*, 14, 25, 2024. https://doi.org/10.1007/s13194-024-00584-7.
2. Biamonte, J., Wittek, P., Pancotti, N., Rebentrost, P., Wiebe, N., Lloyd, S., Quantum machine learning. *Nature*, 549, 7671, 195–202, 2017. https://doi.org/10.1038/nature23474.
3. Preskill, J., Quantum Computing in the NISQ era and beyond. *Quantum*, 2, 79, 2018. https://doi.org/10.22331/q-2018-08-06-79.
4. Benedetti, M., Realpe-Gomez, J., Biswas, R., Perdomo-Ortiz, A., Quantum-assisted learning of hardware-embedded probabilistic graphical models. *Phys. Rev. X*, 7, 4, 041052, 2017. https://doi.org/10.1103/PhysRevX.7.041052.
5. Harrow, A.W., Hassidim, A., Lloyd, S., Quantum algorithm for linear systems of equations. *Phys. Rev. Lett.*, 103, 15, 150502, 2009. https://doi.org/10.1103/PhysRevLett.103.150502.
6. Havlíček, V., Córcoles, A.D., Temme, K., *et al.*, Supervised learning with quantum-enhanced feature spaces. *Nature*, 567, 7747, 209–212, 2019. https://doi.org/10.1038/s41586-019-0980-2.

7. Amin, M.H., Andriyash, E., Rolfe, J., Kulchytskyy, B., Melko, R., Quantum Boltzmann machine. *Phys. Rev. X*, 8, 2, 021050, 2018. https://doi.org/10.1103/PhysRevX.8.021050.
8. Schuld, M., Sinayskiy, I., Petruccione, F., An introduction to quantum machine learning. *Contemp. Phys.*, 56, 2, 172–185, 2015. https://doi.org/10.1080/00107514.2014.964942.
9. Wossnig, L., Zhao, Z., Prakash, A., Quantum linear system algorithm for dense matrices. *Phys. Rev. Lett.*, 120, 5, 050502, 2018. https://doi.org/10.1103/PhysRevLett.120.050502.
10. Zahorodko, P.V., Modlo, Y.O., Kalinichenko, O.O., Selivanova, T.V., Semerikov, S., Quantum enhanced machine learning: An overview. CEUR Workshop Proceedings, 2021.
11. Francesco, T., Chiara, M., Dario, G., Daniele, B., An artificial neuron implemented on an actual quantum processor. *npj Quantum Inf.*, 5, 1, 2019. https://doi.org/10.48550/arXiv.1811.02266.
12. Benedetti, M., Lloyd, E., Sack, S., Fiorentini, M., Parameterized quantum circuits as machine learning models. *Quantum Sci. Technol.*, 4, 4, 043001, 2019. 10.1088/2058-9565/ab4eb5.
13. Schuld, M., Bocharov, A., Svore, K.M., Wiebe, N., Circuit-centric quantum classifiers. *Phys. Rev. A*, 101, 3, 032308, 2020. https://doi.org/10.1103/PhysRevA.101.032308.
14. Lloyd, S., Mohseni, M., Rebentrost, P., Quantum principal component analysis. *Nat. Phys.*, 10, 9, 631–633, 2014. https://doi.org/10.1038/nphys3029.
15. Goertzel, B. and Pennachin, C., *Artificial General Intelligence*, Springer, Springer Berlin, Heidelberg Germany, 2007, https://doi.org/10.1007/978-3-540-68677-4.
16. DeepMind, A Generalist Agent, 2022. [https://www.deepmind.com/publications/a-generalist-agent].
17. Lebedev, M.A. and Nicolelis, M.A., Brain–machine interfaces: past, present and future. *Trends Neurosci.*, 29, 9, 536–546, 2006. https://doi.org/10.1016/j.tins.2006.07.004.
18. Rao, R.P.N., *Brain–Computer Interfacing: An Introduction*, Cambridge University Press, Cambridge, United Kingdom, 2013.
19. Yuste, R., *et al.*, Four ethical priorities for neurotechnologies and AI. *Nature*, 551, 7679, 159–163, 2017. https://doi.org/10.1038/551159a.
20. Elon Musk, Neuralink, An Integrated Brain-Machine Interface Platform With Thousands of Channels. bioRxiv, 2019. https://preprints.jmir.org/preprint/16194.
21. Ienca, M. and Andorno, R., Towards new human rights in the age of neuroscience and neurotechnology. *Life Sciences Soc. Policy*, 13, 1, 5, 2017. https://doi.org/10.1186/s40504-017-0050-1.

22. O'Hara, P.A., Artificial General Intelligence and Autonomous Humanoid Robotics, in: *Principles of Institutional and Evolutionary Political Economy. Springer Texts in Business and Economics*, Springer, Singapore, 2022, https://doi.org/10.1007/978-981-19-4158-0_7.
23. Engelbart, D.C., Augmenting human intellect: A conceptual framework, in: *Augmented Education in the Global Age*, pp. 13–29, Routledge, London, United Kingdom, 2023.
24. Lavazza, A., Freedom of thought and mental integrity: The moral requirements for any neural prosthesis. *Front. Neurosci.*, 12, 82, 2018. https://doi.org/10.3389/fnins.2018.00082.

Index

Also of Interest

Check out these published and forthcoming titles in the "Leading-Edge Breakthroughs in Artificial Intelligence" series from Scrivener Publishing

Affective Artificial Intelligence
Fundamentals, Challenges, and Applications
Published 2026. ISBN 978-1-394-42272-2

Platform Engineering
Concepts, Challenges and Applications
Edited by K. Suresh Kumar, S. Sundaresan, R. Prithiviraj, T. Ananth Kumar and S. Balamurugan
Published 2026. ISBN 978-1-394-39588-0

Responsible AI
Principles and Practices
Edited by Manish Kumar, Nitigya Sambyal, Leena Priyadarshini Singh, Ramasamy V. and S. Balamurugan
Published 2026. ISBN 978-1-394-35544-0

Digital Immune System
Principles and Practices
Edited by Sujata Priyambada Dash, Vaibhav Mishra, Bijeta Shaw, Sandeep Kumar Panda, and S. Balamurugan
Published 2025. ISBN 978-1-394-38375-7

Adaptive AI
Fundamentals, Challenges, and Applications
Edited by P. Pavan Kumar, G. Suresh Kumar, Ajay Kumar Jena, Sandeep Kumar Panda and S. Balamurugan
Published 2025. ISBN 978-1-394-38904-9

AI Trust, Risk and Security Management
Framework, Principles, and Practices
Edited by R. Karthick Manoj, S. Senthilnathan, S. Arunmozhi Selvi, T. Ananth Kumar, S. Balamurugan
Published 2025. ISBN 978-1-394-39299-5

Forthcoming 2026
Neuro-Symbolic AI: Concepts and Applications
Decision Intelligence: Concepts, Applications, and Case Studies
Human-Centered AI: Frameworks and Applications
Composite AI: Fundamentals, Challenges, and Applications
Visual AI: Fundamentals, Challenges, and Applications
Causal AI: Fundamentals and Applications
Embodied AI: Concepts, Challenges, and Applications
Agentic AI: Concepts, Applications, and Case Studies
Data-Centric AI: Concepts, Applications and Case Studies
Sovereign AI: Frameworks, Principles, and Practices
Accelerated AI: Energy Efficient Computing Systems for Large Language Models (LLMs)
Actionable AI: Powered by Large Action Models (LAMs)
AI-Augmented Software Engineering: Principles and Practices
Ambient Invisible Intelligence: Fundamentals, Challenges, and Applications
Artificial Cognition: Technology, Applications, and Challenges
Artificial General Intelligence: Principles and Practices
Autonomous AI Agents: Foundations and Applications
Principles of Confidential Computing
Disinformation Security for Generative AI Applications: Concepts, Methodologies, and Use Cases
Embedded AI: Concepts, Challenges, and Applications
Graph Neural Networks (GNNs): Concepts and Applications
Industry Cloud Platforms: Fundamentals, Applications, and Case Studies
Multi-Modal Generative AI: Frameworks, Principles, and Practices
Objective-Driven AI: Concepts, Applications, and Case Studies
Physical AI for Robotics and Automation: Concepts, Techniques, and Applications
Predictive AI: Concepts, Applications, and Case Studies
Prescriptive AI: Concepts, Applications, and Case Studies
Shadow AI: Concepts, Applications, and Security
Small Language Models (SLMs) for Compact AI: Architectures, Principles, and Applications
www.scrivenerpublishing.com